Hot Quarks 2012: Workshop for Young Scientists on the Physics of Ultrarelativistic Nucleus-Nucleus Collisions

AA002815

Journal of Physics: Conference Series Volume 446

Copamarina, Puerto Rico
14-20 October 2012

ISBN: 978-1-62993-281-1
ISSN: 1742-6588

Printed from e-media with permission by:

Curran Associates, Inc.
57 Morehouse Lane
Red Hook, NY 12571

Some format issues inherent in the e-media version may also appear in this print version.

Printed by Curran Associates, Inc. (2013)

For permission requests, please contact the Institute of Physics at the address below.

Institute of Physics
Dirac House, Temple Back
Bristol BS1 6BE UK

Phone: 44 1 17 929 7481
Fax: 44 1 17 920 0979

techtracking@iop.org

Additional copies of this publication are available from:

Curran Associates, Inc.
57 Morehouse Lane
Red Hook, NY 12571 USA
Phone: 845-758-0400
Fax: 845-758-2634
Email: curran@proceedings.com
Web: www.proceedings.com

TABLE OF CONTENTS

PAPERS

JETS IN THE VACUUM AND IN THE MEDIUM

012001 Jet Quenching at RHIC and the LHC 1
Y-J Lee

012002 Phenomenology of Light Quark Jet Quenching in AdS/CFT 7
A Ficnar, J Noronha, M Gyulassy

012003 Azimuthal Asymmetries From Jets Quenched In Fluctuating Backgrounds 11
R Rodriguez, R J Fries

012004 Jet Measurements in Proton-Proton Collisions with the ALICE Experiment at LHC 15
M Vajzer

012005 Measurement of Jet p_T Spectra in Pb–Pb Collisions at Squareroot(ˢNN)= 2.76 TeV with the ALICE Detector at the LHC 19
S Aiola

012006 Full Jet Reconstruction in 2.76 TeV pp and Pb-Pb Collisions in the ALICE Experiment 23
R Reed

12007 Study of Jet Quenching Using Di-Jet and Photon-Jet Events in PbPb Collisions at 2.76 TeV with CMS 28
Y Kim

012008 Systematic Monte-Carlo Studies of Dijets at RHIC Using the VNI/BMS Parton Cascade 32
C E Coleman-Smith

012009 Jet-Hadron Correlations in pp and Pb-Pb Collisions with ALICE 36
M Connors

012010 Interference Effects Between the Initial and Final State Radiation in a QCD Medium 40
N Armesto, H Ma, M Martinez, Y Mehtar-Tani, C A Salgado

QCD AT HIGH TEMPERATURE / DENSITY AND LATTICE QCD

012011 An Overview of (Selected) Recent Results in Finite-Temperature Lattice QCD 44
A Bazavov

012012 Dipole Asymmetry at Squareroot(ˢNN)= 200 GeV Au+Au Collisions from STAR Experiment at RHIC 50
Y Pandit

012013 Femtoscopy at High m_T in Heavy-Ion Collisions with ALICE 54
H Beck

012014 Can Baryon Stopping Explain the Breakdown of Constituent Quark Scaling and Proposed Signals of Chiral Magnetic Waves at RHIC? 58
J M Campbell, M A Lisa

012015 Beam Energy Scan Results from RHIC 62
P Sorensen

012016 Influence of Quark Masses on the QCD Phase Diagram in the Presence of a Magnetic Field 68
A J Mizher

012017 Measurement of Beam Energy Dependent Nuclear Modification Factors at STAR 72
S P Horvat

012018 'Chemical' Composition of the Quark Gluon Plasma 76
F Scardina, M Colonna, S Plumari, V Greco

012019 Meson Spectra and Thermodynamics in Soft-Wall AdS/QCD 80
S P Bartz, J I Kapusta

RELATIVISTIC HYDRODYNAMIC COLLECTIVE PHENOMENA AND MORE

012020 Overview of Flow and Hydrodynamic Modeling of Nuclear Collisions 84
M Luzum

012021 Initial Conditions with Flow from a McLerran Venugopalan model with Transverse Dynamics 89
G Chen, R J Fries

012022 Transition From Ideal To Viscous Mach Cones In A Partonic Transport Model ... 93
I Bouras, A El, O Fochler, H Niemi, Z Xu, C Greiner

012023 Two-Particle Correlations in 2D p_t Space in Au+Au Collisions at Squareroot(^{s}NN)= 200 GeV at STAR ... 97
E W Oldag

012024 Measurement of the Distribution of Event-by-Event Harmonic Flow in Pb-Pb Collisions at Squareroot(^{s}NN)=2.76 TeV with the Atlas Detector ... 101
S Mohapatra

012025 Transport Approach to Anisotropic Flows from Viscous Hydro Regime to High p_T ... 105
S Plumari, A Puglisi, F Scardina, V Greco

012026 Event Plane Correlation from Single-Shot Hydrodynamics ... 109
D Teaney, L Yan

012027 Charged-Particle Pseudorapidity Density and Anisotropic Flow Over a Wide Pseudorapidity Range Using ALICE at the LHC ... 113
K Gulbrandsen

012028 Measurement of Direct-Photon Elliptic Flow in Pb-Pb Collisions at Squareroot(^{s}NN)= 2.76 TeV ... 117
D Lohner

012029 Fluctuating Hydrodynamics Confronts the Rapidity Dependence of Transverse Momentum Fluctuations ... 121
R Pokharel, S Gavin, G Moschelli

012030 Observation of Near-Side Angular Correlations in pPb Collisions at 5.02TeV with CMS ... 125
D Velicanu

012031 Event-by-Event di-hadron Correlations in Pb-Pb 2.76 TeV Collisions from the ALICE Experiment ... 129
A R Timmins

012032 Effective Model for Deconfinement at High Temperature ... 133
V Skokov

012033 Initial State in RHIC and Ground-State Properties of Nuclei ... 137
P Filip

HEAVY FLAVOR DILEPTONS AND PHOTONS

012034 Heavy-Quark Production in Heavy-Ion Collisions ... 141
A Dainese

012035 Heavy Quark Energy Loss and Angular De-Correlation in a Quark-Gluon Plasma Matter ... 149
S Cao, G-Y Qin, S A Bass, B Müller

012036 Λ_c Analysis with the ALICE Detector at the LHC ... 153
R Romita

012037 Azimuthal Angular Correlations Between Heavy-Flavour Decay Particles and Charged Hadrons in pp Collisions in ALICE ... 157
D Thomas

012038 Measurement of Electrons from Semi-leptonic Heavy-flavour Hadron Decays in Proton-proton and Pb–Pb Collisions with ALICE at the LHC ... 161
M Fasel

012039 D Meson Elliptic Flow in Pb-Pb Collisions at Squareroot(^{s}NN)= 2.76 TeV Measured with ALICE ... 165
G Luparello

012040 D+s Production at Central Rapidity in pp Collisions at Squareroot(S)= 7 TeV and in Pb–Pb Collisions at Squareroot(^{s}NN)= 2.76 TeV with the ALICE Detector ... 169
G M Innocenti

012041 New Measurement of Heavy-Flavor Electron v_2 and Separation of Bottom and Charm Using VTX in PHENIX ... 173
L Ding

012042 J/ψ Production in Pb-Pb Collisions at Squareroot(^{s}NN)=2.76 TeV in the ALICE Experiment ... 177
A Lardeux

012043 Observation of Sequential Υ Suppression in PbPb Collisions ... 181
Z Hu

012044 Open Heavy Flavor and J/psi at RHIC and LHC Within a Transport Model ... 185
J Uphoff, O Fochler, Z Xu, C Greiner

012045 Direct Photon Measurements in pp Collisions at Squareroot(S)= 7 TeV with ALICE ... 189
A Mas

012046 Electroweak Bosons in Heavy-Ion Collisions with the CMS Detector at Squareroot(ˢNN)=2.76 TeV 193
A Florent
012047 Di-Electron Elliptic Flow in Squareroot(ˢNN)= 200 GeV Au+Au Collisions at STAR 197
X Cui
012048 Measurement of Di-electron Continuum in Au+Au Collisions at Squareroot(ˢNN)= 200 GeV by PHENIX Using the Hadron Blind Detector 201
S D Rolnick
012049 Low-Mass Dielectron Measurement in pp and Pb–Pb Collisions in ALICE 206
M K Köhler

EXPERIMENTAL TECHNIQUES AND FUTURE PROGRAMS APPARATUS

012050 The Physics of sPHENIX 210
A M Sickles
012051 The Importance of e+A Collisions at an Electron-Ion Collider 216
M A C Lamont
012052 Physics Performance Studies for the ALICE Inner Tracker Upgrade 220
J Stiller
012053 Understanding the Nuclear Initial State with an Electron Ion Collider 224
T Toll
012054 Muon Physics in ALICE: The MFT Upgrade Project 228
A Uras

BARYONS AND STRANGENESS

012055 Strange and Multi-Strange Particle Production in ALICE 232
D D Chinellato
012056 Hadronic Resonances at ALICE 236
A G Knospe
012057 Vector Meson Production in pp Collision at Squareroot(S)= 7 TeV and at Squareroot(S)= 2.76 TeV Measured with the ALICE Detector 240
E A R Casula
012058 Coherent *J/ψ* Photoproduction with the ALICE Experiment at the LHC 244
J Adam
Author Index

Journal of Physics: Conference Series **446** (2013) 011001 doi:10.1088/1742-6596/446/1/011001

Preface

The 5th edition of the Workshop for Young Scientists on the Physics of Ultrarelativistic Nucleus–Nucleus Collisions (Hot Quarks 2012) was held in Copamarina, Puerto Rico from 14–20 October 2012. As in previous years, this meeting gathered more than 70 participants in the early years of their scientific careers. This issue contains the proceedings of the workshop.

As in the past, the Hot Quarks workshop offered a unique atmosphere for a lively discussion and interpretation of the current measurements from high energy nuclear collisions. Recent results and upgrades at CERN's Large Hadron Collider (LHC) and Brookhaven's Relativistic Heavy Ion Collider (RHIC) were presented. Measurements from the proton-led run at the CERN-LHC were shown for the first time at this meeting. Recent theoretical developments were also extensively discussed, as well as the proposals for future facilities such as the Facility for Antiproton and Ion Research (FAIR) at Darmstadt, the Electron-Ion Collider at Brookhaven, and the LHeC. The conference's goal to provide a platform for young researchers to learn and foster their interactions was successfully met.

We wish to thank the sponsors of the Hot Quarks 2012 Conference, who supported the authors of this volume: Brookhaven National Laboratory (USA), European Laboratory for Particle Physics CERN (Switzerland), European Research Council (EU), ExtreMe Matter Institute EMMI (Germany), Helmholtz International Center for FAIR (Germany), IN2P3/CNRS (France) and the European Research Council via grant #259612, Lawrence Berkeley National Laboratory (USA), Lawrence Livermore National Laboratory (USA), Los Alamos National Laboratory (USA), National Science Foundation (USA), and Netherlands Organization for Scientific Research (Netherlands).

Marcus Bleicher
Goethe-University Frankfurt and HIC4FAIR
Germany
Helen Caines
Yale University
USA
Manuel Calderón de la Barca Sánchez
UC Davis
USA
Rainer J Fries
Texas A&M University
USA
Raphaël Granier de Cassagnac
CNRS-IN2P3 and École polytechnique
France
Boris Hippolyte
CNRS-IN2P3 and Université de Strasbourg
France
André Mischke
Utrecht University and Nikhef Amsterdam
The Netherlands
Ágnes Mócsy
Pratt Institute and Brookhaven National Laboratory
USA
Hannah Petersen
FIAS
Germany
Lijuan Ruan
Brookhaven National Laboratory
USA
Carlos A Salgado
Universidade de Santiago de Compostela
Spain

HQ'12
HOT QUARKS 2012
WORKSHOP FOR YOUNG SCIENTISTS ON
THE PHYSICS OF ULTRARELATIVISTIC
NUCLEUS-NUCLEUS COLLISIONS
OCTOBER 14-20, 2012 COPAMARINA, PUERTO RICO
QCD AT HIGH TEMPERATURE/DENSITY AND LATTICE QCD
INITIAL STATE EFFECTS AND COLOR GLASS CONDENSATE
RELATIVISTIC HYDRODYNAMIC AND COLLECTIVE PHENOMENA
CORRELATIONS AND FLUCTUATIONS
JETS IN THE VACUUM AND IN THE MEDIUM
BARYONS AND STRANGENESS
HEAVY FLAVOR, DILEPTONS AND PHOTONS
APPLICATIONS OF STRING THEORY AND ADS/CFT
EXPERIMENTAL TECHNIQUES AND FUTURE PROGRAMS
Markus Bleicher (Frankfurt(HIC4FAIR), Germany)
Helen Caines (Yale University, USA)
Manuel Calderón de la Barca Sánchez (UC Davis, USA)
Rainer Fries (Texas A&M University, USA)
Raphaël Granier de Cassagnac (Ecole Polytechnique, France)
Boris Hippolyte (IPHC, Strasbourg, France)
Andre Mischke (Utrecht University, The Netherlands)
Ágnes Mócsy (Pratt Institute, Brooklyn, USA)
Hannah Petersen (Goethe University, Frankfurt, Germany)
Lijuan Ruan (BNL, USA)
Carlos Salgado (Universidade de Santiago de Compostela, Spain)
cnrs
CERN
EMMI
Los Alamos
HIC for FAIR
BROOKHAVEN NATIONAL LABORATORY

Journal of Physics: Conference Series **446** (2013) 011002 doi:10.1088/1742-6596/446/1/011002

Peer review statement

All papers published in this volume of *Journal of Physics: Conference Series* have been peer reviewed through processes administered by the proceedings Editors. Reviews were conducted by expert referees to the professional and scientific standards expected of a proceedings journal published by IOP Publishing.

Journal of Physics: Conference Series **446** (2013) 012001 doi:10.1088/1742-6596/446/1/012001

Jet Quenching at RHIC and the LHC

Yen-Jie Lee

CERN, European Organization for Nuclear Research, Geneva, Switzerland

E-mail: Yen-Jie.Lee@cern.ch

Abstract.
Relativistic heavy ion collisions allow one to study Quantum Chromodynamics (QCD) under extreme temperature and density conditions. A new form of matter formed at energy densities above $\sim$ 1 GeV/fm^3, Quark-Gluon Plasma (QGP), is predicted in Lattice QCD calculations. One of the most interesting experimental signatures of QGP formation is jet-quenching due to in-medium energy loss when hard-scattered partons pass through the QGP. Direct jet reconstruction in heavy ion collisions has opened a new era of precision studies of jet-quenching. In this summary talk, recent jet measurements in heavy ion collisions and their implications are reviewed and discussed.

1. Introduction

High-energy collisions of heavy ions allow one to study the fundamental theory of the strong interaction Quantum Chromodynamics (QCD) under extreme temperature and density conditions. A new form of matter [1, 2, 3, 4] formed at energy densities above $\sim$ 1 GeV/fm^3, accessible at high-energy heavy-ion collisions, is predicted in Lattice QCD calculations [5]. This quark-gluon plasma (QGP) consists of an extended volume of deconfined and chirally-symmetric quarks and gluons.

One of the most interesting experimental signatures which are suggested for QGP studies was the suppression of high-transverse-momentum(p_T) particles, often referred to as "jet-quenching", resulting from in-medium energy loss when hard-scattered partons pass through the strongly interacting medium [6]. Results from nucleus-nucleus collisions at the Relativistic Heavy Ion Collider (RHIC) [7, 8, 9, 10] and the Large Hadron Collider [11, 12, 13] have shown evidence for the jet-quenching effect through the suppression of inclusive high-p_T hadron production and modification of high-p_T dihadron angular correlations when compared to proton-proton collisions.

However, high-p_T particle spectra by themselves give limited information on the underlying mechanism. There are three categories of models proposed to describe the observed magnitude of the suppression from RHIC and LHC data. The first group of models explains the parton energy loss by collinear and soft gluon emission which was the main idea of many pre-LHC calculations. The second group of PYTHIA [14] inspired models describe this phenomenon with medium induced branching which may contain hard gluon emissions. The third category of models based on AdS/CFT features wide angle gluon emissions or "QGP heating". Recent reviews on the theoretical progress are available in [15, 16, 17, 18].

Direct jet reconstruction is becoming available at the RHIC and the LHC which opened a new era of the studies of the jet-quenching in heavy ion collisions. In this proceedings, recent

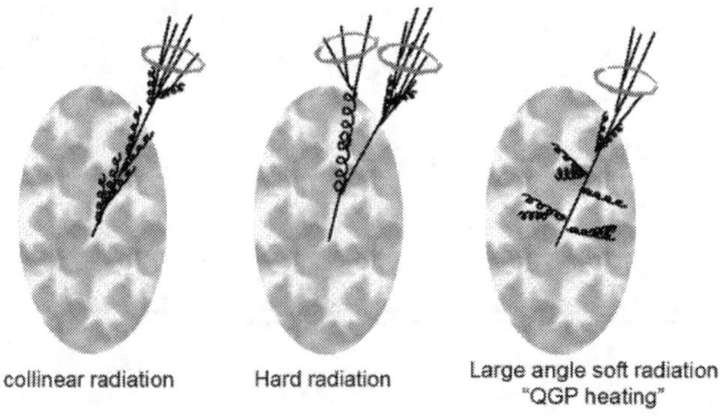

Figure 1. Scenarios of parton energy loss models

results related to jet-quenching measurements using direct jet reconstruction are summarized and discussed.

2. Nuclear modification factors of inclusive jets

To distinguish the scenarios demonstrated in Figure 1, one interesting way is to study the modification of jet p_T spectra using different resolution parameter (R) and check if one can collect the radiated energy back through jet reconstruction. Modifications of measured jet spectra are studied using the nuclear modification factors, the ratio of jet yields measured in the heavy ion collisions to the reference from pp (R_{AA}) or peripheral collisions (R_{CP}), normalized by the number of hard scatterings per collision. The raw jet spectra are typically corrected for jet energy scale and resolution using unfolding techniques. Results from RHIC and the LHC show that the high-p_T jet production is suppressed by a factor of 2-3 in the central CuCu [19], AuAu [20] and PbPb collisions [21, 22] compared to reference from pp collisions, which are smmarized in Figure 2. This implies that a cone of $\Delta R \sim 0.2 - 0.5$ does not collect all the radiated energy back. The R_{CP} of jets reconstructed with R=0.5 is also found to be larger than that with R=0.2 at the same jet p_T [21].

3. Dijet and photon-jet momentum imbalance and azimuthal angle correlation

ATLAS and CMS collaborations has reported increasing dijet momentum imbalance in the central PbPb collisions [23, 24]. This again shows that there is momentum flow out of the jet

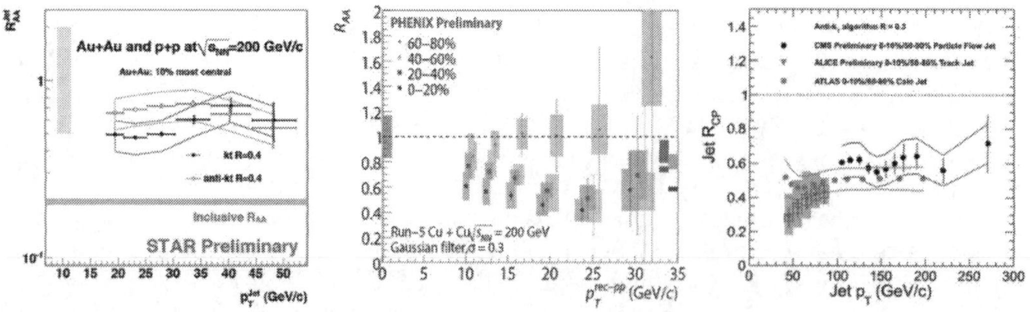

Figure 2. Jet nuclear modification factors measured in AuAu (left panel) [20], CuCu (middle panel) [19] and PbPb collisions (right panel) [21, 22].

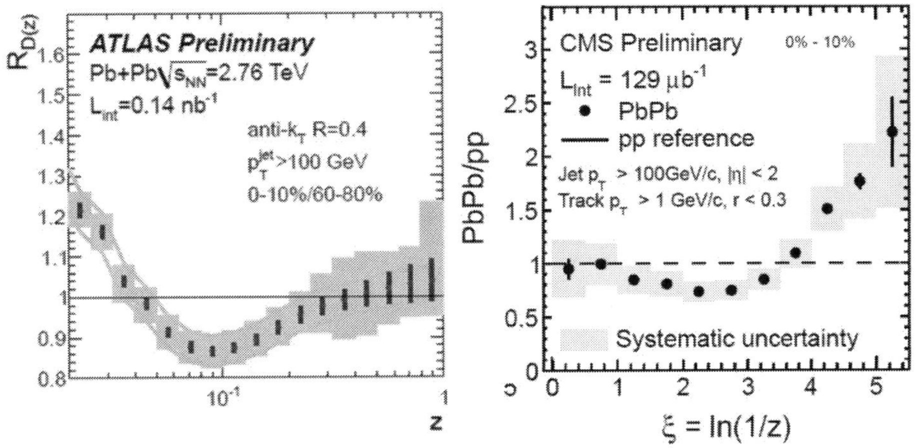

Figure 3. Jet fragmentation function ratios. (Left panel) Ratio of fragmentation function measured in central and peripheral PbPb collisions ($R_{D(z)}$) measured by ATLAS [31]. (Right panel) Ratio of fragmentation function measured in PbPb and pp collisions by CMS [30].

cone. CMS also reported the observed dijet asymmetry persists up to leading jet $p_{\mathrm{T}} \sim 300$ GeV/c and the modification of the mean dijet p_{T} ratio is around 10% [25]. On the other hand, the dijet azimuthal angle correlation is not largely modified, which may provide useful constraints to the hardness of the medium induced radiation.

Triggering on a high p_{T} jet selects a subset of di-jet events which contain a jet passing though a smaller in-medium path-length and losing smaller amount of energy. Therefore, this selection usually biases the position of the hard scattering to the surface of the QGP. It has been shown that the inclusive isolated photons are unmodified in the PbPb collisions with respect to pp reference measurements within the experimental uncertainty [26]. Therefore, requiring a high p_{rmT} isolated photon does not bias the location of the hard scattering inside QGP. Moreover, isolated photons also serve as an unmodified tag for the initial parton energy in the away side in a photon-jet event. CMS has published the first photon-jet analysis [27]. Isolated photons with $p_{\mathrm{T}}^{\gamma} > 60$ GeV/c were correlated with jets with $p_{\mathrm{T}}^{\mathrm{Jet}} > 30$ GeV/c to determine the jet/photon transverse momentum ratio, $x_{J\gamma} = p_{\mathrm{T}}^{\mathrm{Jet}}/p_{\mathrm{T}}^{\gamma}$, and the fraction of photons with an associated jet. The observed momentum ratio $x_{J\gamma}$ is $\sim$ 13% lower than PYTHIA prediction and 20% of the isolated photons lost the associated away-side jet. ATLAS preliminary results [28] on the photon-jet correlation also show a consistent picture as the CMS measurement.

4. Dijet missing p_{T} analysis

Given the large transverse momentum imbalance observed in dijet and photon-jet events, one obvious quest is to measure the distribution of the radiated energy. Information about the overall momentum balance in the dijet events can be obtained using the projection of missing p_{T} of reconstructed charged tracks onto the leading jet axis. For each event, this projection can be calculated as $p_{\mathrm{T}}^{\parallel} = \sum_i -p_{\mathrm{T}}^{i,\mathrm{track}} \cos(\phi_i^{\mathrm{track}} - \phi_{\mathrm{leadingJet}})$ using tracks with $p_{\mathrm{T}} > 0.5$ GeV/c. Studies of $p_{\mathrm{T}}^{\parallel}$ reported by CMS have shown that the overall momentum balance can be recovered if all tracks in the event with $p_{\mathrm{T}} > 0.5$ GeV/c are included in the calculation. In the PbPb data a large fraction of the balancing momentum is carried by tracks having $p_{\mathrm{T}} < 2$ GeV/c. Comparing the momentum balance using only tracks inside and outside of cones of $\Delta R = 0.8$ around the leading and subleading jet axes demonstrates that a large contribution to the momentum balance in data arises from soft particles radiated at $\Delta R > 0.8$ to the jets. [24].

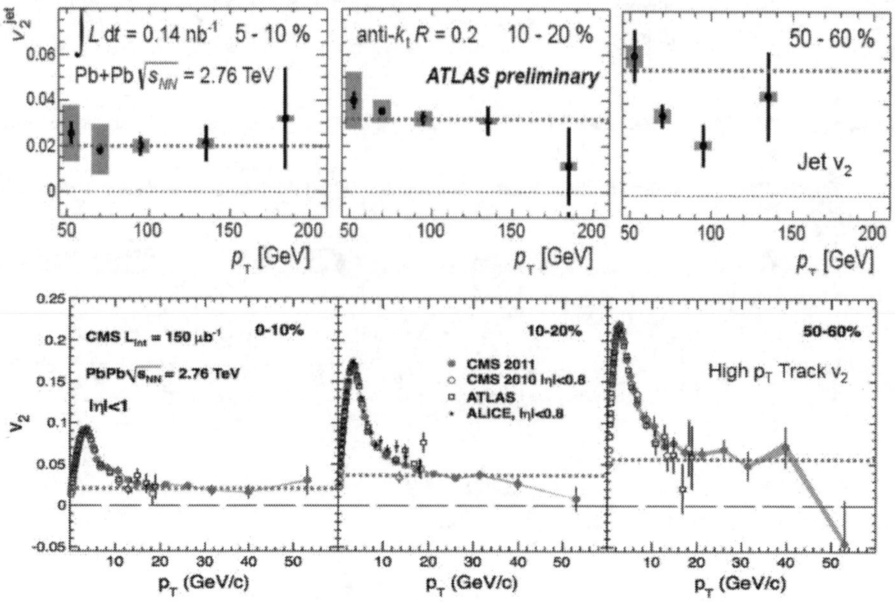

Figure 4. Azimuthal anisotropy measured with jets by ATLAS [33] and high p_T tracks by CMS [34].

Future $p_T^{\|}$ analyses with tracks inside and outside of jet cones with different sizes will give a complete map of the radiated energy and medium response to the hard scattering.

5. Jet shapes and fragmentation functions

Medium induced radiation may redistribute the momenta carried by particles in the jet. The fragmentation function is presented as a function of the variables $z = p_{\|}^{track}/p^{jet}$ and $\xi = -\ln z$. CMS has measured the ratio of jet shapes and fragmentation functions between PbPb and pp data [29, 30], while ATLAS compared the fragmentation function measured in central and peripheral PbPb collisions [31]. As shown in Figure 3, an excess in high ξ (low z) particles in jets with $p_T > 100$ GeV/c is observed in central PbPb collisions. This implies that for central collisions the spectrum of particles in a jet has an enhanced contribution of soft particles compared to one from pp collisions. At low ξ (high z) region, the fragmentation function is not largely modified. ALICE also reported the measurement of jet like near side correlation using charged particles with background subtraction [32]. It is shown that the correlation peak is widened in central collisions.

Detailed studies of jet fragmentation functions and their fluctuations using high statistics dijet and photon-jet events will give a more detailed picture of in-medium jet modification.

6. Jet azimuthal anisotropy

Path length dependence of parton energy loss can be studied by the measurement of jet yields as a function of azimuthal angle with respect to the event plane. The second Fourier harmonics (v_2) of jet azimuthal distribution is measured by ATLAS [33] up to jet $p_T = 200$ GeV/c, remains significantly larger than zero at all measured jet p_T. As shown in Figure 4, the magnitude of the jet v_2 is in agreement with the high p_T track v_2 measurement from CMS [34]. Preliminary results from STAR [35] also show non-zero jet v_2 and the magnitude is similar to results from LHC experiments.

Journal of Physics: Conference Series **446** (2013) 012001 doi:10.1088/1742-6596/446/1/012001

The observed positive jet v_2 at RHIC and the LHC may indicate that high-p_T partons suffer from larger energy loss in the direction of larger transverse path-length in the medium.

7. b-jet fraction

Modification to jets in high-energy heavy-ion collisions is expected to depend on the flavor of the fragmenting parton and the flavor dependence of jet quenching is a crucial test ground for various parton energy loss models. In QCD, quarks carry a smaller color factor compared to gluons, so that the in-medium energy loss for quarks is expected to be smaller than for gluons. Moreover, compared to light quarks, heavy quarks are expected to suffer from smaller radiative energy loss when passing through the medium because gluon radiation is suppressed at angles smaller than the ratio of the quark mass M to its energy E [36], thus further attenuating the effect of the medium.

To disentangle this flavor dependence, jets from heavy quark fragmentation are identified for the first time in heavy ion collisions [37]. Jets are first tagged by their secondary vertices and the contribution from bottom quarks is extracted using template fits to their secondary vertex mass distributions. The bottom quark jet to inclusive jet ratio is measured with the CMS detector from PbPb and pp collisions at a center-of-mass energy of 2.76 TeV per nucleon. This b-jet fraction is measured in the range of $80 < p_T^{\rm Jet} < 200$ GeV/c in PbPb collisions and found to be compatible with the pp b-jet fraction, within sizable uncertainties. The measurement is sufficiently precise to demonstrate that b-jets are subject to jet quenching, although a precise comparison of light and b-jet quenching would require a reduction of the statistical and systematic uncertainties.

8. Summary

The new data from RHIC and the LHC have brought new insights to the jet quenching mechanism. Models which feature soft collinear gluon radiation need mechanism to transport the radiated energy out of the jet cone [38]. Hard gluon radiations and modification of splitting function are constrained by dijet and photon-jet azimuthal angle correlation measurements. Also the distributions of dijet and photon-jet momentum ratio as well as inclusive jet v_2 provide constraints to the path-length dependence of the parton energy loss. The data from RHIC and the LHC can be qualitatively described by models associated with wide angle radiative energy loss. However quantitative comparisons between data and models are needed.

From those studies, we are now in a better position to validate our theoretical understanding of in-medium parton energy loss. It is expected that PbPb collisions at 5.5 TeV will be delivered by the LHC and precision measurements such as $Z^0(W^{\pm})$-jet correlations, b-jet quenching, multi-jet correlations, the azimuthal anisotropy of high p_T jets, and top production may bring us beyond the qualitative observation [39, 40, 41, 42, 43, 44, 45, 46, 47]. Coming implementations of parton energy-loss in realistic event generators, that can be directly compared to the experimental data from RHIC and the LHC, will allow us to go beyond a "mechanistic" understanding of in-medium parton energy loss, and reach our final goal of determining the thermodynamical and transport properties of hot and dense QCD matter via "jet tomography" measurements.

References

[1] E. V. Shuryak, "Theory of Hadronic Plasma", *Sov. Phys. JETP* **47** (1978) 212.

[2] J. C. Collins and M. J. Perry, "Superdense Matter: Neutrons Or Asymptotically Free Quarks?", *Phys. Rev. Lett.* **34** (1975) 1353

[3] N. Cabibbo and G. Parisi, "Exponential Hadronic Spectrum and Quark Liberation", *Phys. Lett.* **B59** (1975) 67

[4] B. A. Freedman and L. D. McLerran, "Fermions and Gauge Vector Mesons at Finite Temperature and Density. 3. The Ground State Energy of a Relativistic Quark Gas", *Phys. Rev.* **D16** (1977) 1169

[5] F. Karsch and E. Laermann, "Thermodynamics and in-medium hadron properties from lattice QCD", in *Quark-Gluon Plasma III, R. Hwa (ed.)*, pp. 1–59. Hackensack, USA, 2003
[6] J. D. Bjorken, "Energy loss of energetic partons in QGP: possible extinction of high pT jets in hadron-hadron collisions", *FERMILAB-PUB-82-059-THY* (1982).
[7] PHENIX Collaboration, "Formation of dense partonic matter in relativistic nucleus nucleus collisions at RHIC: Experimental evaluation by the PHENIX collaboration", *Nucl. Phys.* **A757** (2005) 184.
[8] STAR Collaboration, "Experimental and theoretical challenges in the search for the quark gluon plasma: The STAR collaboration's critical assessment of the evidence from RHIC collisions", *Nucl. Phys.* **A757** (2005) 102.
[9] PHOBOS Collaboration, "The PHOBOS perspective on discoveries at RHIC", *Nucl. Phys.* **A757** (2005) 28
[10] BRAHMS Collaboration, "Quark Gluon Plasma and Color Glass Condensate at RHIC? The perspective from the BRAHMS experiment", *Nucl. Phys.* **A757** (2005) 1
[11] Aamodt K (ALICE Collaboration) 2011 *Phys.Lett.* **B696** 30
[12] Aamodt K (ALICE Collaboration) 2012 Particle-yield modification in jet-like azimuthal di-hadron correlations in Pb-Pb collisions at $\sqrt{s_{NN}} = 2.76$ TeV accepted by *Phys. Rev. Lett.*
[13] S. Chatrchyan *et al.* [CMS Collaboration], Eur. Phys. J. C **72** (2012) 1945 [arXiv:1202.2554 [nucl-ex]].
[14] T. Sjostrand, S. Mrenna and P. Z. Skands, JHEP **0605** (2006) 026 [hep-ph/0603175].
[15] N. Armesto, (ed.), N. Borghini, (ed.), S. Jeon, (ed.), U. A. Wiedemann, (ed.), S. Abreu, V. Akkelin, J. Alam and J. L. Albacete *et al.*, J. Phys. G **35** (2008) 054001 [arXiv:0711.0974 [hep-ph]].
[16] D. d'Enterria, arXiv:0902.2011 [nucl-ex].
[17] N. Armesto, B. Cole, C. Gale, W. A. Horowitz, P. Jacobs, S. Jeon, M. van Leeuwen and A. Majumder *et al.*, Phys. Rev. C **86** (2012) 064904 [arXiv:1106.1106 [hep-ph]].
[18] J. Casalderrey-Solana, H. Liu, D. Mateos, K. Rajagopal and U. A. Wiedemann, arXiv:1101.0618 [hep-th].
[19] Y. -S. Lai [PHENIX Collaboration], Nucl. Phys. A **855** (2011) 295.
[20] M. Ploskon [STAR Collaboration], Nucl. Phys. A **830** (2009) 255C [arXiv:0908.1799 [nucl-ex]].
[21] G. Aad *et al.* [ATLAS Collaboration], Phys. Lett. B **719** (2013) 220 [arXiv:1208.1967 [hep-ex]].
[22] [CMS Collaboration], CMS-PAS-HIN-12-004.
[23] G. Aad *et al.* [Atlas Collaboration], Phys. Rev. Lett. **105** (2010) 252303 [arXiv:1011.6182 [hep-ex]].
[24] S. Chatrchyan *et al.* [CMS Collaboration], Phys. Rev. C **84** (2011) 024906 [arXiv:1102.1957 [nucl-ex]].
[25] S. Chatrchyan *et al.* [CMS Collaboration], Phys. Lett. B **712** (2012) 176 [arXiv:1202.5022 [nucl-ex]].
[26] S. Chatrchyan *et al.* [CMS Collaboration], Phys. Lett. B **710** (2012) 256 [arXiv:1201.3093 [nucl-ex]].
[27] S. Chatrchyan *et al.* [CMS Collaboration], Phys. Lett. B **718** (2013) 773 [arXiv:1205.0206 [nucl-ex]].
[28] [ATLAS Collaboration], ATLAS-CONF-2012-121.
[29] S. Chatrchyan *et al.* [CMS Collaboration], JHEP **1210** (2012) 087 [arXiv:1205.5872 [nucl-ex]].
[30] [CMS Collaboration], CMS-PAS-HIN-12-013.
[31] [ATLAS Collaboration], ATLAS-CONF-2012-115.
[32] A. Morsch [ALICE Collaboration], [arXiv:1207.7187 [nucl-ex]].
[33] [ATLAS Collaboration], ATLAS-CONF-2012-116.
[34] S. Chatrchyan *et al.* [CMS Collaboration], Phys. Rev. Lett. **109**, 022301 (2012) [arXiv:1204.1850 [nucl-ex]].
[35] A. Ohlson [STAR Collaboration], arXiv:1210.8112 [nucl-ex].
[36] Y. L. Dokshitzer and D. E. Kharzeev, Phys. Lett. B **519**, 199 (2001) [hep-ph/0106202].
[37] [CMS Collaboration], CMS-PAS-HIN-12-003.
[38] J. Casalderrey-Solana, J. G. Milhano and U. A. Wiedemann, J. Phys. G **38**, 035006 (2011) [arXiv:1012.0745 [hep-ph]].
[39] X. -N. Wang and Y. Zhu, arXiv:1302.5874 [hep-ph].
[40] G. -L. Ma, arXiv:1302.5873 [nucl-th].
[41] I. Vitev, arXiv:1210.1264 [hep-ph].
[42] W. Dai, I. Vitev and B. -W. Zhang, arXiv:1207.5177 [hep-ph].
[43] K. C. Zapp, F. Krauss and U. A. Wiedemann, JHEP **1303**, 080 (2013) [arXiv:1212.1599 [hep-ph]].
[44] T. Stavreva, F. Arleo and I. Schienbein, JHEP **1302**, 072 (2013) [JHEP **1302**, 072 (2013)] [arXiv:1211.6744 [hep-ph]].
[45] G. -Y. Qin, arXiv:1210.6610 [hep-ph].
[46] J. Casalderrey-Solana, Y. Mehtar-Tani, C. A. Salgado and K. Tywoniuk, arXiv:1210.7765 [hep-ph].
[47] L. Bhattacharya, K. Ghosh and K. Huitu, arXiv:1210.0116 [hep-ph].

Phenomenology of light quark jet quenching in AdS/CFT

Andrej Ficnar[1], **Jorge Noronha**[2] **and Miklos Gyulassy**[1]

[1] Department of Physics, Columbia University, New York, NY 10027, USA
[2] Instituto de Física, Universidade de São Paulo, 05315-970 São Paulo, Brazil

E-mail: `aficnar@phys.columbia.edu`

Abstract. We explore phenomenological signatures of light quark jet quenching within the AdS/CFT correspondence. Firstly, we note that the numerical studies indicate a linear path dependence of the instantaneous energy loss of light quarks modeled as falling strings. Secondly, we propose a phenomenological model for generic description of their energy loss and use it to compute the nuclear modification factor R_{AA} for light quarks in an expanding plasma with Glauber initial conditions. Comparing with the light hadron R_{AA} data at the LHC, a qualitative agreement is shown. Thirdly, we show how the observed quantitative disagreement can be partially alleviated by including the effects from higher derivative corrections to AdS_5.

1. Introduction

The AdS/CFT correspondence [1, 2, 3], a conjectured duality between $\mathcal{N} = 4$ $SU(N_c)$ super-Yang-Mills theory and type IIB string theory on $AdS_5 \times S^5$ space, has been a very useful tool in the study of many properties of the strongly-coupled quark-gluon plasma [4] created in heavy ion collisions at RHIC and LHC. The correspondence allows one to study this strongly coupled gauge theory in the $N_c \gg \lambda \gg 1$ limit by performing classical, two-derivative (super)gravity calculations.

One of the important applications of the AdS/CFT correspondence has been the study of jet quenching in strongly-coupled systems [5]. In the light of the new LHC results on the suppresion of light hadrons in AA collisions [6, 7], a more consistent treatment of energy loss of light quarks in gauge/gravity duality has become necessary, in order to be able to compute jet quenching observables such as the nuclear modification factor R_{AA}. Here we propose a possible way to compute the R_{AA} using the input from the numerical studies of the energy loss of light quarks in AdS/CFT and analyze its features.

2. Light quarks in AdS/CFT and their energy loss

In AdS/CFT, degrees of freedom of mass m_Q in the fundamental representation in the boundary theory ('quarks') at a finite temperature T correspond to a setup with a D7-brane that spans from the boundary ($r = 0$) to some $r_m \propto 1/m_Q$ [8] in the presence of a black brane in the AdS_5 geometry with an event horizon at $r_h \propto 1/T$ [2]. For light quarks, since the D7-brane fills the entire geometry between the boundary and the black hole, one can study their energy loss by studying the free motion of the strings that have both of their endpoints on the D7-brane (representing dressed $q\bar{q}$ pairs), the so-called falling strings [9].

Journal of Physics: Conference Series **446** (2013) 012002 doi:10.1088/1742-6596/446/1/012002

One of the first results on the energy loss of light quarks in AdS/CFT was their maximum stopping distance in a strongly coupled $\mathcal{N} = 4$ SYM plasma, which was shown in [9] to scale with energy as $\Delta x_{max} \sim E^{1/3}$. However, to compute observables such as R_{AA}, one needs the knowledge of the instantaneous energy loss, for which the spacetime momentum currents Π^a_μ on the string worldsheet need to be analyzed.

In case of a non-stationary environment of falling strings, the spacetime momentum currents become non-trivial, time-dependent quantities and the details of the geometry on the worldsheet become important in connecting them to the energy loss. This was examined in [10], where a general expression was derived for calculating the instantaneous energy loss in time-dependent string configurations, which shows that the energy loss, in general, receives a correction to the simple Π^σ_μ component of the flux.

If one defines a light quark jet as a part of the falling string within a certain fixed $\Delta x \sim 1/(\pi T)$ distance from the endpoint (as in [9]), it was demonstrated [10] that this correction becomes especially important at late times and substantially decreases the magnitude of the Bragg-like peak reported in [9] (see Figure 1). We also note that, although the early time behavior of the energy loss is sensitive to the initial conditions, numerical studies suggest that it is linear in time, $dE/dt \sim t$ [10].

3. Computing the light quark R_{AA}

In order to compute the nuclear modification factor of light quarks in AdS/CFT, we need to have a definite temperature T, initial energy E_0 and path length x dependence of the instantaneous energy loss of a light quark moving through an $\mathcal{N} = 4$ SYM plasma. For that purpose, motivated by [11], we model the phenomenologically relevant part of the energy loss in the following way:

$$\frac{dE}{dx}(E_0, T, x) = -c(E_0, T)x^1\Theta\left[L_s(E_0, T) - x\right] . \tag{1}$$

Here L_s is the stopping distance, Θ is the step function and we have also utilized the conjectured linear path dependence from the previous section. The initial energy and temperature dependence has been packed into an unknown function $c(E_0, T)$ which will be determined from the dependence of the stopping distance on the initial energy and temperature [9]:

$$L_s(E_0, T) = \frac{\kappa}{T}\left(\frac{E_0}{\sqrt{\lambda}T}\right)^{1/3} . \tag{2}$$

Here λ is the 't Hooft coupling and κ is a numerical factor that in general depends on the initial string configuration (for the maximum stopping distance it was shown to be ≈ 0.5 [9]). Combining (1) and (2), we can fix the unknown function $c(E_0, T)$ and finally obtain:

$$\frac{dE}{dx} = -\chi E_0^{1/3} x^1 T^{8/3} , \tag{3}$$

where we have defined an effective coupling $\chi \equiv 2\lambda^{1/3}/\kappa^2$, which determines the overall magnitude of the energy loss and is the only free parameter in this formula. For $\kappa \approx 0.5$ [9] and an unphysically small $\lambda = 1$, we have $\chi \approx 8$. As noted in [10], formula (3) has a surprising similarity to the typical qualitative behavior of energy loss of light quarks in pQCD in the strong LPM regime [11].

With this formula, we can compute the R_{AA} for light quarks following the procedure in [12]: we use Glauber initial conditions for determining the temperature profile of the expanding plasma and average over the jet azimuthal directions and production points in the transverse plane. The results are shown in in Figure 2, where we see that the value of $\chi = 8$ gives an

R_{AA} of a rather low magnitude, indicating strong quenching. However, using a lower value of $\chi = 1$ we see that R_{AA} has the correct qualitative behavior as displayed by the LHC data on the suppression of light hadrons in AA collisions [7]. This suggests that the main problem could be simply in the low magnitude of R_{AA}, or, equivalently, too strong quenching.

4. Higher derivative corrections

To make our setup more realistic, in the hope of better fitting the R_{AA} data (i.e. decreasing the strength of the quenching while keeping the correct qualitative behavior), one can include higher derivative R^2 corrections to the gravity sector of AdS_5, which are the leading $1/N_c$ corrections in the presence of a $D7$-brane. These we will model by a Gauss-Bonnet term, i.e. we will consider the action of the form:

$$S = \frac{1}{2\kappa_5^2}\int d^5x\sqrt{-G}\left[R + \frac{12}{L^2} + L^2\frac{\lambda_{GB}}{2}\left(R^2 - 4R_{\mu\nu}^2 + R_{\mu\nu\rho\sigma}^2\right)\right], \tag{4}$$

where λ_{GB} is a dimensionless parameter, constrained to be $-\frac{7}{36} < \lambda_{GB} \leq \frac{9}{100}$ to avoid causality violations [13] and to have positive energy density on the boundary [14]. A black brane solution in this case is known analytically [15]. Following a similar procedure as in [9], by analyzing null geodesics in geometry given by (4) and relating its parameters to the energy of the string, we can estimate the maximum stopping distance of falling strings up to linear order in λ_{GB} [16]:

$$\Delta x_{max} = \frac{\mathcal{C}}{T}\left(\frac{E}{T\sqrt{\lambda}}\right)^{1/3}(1 - \mathcal{F}\lambda_{GB}) + \mathcal{O}(\lambda_{GB}^2), \tag{5}$$

where $\mathcal{C} \sim 1/2$ and $\mathcal{F}$ is a numerical factor > 1, for which our preliminary estimate gives $\mathcal{F} = 11/6$. We see that the $\sim E^{1/3}$ scaling is still present and that for negative values of λ_{GB} we can increase the stopping distance by up to $\sim 30-40\%$, compared to the case of pure AdS_5 with no higher derivative corrections. In addition to the $\sim E^{1/3}$ scaling, there are indications that, at linear order in λ_{GB}, we can also have an additional new energy scaling (though numerically suppressed), details of which, together with the precise value of the parameter $\mathcal{F}$ in (5), will be presented in a future publication [16].

From (5) we see that the Gauss-Bonnet parameter λ_{GB} affects the κ coefficient in (2) and hence the effective coupling χ, which governs the strength of the quenching. For $\lambda_{GB} = -1/5$, this gives $\kappa = 0.65$ instead of ≈ 0.5 in the pure AdS case and, as we can see from Figure 3, this has the effect of increasing the magnitude of R_{AA} by almost 100%, while preserving the correct qualitative behavior.

5. Conclusions and outlook

In [10] it was shown that, defining the light quark jet in AdS/CFT as a part of the falling string within a certain fixed Δx distance from the endpoint, its instantaneous energy loss seems to display a linear path dependence. Using this in a phenomenological model (3) for the energy loss, we have computed the nuclear modification factor R_{AA} for light quarks in an expanding plasma with Glauber initial conditions [12]. While the shape of our model calculations for R_{AA} seems to qualitatively agree with the LHC light hadron R_{AA} data [6, 7], the overall magnitude is too low. We have also shown that taking into account higher derivative R^2 corrections to the gravity sector of AdS_5 can increase the stopping distance (5) and hence increase the magnitude of R_{AA} while preserving its qualitative behavior.

An interesting next step in a more realistic modeling of light quark energy loss would be to study the effect of higher derivative corrections in a non-conformal plasma (the energy loss of heavy quarks in simple non-conformal models was studied in [17, 12]), as well as to examine the falling strings in the Janik-Peschanski metric [18], dual to perfect fluid hydrodynamics, to take into account realistic effects of an expanding plasma [16].

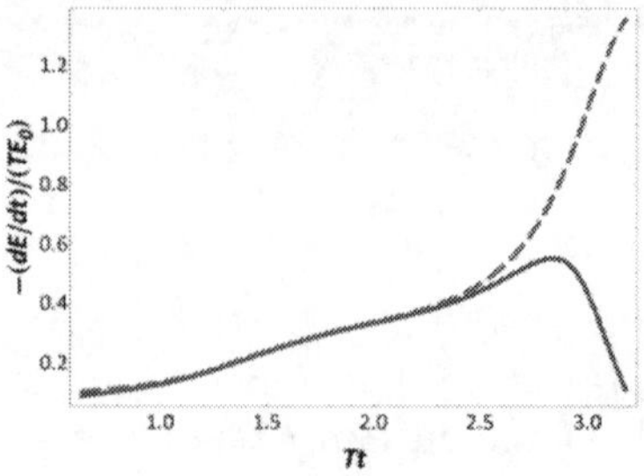

Figure 1. Comparison of the (normalized) instantaneous energy loss as a function of time (i.e. with the correction from [10]; solid blue curve) with the apparent energy loss Π_0^σ (dashed red curve), for a light quark jet defined as a part of the falling string within a certain fixed Δx distance from the endpoint.

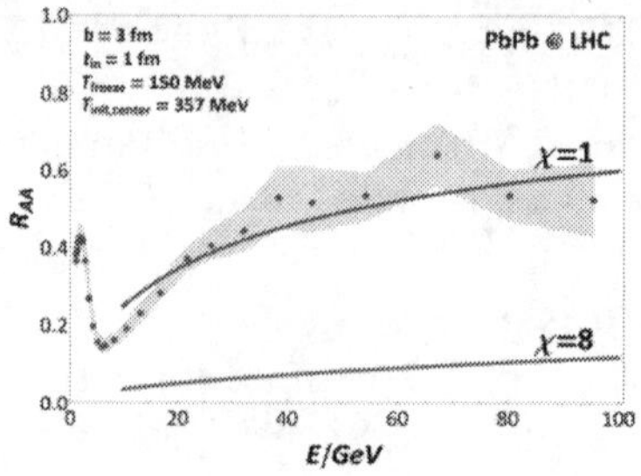

Figure 2. Nuclear modification factor R_{AA} for light quarks as a function of the final parton energy E for two different values of the effective coupling χ, compared to the LHC data from [7]. The initial jet production time t_i, the freezout temperature T_{freeze} and the initial temperature at the center of the plasma $T_{init,center}$ are indicated in the plot.

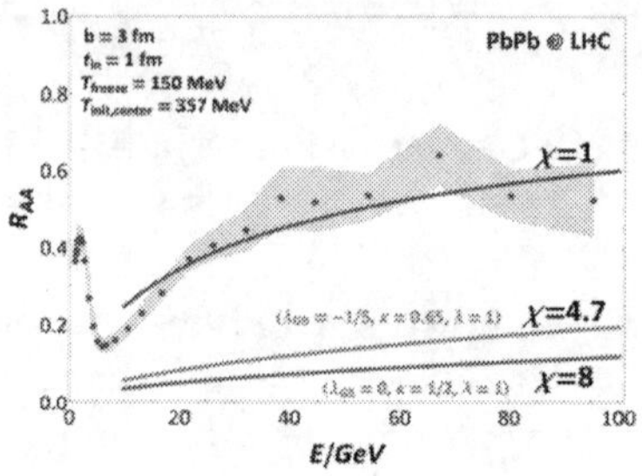

Figure 3. Comparison of the light quark R_{AA} at LHC as a function of the final parton energy E, computed as in [12], for different values of the effective coupling χ with and without Gauss-Bonnet higher derivative corrections (4), compared to the LHC data from [7].

Acknowledgments

We thank W. Horowitz, A. Buzzatti, G. Torrieri and S. S. Gubser for helpful discussions. A.F. and M.G. acknowledge support by US-DOE Nuclear Science Grant No. DE-FG02-93ER40764. J.N. is supported by the Brazilian funding agencies FAPESP and CNPq.

References

[1] Maldacena J M 1998 *Adv. Theor. Math. Phys.* **2** 231-252 (*Preprint* hep-th/9711200)
[2] Witten E 1998 *Adv. Theor. Math. Phys.* **2** 253-291 (*Preprint* hep-th/9802150)
[3] Gubser S S, Klebanov I R and Polyakov A M 1998 *Phys. Lett. B* **428** 105-114 (*Preprint* hep-th/9802109)
[4] Gyulassy M and McLerran L 2005 *Nucl. Phys. A* **750** 30-63 (*Preprint* nucl-th/0405013)
[5] Gubser S S 2006 *Phys. Rev. D* **74** 126005 (*Preprint* hep-th/0605182); Herzog C P, Karch A, Kovtun P, Kozcaz C and Yaffe L G 2006 *JHEP* **0607** 013 (*Preprint* hep-th/0605158); Casalderrey-Solana J and Teaney D 2006 *Phys. Rev. D* **74** 085012 (*Preprint* hep-ph/0605199); Noronha J, Gyulassy M and Torrieri G 2010 *Phys. Rev. C* **82** 054903 (*Preprint* 1009.2286 [nucl-th])
[6] Aamodt K et al. (ALICE Collaboration) 2011 *Phys. Lett. B* **696** 30 (*Preprint* 1012.1004 [nucl-ex])
[7] CMS Collaboration 2012 *Eur. Phys. J. C* **72** 1945 (*Preprint* 1202.2554 [nucl-ex])
[8] Karch A and Katz E 2002 *JHEP* **0206** 043 (*Preprint* hep-th/0205236)
[9] Chesler P M, Jensen K, Karch A and Yaffe L G 2009 *Phys. Rev. D* **79** 125015 (*Preprint* 0810.1985 [hep-th])
[10] Ficnar A 2012 *Phys. Rev. D* **86** 046010 (*Preprint* 1201.1780 [hep-th])
[11] Horowitz W A and Gyulassy M 2011 *Nucl. Phys. A* **872** 265 (*Preprint* 1104.4958 [hep-ph])
[12] Ficnar A, Noronha J and Gyulassy M 2011 *J. Phys. G: Nucl. Part. Phys.* **38** 124176 (*Preprint* 1106.6303 [hep-ph])
[13] Brigante M, Liu H, Myers R C, Shenker S and Yaida S 2008 *Phys. Rev. D* **77** 126006 (*Preprint* 0712.0805 [hep-th])
[14] Hofman D M and Maldacena J 2008 *JHEP* **0805** 012 (*Preprint* 0803.1467 [hep-th])
[15] Cai R-G 2002 *Phys. Rev. D* **65** 084014 (*Preprint* hep-th/0109133)
[16] Ficnar A, Noronha J and Gyulassy M, to appear.
[17] Ficnar A, Noronha J and Gyulassy M 2011 *Nucl. Phys. A* **855** 372-375 (*Preprint* 1012.0116 [hep-ph])
[18] Janik R A and Peschanski R 2006 *Phys. Rev. D* **73** 045013 (*Preprint* hep-th/0512162)

Azimuthal Asymmetries From Jets Quenched In Fluctuating Backgrounds

R Rodriguez[1], **R J Fries**[2]

[1] Department of Mathematics and Physics, Ave Maria University, Ave Maria FL 34142, USA
[2]Cyclotron Institute and Department of Physics and Astronomy, Texas A&M University, College Station, TX 77843, USA

E-mail: ricardo.rodriguez@avemaria.edu, rjfries@comp.tamu.edu

Abstract. High momentum jets and hadrons are important probes for the quark gluon plasma (QGP) formed in nuclear collisions at high energies. We investigate how fluctuations in the background density of the QGP and fluctuations in the spatial distribution of the hard process create azimuthal asymmetries of the high momentum hadron spectrum, described by the Fourier coefficients v_n, $n > 0$. We estimate the coefficients up to v_6 in a simple energy loss model tuned to single inclusive hadron suppression.

With the study of the QGP created in collisions of relativistic heavy ions moving into an increasingly accurate quantitative phase it has been found important to include fluctuations in the space-time structure of the fireball into calculations of bulk quantities [1, 2]. These can emerge from fluctuations in the initial energy density $\epsilon(x, y)$ in the plane transverse to the beam axis and leave signature effects on bulk quantities like the azimuthal asymmetry coefficients v_n. For example fluctuations can lead to a sizeable triangular flow v_3 which would be vanishing in an averaged fireball due to the overall geometry of the nuclear overlap. [2]. At large transverse momentum p_T fluctuations in the position of the hard process can also affect observables accessible in current heavy ion experiments.

Here we have explored the role of such fluctuations on the suppression of high-p_T hadrons and their generalized azimuthal asymmetry coefficients v_n. The v_n are defined as a Fourier decomposition of the azimuthal angle dependent spectrum

$$\frac{dN}{dp_T^2 d\Phi} = \frac{dN}{2\pi p_T dp_T}\left[1 + 2\sum_{n>0} v_n(p_T)\cos(n\Phi + \delta_n)\right] \quad (1)$$

where the angle Φ is measured with respect to the reaction plane defined below and the δ_n are phases that encode a misalignment with the reaction plane. We note that for smooth, non-fluctuating fireballs we expect all odd coefficients v_1, v_3 etc. at midrapidity to vanish for symmetry reasons.

However, in any given single event the initial energy density will typically exhibit a non-vanishing triangular eccentricity ϵ_3 which could in turn lead to a non-vanishing v_3. The event-by-event fluctuations in initial energy density are driven by fluctuations of the positions of nucleons in the initial nuclei and in the amount of energy deposited around midrapidity for every nucleon-nucleon collision [3]. We also expect that the triangular eccentricity is not correlated with the

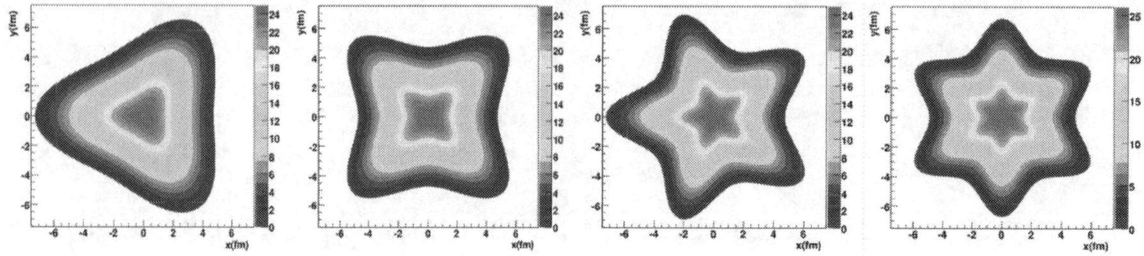

Figure 1. Engineered events with $n = 3, 4, 5, 6$

reaction plane, i.e. δ_3 should appear random. Similar arguments can be made for other $n > 0$ and we expect all odd v_n to acquire non-vanishing values with realistic fluctuations. While all of this has first been discussed for the bulk of the fireball [3] these statements easily transfer to hard probes. Fluctuations in the energy density lead to fluctuations in energy loss. In addition, the position of a hard process which creates a hard probe is subject to fluctuations.

Systematic measurements of v_n at large momentum could lead to further constraints on the type of energy loss prevalent in QGP, and on the size of the transport coefficient $\hat{q}$. It can also give an independent handle on the size and granularity of initial state fluctuations. Here we report on a quantitative study of high momentum azimuthal coefficients using a simple energy loss model with realistic fluctuations.

We calculate high momentum hadron spectra using our simulation package PPM [4, 5]. It samples initial momentum distributions of quark and gluon jets from a perturbative calculation and propagates leading partons through a given background fireball. Different energy loss models can be employed. Here we will show results from a simple LPM-inspired (sLPM) deterministic energy loss model $dE/dx \sim \hat{q}x$ where $\hat{q}$ scales with the 3/4th power of the local energy density [4]. As a cross check we will sometimes also use the non-deterministic Armesto-Salgado-Wiedemann (ASW) model [6, 7]. Fitted to the same experimental data on single hadron suppression these models cover a wide range of values for $\hat{q}$. PPM will eventually fragment leading partons into hadrons and all results here will be shown for pions.

We have used the Glauber Monte Carlo generator GLISSANDO [8] to produce an ensemble of Au+Au events at top RHIC energy using the three different centralities $b = 3.2$, 7.4 and 11 fm. However, first we check the general relation between the spatial eccentricities ϵ_n [9]

$$\epsilon_n = \frac{\sqrt{\langle r^n \cos(n\phi)\rangle^2 + \langle r^n \sin(n\phi)\rangle^2}}{\langle r^n \rangle} \tag{2}$$

and the azimuthal asymmetry coefficients v_m using "engineered" events with particular fixed eccentricities. These are created with the energy density modeled as simple Gaussians in the transverse plane with a $\cos n\phi$ modulation of the mean square radius, as shown in the examples in Fig.1. Observe that r is the distance to the origin of the participant plane.

First we scanned the space (v_n, ϵ_m), $n, m = 1, 6$ in search of correlations. As expected we find non-zero v_n for a given ϵ_m only if $n = im$ where $i > 0$ is an integer. Next we explored the scaling of v_n with the size of the eccentricity ϵ_n. We expected a monotonically increasing function $v_n(\epsilon_n)$ which can be seen confirmed in Fig. 2. Deviations from monotonic behavior only occur for unrealistically large eccentricities. These basic results should hold if realistic fluctuations are considered. Fig. 3 shows the correlation between v_2 and ϵ_2 for our ensemble of GLISSANDO events for all 3 impact parameters for Au+Au collisions at top RHIC energies. The basic linear correlation persists for $n = 2$ but is washed out. Correlations for $n > 2$, not shown here, are much

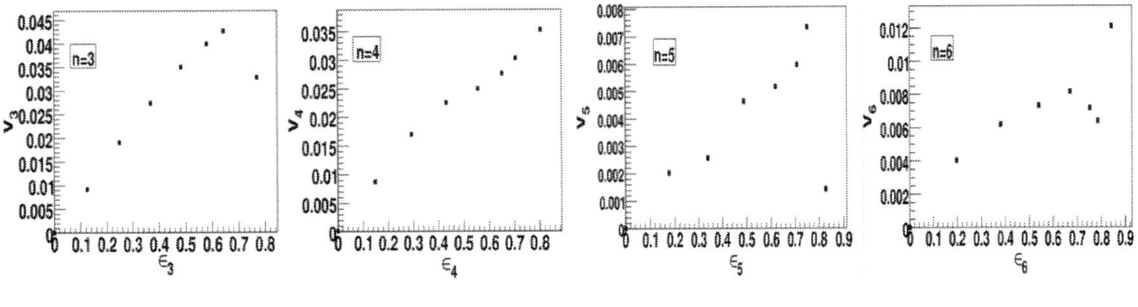

Figure 2. Azimuthal asymmetry v_n vs eccentricity ϵ_n in engineered events for $n = 3, 4, 5, 6$.

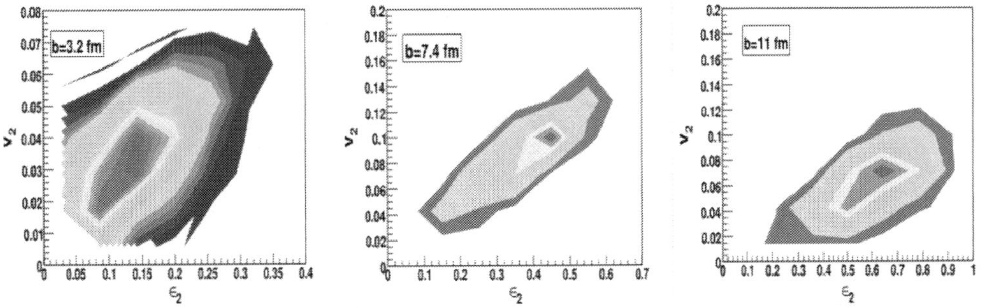

Figure 3. Correlation between v_2 and ϵ_2 for an ensemble of Au+Au collisions at top RHIC energy in GLISSANDO for three impact parameters 3.2, 7.4 and 11 fm.

more weakened to a point that makes it hard to predict v_n for a known ϵ_n. All results shown are for sLPM energy loss but the corresponding result for ASW show no noticable difference, making the conclusions rather robust against large variations in the microscopic origin of energy loss.

We determined the phases δ_n for high momentum pions in our ensemble of GLISSANDO events for three different impact parameters. In Fig. 4 we plot those phases relative to the reaction plane determined by the eccentricity ϵ_2. In other words our definition of a reaction plane is given by the fundamental initial ellipticity, which in general deviates from the plane defined by beam axis and impact vector. We observe that δ_3 and δ_5 are randomly distributed, so there is no correlation between the reaction plane and the fluctuations which create ϵ_3 or ϵ_5. This had also been found for the bulk azimuthal asymmetries before [9]. δ_4 and δ_6 on the other hand show a correlation with the reaction plane which is consistent with v_4 and v_6 receiving contributions from ϵ_2.

The transverse momentum dependence of the coefficients v_n for pions for our ensemble of GLISSANDO Au+Au events is shown in Fig. 5. We observe two hierarchies of coefficients, one being $v_2 > v_4 > v_6$ and the second one being $v_1 > v_3 > v_5$. v_2 is always the largest coefficient, even in the most central events. Interestingly v_1 is non-zero and the second largest coefficient, beating v_3 and v_4 by more than a factor 2. Momentum conservation dicatates a sum rule for v_1 integrated over p_T. The recoil of the medium in which energy is lost would lead to a negative v_1 at lower momentum. Such a back reaction is not included in this calculation. v_1 at intermediate and large p_T could be sensitive to the mechanism of medium recoil. Generally we point out that our results start to become unreliable below 4 to 6 GeV/c since transverse expansion was not included in the calculation. Fig. 5 shows the results for both sLPM and ASW energy

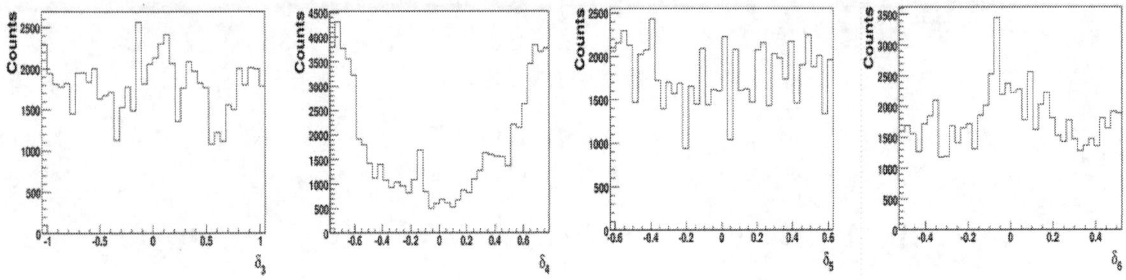

Figure 4. Distribution of the phases δ_n for $n = 3, 4, 5, 6$ in our ensemble of Au+Au events.

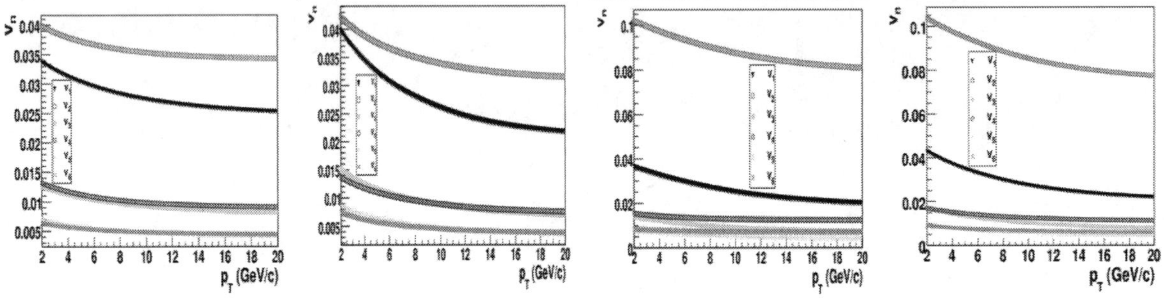

(a) sLPM, $b = 3.2$ fm (b) ASW, $b = 3.2$ fm (c) sLPM, $b = 7.4$ fm (d) ASW, $b = 7.4$ fm

Figure 5. Pion v_n vs p_T for two impact parameters for Au+Au collisions at top RHIC energy, calculated with either sLPM or ASW energy loss.

loss. We have to conclude that azimuthal asymmetry coefficients are not particularly useful to discriminate between energy loss models. We also find that the p_T dependence of the coefficients becomes rather weak at large momenta.

To summarize, in this study we have explored higher order azimuthal asymmetry coefficients at large momentum in heavy ion collisions. We find that in general v_n rises with ϵ_n but this correlation weakens for larger n. We also find that there are only a few cross correlations between v_n and ϵ_m. We have also classified the preferred angular orientation of the azimuthal asymmetries as given by the phases δ_n with respect to the second order participant plane. We find a decorrelation with the reaction plane for all odd n. Finally we have made predictions for v_n as a function of p_T in two different energy loss models. We find mostly consistent results between those two models with v_2 being the largest coefficient followed by v_1. In general the v_n carry important geometrical information and measurements at large momentum can be complementary to those for bulk observables, but they seem less useful to distinguish different energy loss models.

References

[1] B. Alver *et al.* Phys. Rev. C **77**, 014906 (2008)
[2] B. Alver and G. Roland, Phys. Rev. C **81** (2010) 054905 [Erratum-ibid. C **82** (2010) 039903]
[3] G. -Y. Qin, H. Petersen, S. A. Bass and B. Muller, Phys. Rev. C **82**, 064903 (2010)
[4] R. Rodriguez, R. J. Fries and E. Ramirez, Phys. Lett. B **693**, 108 (2010)
[5] R. J. Fries and R. Rodriguez, Nucl. Phys. A **855**, 424 (2011)
[6] C. A. Salgado and U. A. Wiedemann, Phys. Rev. D **68**, 014008 (2003)
[7] A. Dainese, C. Loizides and G. Paic, Eur. Phys. J. C **38**, 461 (2005)
[8] W. Broniowski, M. Rybczynski and P. Bozek, Comput. Phys. Commun. **180**, 69 (2009)
[9] H. Petersen, G. -Y. Qin, S. A. Bass and B. Muller, Phys. Rev. C **82**, 041901 (2010)

Jet measurements in proton-proton collisions with the ALICE experiment at LHC

Michal Vajzer[1,2] on behalf of the ALICE collaboration
[1]Nuclear Physics Institute – Prague, Czech Republic
[2]Faculty of Nuclear Sciences and Physical Engineering, CTU – Prague, Czech Republic

E-mail: michal.vajzer@cern.ch

Abstract. The study of jets, collimated sprays of particles associated with hard partons, is an important tool in testing perturbative quantum chromodynamics (pQCD) and probing hot and dense nuclear matter created in high energy heavy-ion collisions. Jets enable the study of hard scatterings, fragmentation and hadronisation and their modification in the presence of a nuclear medium with respect to baseline vacuum measurements, which is acquired from jet measurements in proton-proton collisions.

We have analysed data from proton-proton collisions at $\sqrt{s}$= 7 TeV measured by the ALICE experiment at the LHC and reconstructed the inclusive spectra of charged particle jets at mid-rapidity using anti-k_T clustering algorithm. We present the jet spectra corrected for detector effects using several unfolding methods. Furthermore, we examine various properties of jets, such as their charged particle multiplicity and jet shapes.

1. Introduction

Jets are collimated sprays of particles created from an energetic parton (quark or gluon), which is produced in the initial stage of the collision. They are an excellent tool to test pQCD and study fragmentation and hadronisation in proton-proton collisions and their modification in collisions of heavy ions. Here, they also serve as probe of hot and dense nuclear matter with an aim of exploring quark-gluon plasma (QGP), which is a challenging task due to an elusive characteristics that requires a combination of various signatures. A Large Ion Colliding Experiment (ALICE) has excellent tracking capabilities and can reconstruct and identify charged particles with transverse momenta down to 150 MeV/c. We report measurements of the charged jet spectra and jet shape observables, such as charged track multiplicities and radial momentum density distributions [1]. These observables are studied at centre of mass energy $\sqrt{s}$= 7 TeV and compared to predictions obtained from Monte-Carlo generators.

2. Data analysis

The results presented in this paper are based on an analysis of the data from proton-proton collisions at the LHC taken in 2010 by the ALICE. The online event selection is done using the V-ZERO (V0) detector and Inner Tracking System (ITS), selecting 160M minimum bias events. From these events, only those with reconstructed primary vertex within 10 cm from the nominal interaction point are further analysed.

For charged jet reconstruction, tracks from the ITS and Time Projection Chamber (TPC) are used. These tracks are selected to be within pseudorapidity $|\eta| \leq 0.9$ and transverse momenta

(p_T) greater than 0.15 GeV/c. They are used as input for FastJet [2] to reconstruct jets using anti-k_T algorithm. It belongs to sequential recombination jet algorithms. Anti-k_T merges hard particles first while k_T starts with soft particles and is therefore more suitable for determination of background. Resolution parameters $R = \sqrt{\Delta\eta^2 + \Delta\varphi^2}$, used in this analysis are 0.2, 0.4 and 0.6. For further analysis, jets had to have p_T of at least 5 GeV/c and be within $|\eta_{jet}| \leq 0.9 - R$.

In order to obtain jet shape observables and charged jet spectra at particle level we have to correct effects of detector reconstruction. We are using PYTHIA [3] (Perugia0 tune) simulated data to acquire spectra on a particle level and a full GEANT simulation of detector to extract detector response matrices linking generator level jet spectra with recontructed spectra. We use 3 different correction procedures.

As PYTHIA reproduces the measured jet observables such as spectra and jet shapes, reasonably well, as seen in [1], we adopted a bin-by-bin correction procedure using simulated events as first method. In this technique, the correction factors (CF) are computed for each jet p_T bin as the ratio between the observable at particle level and the same observable at a detector level in the same p_T bin,

$$\mathrm{CF}\,(p_\mathrm{T}) = \left.\frac{\mathrm{d}\sigma^{\mathrm{ch}}_{\mathrm{jet}}}{\mathrm{d}p_\mathrm{T}\mathrm{d}\eta}\right|^{\mathrm{part}}_{\mathrm{MC}} \Bigg/ \left.\frac{\mathrm{d}\sigma^{\mathrm{ch}}_{\mathrm{jet}}}{\mathrm{d}p_\mathrm{T}\mathrm{d}\eta}\right|^{\mathrm{det}}_{\mathrm{MC}}. \quad (1)$$

This correction is subsequently applied to a raw jet spectrum.

Second method, we used to find an inverse matrix to the response matrix, is Bayesian unfolding [4]. This is possible because the response matrix is composed of a probability distribution of finding given a cause – generated jet of given $p_{\mathrm{T,jet}}^{\mathrm{ch,MC-gen}}$ – as an effect – reconstructed jet with any value $p_{\mathrm{T,jet}}^{\mathrm{ch,MC-rec}}$.

Third method and another regularized unfolding procedure used as a cross-check in jet spectrum analysis is the Singular Value Decomposition of the response matrix [5]. This method decomposes the response matrix into two square orthogonal matrices and a diagonal singular matrix. Subsequently it converts the problem to a system of linear equations and solves it.

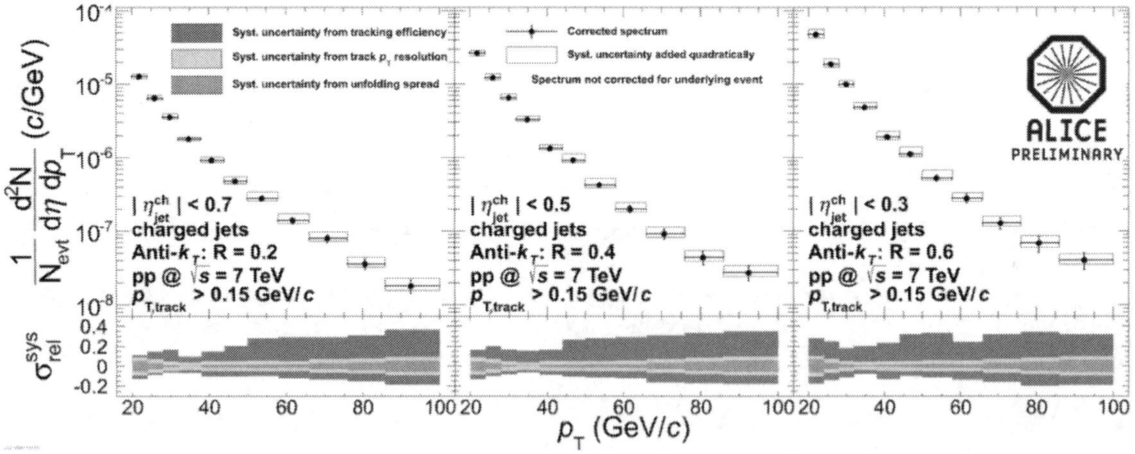

Figure 1. (Color online) Charged jet spectra corrected for detector effects reconstructed using anti-k_T algorithm with resolution parameter R (left) 0.2; and (middle) 0.4; and (right) 0.6; from pp collisions at $\sqrt{s}$= 7 TeV. Error bars indicate statistical uncertainties. In bottom part are relative systematic uncertainties obtained from different sources for the anti-k_T algorithm with corresponding R.

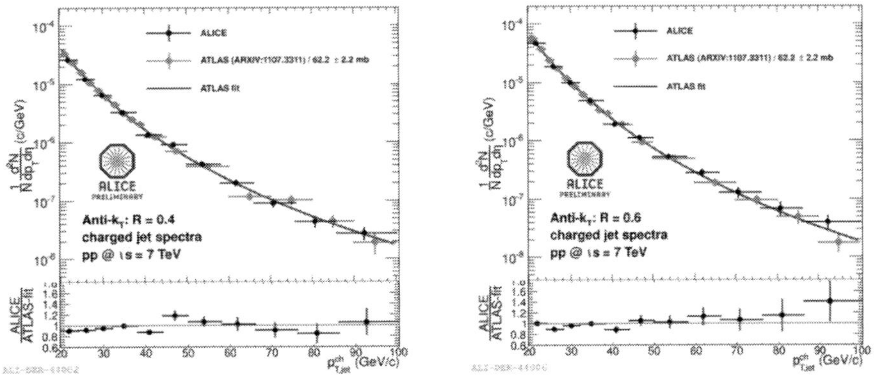

(a) Anti-k_{T}; R = 0.4 (b) Anti-k_{T}; R = 0.6

Figure 2. (Color online) Charged jet spectra from anti-k_{T} algorithm with resolution parameter R of (a) 0.4; and (b) 0.6; from pp collisions at $\sqrt{s}$= 7 TeV measured by the ALICE (black points) and the ATLAS (red points) collaborations. The line is a Tsallis fit of the ATLAS results. Error bars indicate statistical uncertainties. In bottom part is ratio of the ALICE charged jet spectrum to the ATLAS results.

3. Results

In Fig. 1 we present the unfolded charged jet spectra for anti-k_{T} algorithm with resolution parameters R of 0.2, 0.4 and 0.6 in top panel, with the relative systematic uncertainties in the bottom panel of each plot, reconstructed from proton-proton collisions at $\sqrt{s}$= 7 TeV using the ALICE detector system. The most significant uncertainties are obtained from varying track reconstruction effiency, track p_{T} resolution and simulation with parametrised detector response. The final systematic uncertainties are obtained by adding above mentioned uncertainties in quadrature.

The results presented here are consistent with results obtained by the ATLAS collaboration [6], as shown in the Fig 2, obtained from charged tracks in $|\eta| \leq 0.5$ with $p_{\mathrm{T}} \geq 0.3$ GeV/c.

Fig. 3 shows jet shape variables. The charged track multiplicity of a jet, defined as a mean number of charged tracks in the leading jet, i.e. one with the largest transverse momentum in the event, belonging to given $p^{\mathrm{ch}}_{\mathrm{T,jet}}$ bin, is obtained from charged tracks used for the jet reconstruction within a jet cone. The mean track multiplicity in such jets obtained from the analysed data is shown in Fig 3(a), with respect to a charged jets' transverse momentum bin. Increasing value of average number of tracks is observed with increasing jet p_{T} bin, as previously measured by the CDF collaboration [8]. Also, predictions obtained from the PYTHIA (Perugia0 tune), PYTHIA (Perugia11 tune) and PHOJET [7] are shown here. Results from data differ by 10% from values obtained from Monte-Carlo predictions.

The radial momentum density is defined for jets of given $p^{\mathrm{ch}}_{\mathrm{T,jet}}$ bin as a sum of momenta of charged tracks $p^{\mathrm{ch}}_{\mathrm{T,track}}$ belonging to the leading jet with respect to their distance from the jet axis in $\eta - \varphi$ plane, normalized by number of jets,

$$\frac{\mathrm{d}p^{\mathrm{sum}}_{\mathrm{T}}}{\mathrm{d}r}(r) = \frac{1}{2\Delta r}\frac{1}{\mathrm{N_{jet}}}\sum_{\mathrm{jets}} p^{\mathrm{ch}}_{\mathrm{T,jet}}(r\text{ - }\Delta r,\ r + \Delta r)\,, \tag{2}$$

where $p^{\mathrm{ch}}_{\mathrm{T,jet}}$ (r -Δ r , r +Δ r) denotes the sum of $p^{\mathrm{ch}}_{\mathrm{T,track}}$ of all charged tracks which are displaced from jet axis by $r \pm \Delta r$ of given jet. N_{jet} is number of leading jets in analysis.

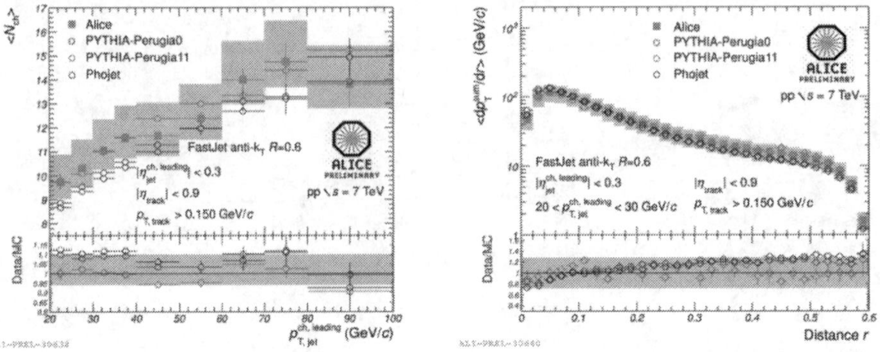

(a) Charged track multiplicity (b) Radial momentum distribution

Figure 3. (Color online) (a) Mean charged track multiplicities of leading jet as a function of $p^{\text{ch}}_{\text{T,jet}}$. (b) The radial momentum distribution from a jet axis in leading charged jets in $20 \leq p^{\text{ch}}_{\text{T,jet}} \leq 30$ GeV/c momentum bin. Results from pp collision at $\sqrt{s}=$ 7 TeV using the ALICE (magenta squares) and predictions from PYTHIA (Perugia0 tune) (blue circles), PYTHIA (Perugia11 tune) (red circles) and PHOJET (black circles) are shown with thes ratio of data and predictions from Monte-Carlo simulations in the bottom part. Error bars indicate statistical uncertainties and gray bands systematic uncertainties.

In the Fig 3(b) we observe a decreasing radial momentum density with decreasing distance from a jet axis. This suggests that most of the jet energy is concentrated around a jet axis. Comparison of the slopes of two different jet p_{T} bins at a small distances from jet axis indicate that high energy jets are more collimated than jets with smaller energy.

4. Conclusions

We reported measurements of inclusive charged jet spectra in pp collisions at midrapidity, charged track multiplicities within leading jet and the radial momentum density for jets reconstructed using anti-k_{T} algorithm. The reconstructed jet spectra are consistent with results obtained from the ATLAS collaboration and jet shape observables follow the trend expected from previous results of the CDF collaboration and are in reasonable agreement with predictions from PYTHIA and PHOJET calculations.

Acknowledgements

This work was supported by the Grant Agency of the Czech Technical University in Prague, grant No. SGS10/292/OHK4/3T/14 and by Ministry of education, youth and sports of Czech Republic, grant No. LA08015.

References

[1] S. K. Prasad [ALICE Collaboration], J. Phys. Conf. Ser. **389**, 012005 (2012).
[2] M. Cacciari, G. P. Salam and G. Soyez, Eur. Phys. J. C **72** (2012) 1896.
[3] T. Sjstrand, S. Mrenna and P. Skands, JHEP05 (2006) 026.
[4] G. D'Agostini,NIM A 362 (1995) 487.
[5] A. Höcker and V. Kartvelishvili, NIM A 372 (1996) 469.
[6] G. Aad *et al.* [ATLAS Collaboration], Phys. Rev. D **84** (2011) 054001
[7] S. Roesler, R. Engel and J. Ranft, hep-ph/0012252.
[8] T. Affolder *et al.* [CDF Collaboration], Phys. Rev. D 65, 092002 (2001).

Measurement of jet p_T spectra in Pb–Pb collisions at $\sqrt{s_{NN}} = 2.76$ TeV with the ALICE detector at the LHC

Salvatore Aiola[1,2]

[1] University of Catania and INFN Sezione di Catania, 95123 Catania, Italy
[2] Lawrence Berkeley National Laboratory, Berkeley, CA 94720

E-mail: Salvatore.Aiola@cern.ch

Abstract. Reconstruction of jets in high-energy heavy-ion collisions is challenging due to the large and fluctuating background coming from the underlying event. We report results on full jet reconstruction, obtained from data collected in 2011 by the ALICE detector at LHC for Pb–Pb collisions at $\sqrt{s_{NN}} = 2.76$ TeV. The analysis makes use of the tracking system and the electromagnetic calorimeter. Signal jets, which come from hard scattered partons, are reconstructed using the anti-k_T jet finder algorithm. The average background is subtracted on a jet-by-jet basis to reduce the contribution to the jet reconstructed energy coming from the underlying event. The jet spectrum is corrected to account for fluctuations in the background momentum density and detector effects through unfolding.

1. Introduction

QCD jets are produced in high-energy particle collisions as a result of the fragmentation of a high momentum scattered parton. Experimentally they are reconstructed using a well-defined algorithm, which acts as a working definition of a jet, to be used consistently also in phenomenological models. In heavy-ion collisions, hard scattered partons are produced in the early stages of the collision, so that they propagate through (and potentially are affected by) the hot and dense nuclear medium. The interactions suffered by the parton can result in energy loss, widening and/or complete absorption of jets. These phenomena go under the name of "jet quenching". RHIC and LHC experiments have already collected convincing evidence of jet quenching in a number of measurements, such as the suppression of high-p_T particles relative to a pp baseline [1, 2, 3, 4], hadron-hadron correlations [5, 6] and jet-jet correlations at high-p_T [7, 8]. However, performing a full jet reconstruction in the low and intermediate p_T region is particularly challenging due to the overwhelming soft particle background coming from the underlying event (UE). ALICE has recently reported on a first detailed study of the background for jet reconstruction in the heavy-ion environment [9].

In these proceedings the analysis techniques utilized to fully reconstruct jets in Pb–Pb collisions with the ALICE experiment are presented. The results shown here come from data collected by ALICE in fall 2011 at $\sqrt{s_{NN}} = 2.76$ TeV.

2. Inputs to the jet finder

The ALICE tracking system benefits of the Inner Tracking System, a six-layer silicon detector which provides a precise measurement of the primary vertex together with the first points of the

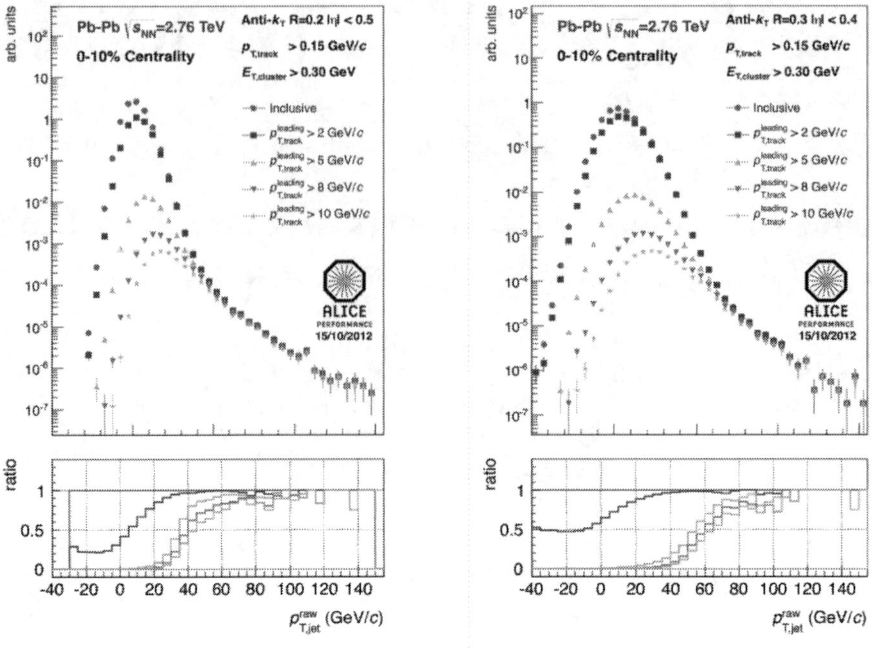

(a) $R = 0.2$. (b) $R = 0.3$.

Figure 1: Raw jet p_{T} spectra in Pb–Pb collisions at $\sqrt{s_{\mathrm{NN}}} = 2.76$ TeV in the 0-10% centrality range at mid-rapidity: inclusive is shown in full circles and with increasing minimum leading hadron p_{T} requirement in the other symbols (color online). Jets are reconstructed using the anti-k_{T} algorithm with two different resolution parameters, $R = 0.2$ (left) and $R = 0.3$ (right).

tracks, and of a large Time Projection Chamber [10]. Tracks are reconstructed at mid-rapidity ($|\eta| < 0.9$) and in full azimuth. The ALICE electromagnetic calorimeter [11] is a Pb-scintillator sampling calorimeter, which covers mid-rapidity ($|\eta| < 0.7$) and partial azimuth ($\Delta\varphi = 100°$). The EMCal measures photons, e.g. from π^0 decays, which are included in the jet finder input. The shift of the jet energy scale due to unreconstructed particles, such as K^0_L and neutrons, has been studied in pp simulations and accounted for in the final result. A full description of the ALICE experiment is available in Ref. [10].

3. Jet finding and average background

The anti-k_{T} [12] jet finding algorithm has been employed in its `FastJet` [13] implementation. This sequential recombination algorithm has the advantage of being "soft-resilent", namely it is little affected by the soft background [14]. Anti-k_{T} jets are pretty regular, cone-like shaped around some high-p_{T} particle. The uncorrelated energy density from soft processes is large in heavy-ion events and needs to be subtracted on an event-by-event basis. The main problem in the estimation of the average background is the separation of the UE from the hard scattering. The method used here has been proposed in Ref. [15]. The average background ρ is calculated, event-by-event, as the median of the p_{T} density (jet p_{T} over jet area) of the k_{T} algorithm [16] reconstructed jets. A jet-by-jet subtraction is performed: $p^{\mathrm{raw}}_{\mathrm{T,jet}} = p^{\mathrm{uncorr}}_{\mathrm{T,jet}} - \rho \times A_{\mathrm{jet}}$, where $p^{\mathrm{uncorr}}_{\mathrm{T,jet}}$ and A_{jet} are respectively the transverse momentum and the area of the jet. Combinatorial jets, i.e. jets reconstructed out of the soft background, are efficiently removed by requiring a minimum leading hadron p_{T}. Figure 1 shows the anti-k_{T} raw jet p_{T} spectra obtained applying different

Journal of Physics: Conference Series **446** (2013) 012005 doi:10.1088/1742-6596/446/1/012005

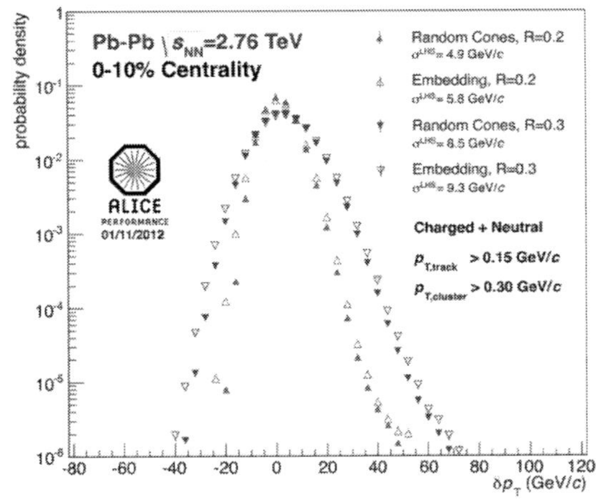

Figure 2: δp_{T} distributions for Pb–Pb collisions at $\sqrt{s_{\mathrm{NN}}} = 2.76$ TeV in the 0-10% centrality class. Two different methods have been implemented: "random cones", shown with full symbols, and single particle embedding, shown with open symbols (see text for details); the distributions for $R = 0.2$ are narrower w.r.t. $R = 0.3$ (color online).

minimum leading hadron p_{T} requirement, from 0 (inclusive) to 10 GeV/c, for two different jet resolution parameters $R = \sqrt{\Delta\eta^2 + \Delta\varphi^2}$.

4. Unfolding

The spectra shown in Fig. 1 are not directly comparable with model predictions. In fact the measured values of the observable, the jet p_{T}, are subject to random fluctuations, due to region-to-region differences in the background momentum density and to the detector response. This means that each observation is characterized by a true (and unknown) value $p_{\mathrm{T,jet}}^{\mathrm{true}}$, and by a measured value $p_{\mathrm{T,jet}}^{\mathrm{meas}}$. The histograms for $p_{\mathrm{T,jet}}^{\mathrm{true}}$ and $p_{\mathrm{T,jet}}^{\mathrm{meas}}$ are related by a convolution through the response matrix $\mathrm{RM_{tot}} = \mathrm{RM_{det}} \times \mathrm{RM_{bkg}}$, where $\mathrm{RM_{det}}$ parametrizes the detector reponse whereas $\mathrm{RM_{bkg}}$ describes background fluctuations. Unfolding is the numerical procedure that allows to get back the true distribution given the measured one and the response matrix $\mathrm{RM_{tot}}$.

Background fluctuations have been estimated in two different ways [9], namely using random cones (scalar sum of the p_{T} of all particles found in a cone randomly placed in the event) and single particle embedding (with the anti-k_{T} algorithm). The residual p_{T} differences due to region-to-region fluctuations are calculated as: $\delta p_{\mathrm{T}} = p_{\mathrm{T,jet}} - p_{\mathrm{T,probe}} - \rho \times A_{\mathrm{jet}}$, where $p_{\mathrm{T,probe}}$ is the p_{T} of the embedded probe ($p_{\mathrm{T,probe}} = 0$ for random cones). The δp_{T} distributions, shown in Fig. 2, tell us how much the jet p_{T} is smeared due to background fluctuations.

The detector response to jet reconstruction has been studied with pp simulated events, using the PYTHIA6 [17] generator and the GEANT3 [18] transport code. Jets are reconstructed both at generator level and at detector level. The generator-level and detector-level jets are matched following a geometrical criterion.

5. Results

This measurement makes use of the unfolding iterative method proposed by D'Agostini [19], which contains elements of Bayesian statistics. In Bayesian unfolding the number of iterations plays the role of the regularization parameter. Based on the evolution of the covariance matrix and the converging of the solution itself, a number of four iterations has been chosen as default. Indeed after three/four iterations the procedure starts to converge. However, with more iterations a characteristic fluctuation pattern does appear along with larger variances and (anti-)correlations between far bins (indicating under-regularization). Systematic uncertainties have been estimated taking into account variations of the solution for ± 1 iterations ($\sim 5\%$).

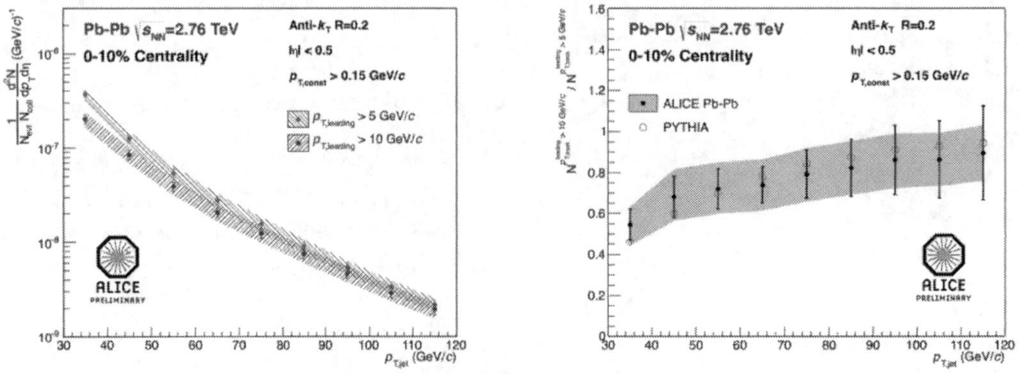

Figure 3: On the left: spectra of reconstructed jets in Pb–Pb collisions at $\sqrt{s_{\mathrm{NN}}} = 2.76$ TeV in the 0-10% centrality class with two different minimum leading hadron p_{T} requirements. On the right: ratio of the two spectra (solid circles) and comparison with a PYTHIA pp expectation (open circles). The bands represent the systematic uncertainty; statistical uncertainties are shown with bars, when they are larger than markers (colors online).

To better understand the effect of the leading hadron requirement in the jet p_{T} spectrum, the analysis was performed for $p_{\mathrm{T,leading}} > 5,\ 10$ GeV/c. Figures 3 show the two spectra with the two different leading hadron thresholds, and the ratio, compared to a PYTHIA expectation.

6. Conclusions

We have reported on the analysis techniques utilized to reconstruct jets in the 10% most central Pb–Pb collisions at $\sqrt{s_{\mathrm{NN}}} = 2.76$ TeV recorded by ALICE in 2011. The raw jet p_{T} spectra are corrected for the average background and biased requiring a minimum leading hadron p_{T}. Corrections for background fluctuations and detector effects are applied via Bayesian unfolding.

The effect of the leading hadron requirement was studied for two different thresholds, 5 GeV/c and 10 GeV/c. The ratio between the two corrected spectra is in reasonable agreement with a PYTHIA simulation, which indicates a vacuum-like fragmentation of the jet core.

References

[1] Adams J *et al.* (STAR Collaboration) 2003 *Phys. Rev. Lett.* **91** 172302
[2] Adler S *et al.* (PHENIX Collaboration) 2004 *Phys. Rev.* C **69** 034910
[3] Aamodt K *et al.* (ALICE Collaboration) 2011 *Physics Letters* B **696** 30 ISSN 0370-2693
[4] Chatrchyan S *et al.* (CMS Collaboration) 2012 *The European Physical Journal* C **72**(3) 1–22 ISSN 1434-6044
[5] Adams J *et al.* (STAR Collaboration) 2006 *Phys. Rev. Lett.* **97** 162301
[6] Aamodt K *et al.* (ALICE Collaboration) 2012 *Phys. Rev. Lett.* **108**(9) 092301
[7] Aad G, *et al.* (ATLAS Collaboration) 2010 *Phys. Rev. Lett.* **105**(25) 252303
[8] Chatrchyan S *et al.* (CMS Collaboration) 2011 *Phys. Rev.* C **84**(2) 024906
[9] Abelev B *et al.* (ALICE Collaboration) 2012 *JHEP* **1203** 053
[10] Aamodt K and other (ALICE Collaboration) 2008 *Journal of Instrumentation* **3** S08002
[11] Bellwied R (for the ALICE EMCal Collaboration) 2010 *Arxiv preprint arXiv:1008.0413*
[12] Cacciari M, Salam G and Soyez G 2008 *JHEP* **04** 063
[13] Cacciari M, Salam G and Soyez G 2011 *Arxiv preprint arXiv:1111.6097*
[14] Cacciari M, Rojo J, Salam G and Soyez G 2011 *The European Physical Journal* C **71** 1–21
[15] Cacciari M and Salam G 2008 *Physics Letters* B **659** 119–126
[16] Ellis S D and Soper D E 1993 *Phys. Rev.* D **48** 3160
[17] Sjöstrand T, Mrenna S and Skands P 2006 *Journal of High Energy Physics* **5** 1–581
[18] Geant3 webpage, http://wwwasd.web.cern.ch/wwwasd/geant/
[19] D'Agostini G 1995 *NIM* A **362** 487

Journal of Physics: Conference Series **446** (2013) 012006 doi:10.1088/1742-6596/446/1/012006

Full Jet Reconstruction in 2.76 TeV pp and Pb-Pb collisions in the ALICE experiment

Rosi Reed
Yale University, New Haven, CT 06520
E-mail: Rosi.Reed@Yale.edu

Abstract. Measurements of the suppression of high-p_T particles and the away-side jets from heavy-ion collisions at RHIC have shown that medium-induced energy loss affects partons produced in the early stage of a heavy-ion collision. At LHC energies the initial production cross-section is much higher, which allows jets to be reconstructed with a wide kinematic range. Measuring fully reconstructed jets by taking advantage of the ALICE Electromagnetic Calorimeter allows for a more differential investigation of the parton energy loss. Parton energy loss will allow us to access key observables of the hot, dense matter created in heavy ion collisions. The data presented was collected during the 2.76 TeV Pb-Pb runs, as well as baseline measurements from the 2.76 TeV pp run. The procedures used to reconstruct jets and extract them from a fluctuating background will be discussed. The procedure for quantifying the background with a limited acceptance will also be discussed. These results are compared to pp measurements and simulations.

1. Introduction

One of the aims of studying heavy-ion collisions is to investigate how partons lose energy in the quark gluon plasma (QGP) that is formed in these collisions. Hard scattered partons will fragment into a spray of hadrons, which is a jet. The cross-section and fragmentation of jets has been extensively studied in proton-proton collisions so that these measurements can be a reference for heavy-ion collisions. In a heavy-ion collision, hard-scattered partons are produced early in the collision, which makes them an ideal probe of the QGP and for studying energy loss within the medium. This parton energy loss will be reflected in the measured heavy-ion jet spectrum so that it is suppressed relative to a pp reference, which we call jet quenching [1]. High-p_T particle measurements, a proxy for jets, as well as other jet measurements have shown that the jet spectrum in heavy-ion collisions does deviate from what would be expected if the heavy-ion collision could be treated as a superposition of independent pp collisions [2, 3, 4, 5, 6, 7, 8].

With the added kinematic range at LHC energies and with improvements in jet finding techniques, it is now possible to fully reconstruct jets in heavy-ion collisions, including the charged component and the neutral components from π^0s, which should correspond better to the kinematics of the scattered parton than either charged jets or single particle measurements can. The results reported in these proceedings are from data collected by the ALICE experiment in 2011 at an energy of 2.76 TeV per nucleon pair. Charged particles were reconstructed with the Time Projection Chamber (TPC) and the Inner Tracking System (ITS), while the neutral energy component from π^0s was reconstructed with the Electromagnetic Calorimeter (EMCal).

The Pb-Pb results are from the top 10% most central events where the modification of the jet spectrum due to the QGP is at a maximum.

2. Jets at ALICE

The cells of the EMCal were clustered prior to inclusion in the jet finder by an algorithm that requires that each cluster only have a single local maximum. Since charged particles, such as electrons or charged pions, will leave an energy signature in the EMCal the clusters were corrected for the hadronic contamination in order to prevent double counting of the charged energy. All tracks with $p_{\mathrm{T}}> 0.15$ GeV/c were propagated to the EMCal surface and then matched to clusters with $E_{\mathrm{T}}> 0.15$ GeV in the window $|\Delta\eta| <0.015$ and $|\Delta\varphi| < 0.03$. Each track was only matched to at most one cluster, though each cluster could have multiple tracks matched to it. Then 100% of the sum of the momenta of all matched tracks was removed from the cluster energy and then those clusters with $E_{\mathrm{T}}< 0.30$ GeV after hadronic correction were removed from the sample [9].

The collection of tracks and corrected EMCal clusters were then assembled into jets using the anti-k_{T} or the k_{T} algorithms [10, 11, 12, 13] with a resolution parameter of $R = 0.2$. Only those jets that were at least R away from the EMCal boundaries of $|\eta| < 0.7$ and $1.4 < \phi < \pi$ and thus fully contained within the EMCal acceptance were used in the analyses described here. The jets found by the anti-k_{T} algorithm were used to determine the signal jets in both the pp and Pb-Pb analyses, the jets found by the k_{T} algorithm were used to quantify the underlying event density in the Pb-Pb analysis.

3. pp

In order to fully quantify the modification of the jet spectrum in heavy-ion collisions due to the presence of the hot and dense QGP medium, the differential cross-section of fully reconstructed jets in pp collisions at 2.76 TeV was measured. Two different triggers were used to measure this spectrum, a minimum bias (MB) trigger and a EMCal trigger that required a single shower with $E_{\mathrm{T}} > 3$ GeV. The combination allowed the pp jet spectrum to be measured over a momentum range of 20 to 125 GeV/c. The details of this trigger scheme can be found in [14].

The measured jets were corrected back to the particle level utilizing a simulation based bin-by-bin technique. This was done by comparing the particle level jet spectra from a PYTHIA simulation to that the detector level spectrum, where the detector corrections were applied by use of GEANT. The contributions to the correction can be divided into two categories: effects that smear the jet energy and effects that shift the jet energy. The effects that shift the jet energy are the hadronic correction, the tracking efficiency and unmeasured neutral particles. The effects that smear the jet energy are the event-by-event fluctuations of the jet energy shift and the resolution of the detectors in measuring the jet constituents. In this analysis, the jet energy shift was approximately 20% with an uncertainty less than 4% [14]. Figure 1 shows the inclusive differential jet cross section obtained with $R = 0.2$ compared to a pQCD calculation at NLO and a PYTHIA8 prediction [14]. The calculations and simulations agree well with the measured spectrum which indicate that the process of jet formation is well understood in collisions where a QGP is not formed.

4. Pb-Pb

One of the main experimental challenges in heavy-ion collisions is removing the contribution from the underlying event from the jet spectrum. This is done by determining the average background energy density, $\rho_{\mathrm{ch+em}}$, by finding the median of $p_{T,jet}/A_{jet}$, where A_{jet} is the area and $p_{T,jet}$ is the uncorrected momentum of the k_{T} jets. This background was subtracted from the reconstructed momentum of the anti-k_{T} jets using the formula $p_{\mathrm{T},jet} = p_{\mathrm{T}}^{rec} - \rho \cdot A_{jet}$ [12, 13]. However this does not account for the effect of the point-to-point background fluctuations which smear the jet energy. These fluctuations are quantified by $\delta p_{\mathrm{T}} = p_{\mathrm{T}}^{rec} - \rho \cdot A_{jet} - p_{\mathrm{T}}^{probe}$ [15]. The width of the distribution was determined to be 6.1 GeV/c using a random cone method, and the effect of the smearing of the jet energy due to these background fluctuations is corrected through unfolding [9, 16].

Unfolding will also correct the measured background subtracted jet spectrum for detector effects as outlined in Section 3. In order to have a stable result it was necessary to bias the collection of measured jets by requiring that the jets have a track with $p_{\mathrm{T}} > 5$ GeV/c. This reduces the contribution to the measured spectrum due to the combinatorial background. The resulting spectrum is shown in Figure 2. The systematic uncertainty is p_{T} dependent with a value of 18% at 60 GeV/c, and dominated by the hadronic correction, the tracking efficiency and the unfolding process itself. Also see [16] for more details.

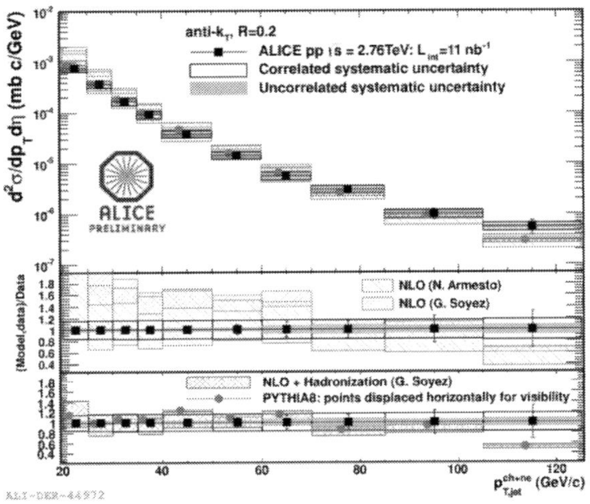

Figure 1. The solid boxes indicate the full inclusive jet differential cross section obtained with resolution parameter $R = 0.2$ measured in pp collisions at 2.76 TeV data [14]. The solid circles represent a PYTHIA8 prediction. The solid boxes are the uncorrelated systematic uncertainties on the cross-section, the open boxes are the correlated systematic uncertainties. The hashed boxes indicate the NLO calculations (color online).

5. R_{AA}

The nuclear modification factor, R_{AA}, is defined as the ratio of the jet spectrum from Pb-Pb collisions as shown in Figure 2 to the differential cross-section from pp scaled by T_{AA}, where T_{AA} is the ratio of the number of binary collisions over the total inelastic pp cross-section. R_{AA} is constructed so that it would be equal to one if there is no suppression or enhancement of a signal in heavy ion collisions as compared to pp collisions. However since a bias was necessary to measure the jet spectrum in Pb-Pb collisions, if the unbiased reference spectra shown in Figure 1 was used, R_{AA} could be constructed as less than 1 even if there is no modification of the heavy ion jet spectrum due to nuclear effects. In order to avoid this, a biased pp reference spectrum was constructed by applying the same method outlined in [14]. The resulting R_{AA} can be seen in Figure 2. The systematic and statistical uncertainties from the Pb-Pb and pp analysis were added in quadrature. We have found that fully reconstructed jets are suppressed in the 10% most central heavy-ion collisions in a p_{T} dependent manner. This measurement is not inconsistent with conservation of energy as the energy lost by the jets could be recovered elsewhere, for

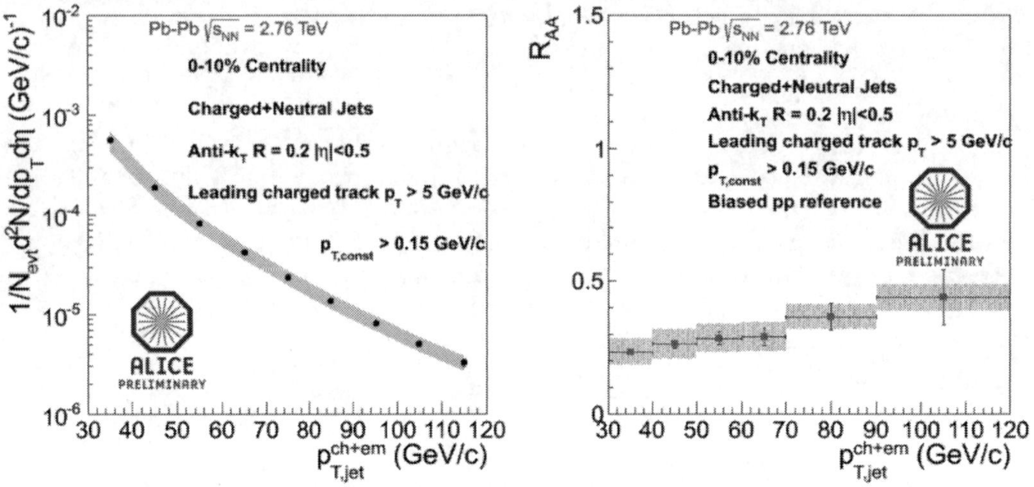

Figure 2. On the left is the fully reconstructed jet spectrum from 0-10% central events in Pb-Pb collisions measured by ALICE at $\sqrt{S_{NN}} = 2.76$ TeV. The jets were constrained to the ALICE EMCal and biased so that their leading track had $p_T > 5$ GeV/c. The band around the data points represents the systematic uncertainty. The R_{AA} for $R = 0.2$ biased jets in 0-10% central events is on the right. The statistical and systematic uncertainties from the pp and Pb-Pb analyses are added in quadrature. The combined systematics are shown as dotted boxes.

instance at a larger R or at a lower momentum. These results agree well the the previous charged only results [17], though the energy scale is not the same in both cases so a direct comparison is not possible.

6. Conclusions

In this proceedings we have reported on a corrected fully reconstructed jet spectrum from the 2011 Pb-Pb data from the 10% most central evens in ALICE. The spectra combined with the differential cross section measured in pp collisions allowed us to determine the nuclear modification factor. We have seen that $R = 0.2$ jets are suppressed in these most central events and that this suppression has a p_T dependence. This result is still consistent with conservation of energy laws as the lost energy may be recoverable at larger angles or lower momentum than what are measuring in this analysis. Additional measurements, such as determining the dependence of the nuclear modification factor with centrality, event plane and R, will be necessary to fully quantify the measured jet quenching. These measurement can then be compared to available energy loss models. These results are consistent with the CMS $R = 0.2$ results from the 5% most central events [18], though there are variations between the analyses such as the background determination. Jet analyses in heavy-ion collisions are experimentally difficult, in particular the background subtraction and correction of fluctuations can be performed in different ways. Therefore it is important for all of the LHC heavy-ion experiments to converge upon a single consistent jet measurement.

References

[1] U. A. Wiedemann, arXiv:0908.2306 [hep-ph].
[2] J. Adams *et al.* [STAR Collaboration], Phys. Rev. Lett. **91**, 172302 (2003) [nucl-ex/0305015].
[3] K. Adcox *et al.* [PHENIX Collaboration], Phys. Lett. B **561**, 82 (2003) [nucl-ex/0207009].
[4] S. Chatrchyan *et al.* [CMS Collaboration], Eur. Phys. J. C **72**, 1945 (2012) [arXiv:1202.2554 [nucl-ex]].

[5] C. Adler *et al.* [STAR Collaboration], Phys. Rev. Lett. **90**, 082302 (2003) [nucl-ex/0210033].
[6] K. Aamodt *et al.* [ALICE Collaboration], Phys. Lett. B **696**, 30 (2011) [arXiv:1012.1004 [nucl-ex]].
[7] A. Adare *et al.* [PHENIX Collaboration], Phys. Rev. Lett. **98**, 232302 (2007) [nucl-ex/0611019].
[8] G. Aad *et al.* [ATLAS Collaboration], arXiv:1208.1967 [hep-ex].
[9] R. Reed for the ALICE Collaboration, arXiv:1211.5016 [nucl-ex].
[10] M. Cacciari, G. P. Salam and G. Soyez, JHEP **0804**, 063 (2008) [arXiv:0802.1189 [hep-ph]].
[11] M. Cacciari and G. P. Salam, Phys. Lett. B **641**, 57 (2006) [hep-ph/0512210].
[12] M. Cacciari and G. P. Salam, Phys. Lett. B **659**, 119 (2008) [arXiv:0707.1378 [hep-ph]].
[13] M. Cacciari, J. Rojo, G. P. Salam and G. Soyez, Eur. Phys. J. C **71**, 1539 (2011) [arXiv:1010.1759 [hep-ph]].
[14] R. Ma [ALICE Collaboration], arXiv:1208.4537 [nucl-ex].
[15] B. Abelev *et al.* [ALICE Collaboration], JHEP **1203**, 053 (2012) [arXiv:1201.2423 [hep-ex]].
[16] S. Aiola for the ALICE Collaboration, in these proceedings
[17] M. Verweij [ALICE Collaboration], arXiv:1208.6169 [nucl-ex].
[18] The CMS Collaboration, CMS-PAS-HIN-12-004

Hot Quarks 2012 IOP Publishing
Journal of Physics: Conference Series **446** (2013) 012007 doi:10.1088/1742-6596/446/1/012007

Study of jet quenching using di-jet and photon-jet events in *PbPb* collisions at 2.76 TeV with CMS

Yongsun Kim

24-519 Heavy Ion Group, 77 Massachusetts Avenue, Cambridge, MA, USA

E-mail: jazzitup@mit.edu

Abstract. The first measurement of the transverse momentum imbalance of isolated-photon-jet pairs and the modification of fragmentation function of the inclusive jets in relativistic heavy ion collisions are reported. The analysis uses data from *PbPb* collisions at a center-of-mass energy of 2.76 TeV per nucleon pair and corresponding to an integrated luminosity of 150 μb^{-1} recorded by the CMS experiment at the LHC in 2011. For events containing an isolated photon with transverse momentum $p_T > 60$ GeV/c and an associated jet with $p_T > 30$ GeV/c, the photon-jet p_T imbalance is studied as a function of collision centrality and compared to *pp* data and PYTHIA calculations at the same center-of-mass energy. Using the p_T of the isolated photon as an estimate of the energy of the associated parton at production, this measurement allows an unbiased characterization of the in-medium parton energy loss. In addition, further study on the modification of jet fragmentation function was studied using inclusive jets with $p_T >$ 100 GeV/c. Both show the gradual centrality dependence of jet energy loss and fragmentation function modification.

1. Introduction

These proceedings can be divided into two subjects. The first half shows the study of the jet energy loss using isolated-photon-jet pairs from *PbPb* data taken by CMS detector [1]. The result was compared to *pp* reference dataset with $\int \mathcal{L}dt = 230\ \mathrm{nb}^{-1}$ at $\sqrt{s} = 2.76$ TeV which was obtained in 2011. The properties of isolated-photon-jet pairs are studied via the azimuthal angular correlation in$\Delta\ \phi_{J\gamma} = |\phi - \phi^{\gamma}|$ and the transverse momentum ratio given by $x_{J\gamma} = p_T^{Jet}/p_T^{\gamma}$. Photons with transverse momentum of $p_T^{\gamma} > 60$ GeV/c are selected in a pseudorapidity range of $|\eta^{\gamma}| < 1.44$, using isolation criteria. These photons are then correlated with jets having $p_T > 30$ GeV/c and $|\eta| < 1.6$ [2].

In the second half, the measurement of fragmentation function for inclusive jets above 100 GeV/c is discussed [3]. The fragmentation properties of jets connect the perturbatively calculable production of high-p_T quarks and gluons with the hadronized final state particles. The study of jet production via final state particles relies on hadronization models to determine e.g. the non-perturbative hadronization corrections of the production cross section, and also the experimental jet energy scale. It also provides an important experimental check on the validity of the assumed jet fragmentation.

2. Photon-jet correlations

2.1. Azimuthal angle correlations

Possible medium effects on the back-to-back alignment of the photon and recoiling jet can be studied using the distribution of the number of photon-jet pairs, $N_{J\gamma}$, as a function of the relative azimuthal angle,$\Delta\ \phi_{J\gamma}$, normalized by the total number of pairs, $(N_{J\gamma})^{-1}\mathrm{dN_{J\gamma}}/\mathrm{d}\Delta\phi_{J\gamma}$. The *PbPb* data are compared to PYTHIA + HYDJET simulation and *pp* data. For both *PbPb* data and MC distributions, the jet is found to be well aligned opposite to the photon direction, with a clear peak at$\Delta\ \phi_{J\gamma} = \pi$. The shape of the$\Delta\ \phi_{J\gamma}$ correlation peak is similar in *PbPb* data and MC. An excess in the tail of the 0–10% centrality bin was found. However, further investigation showed that it was not inconsistent with PYTHIA + HYDJET within the given statistics.

To study the centrality evolution of the shape, the distributions are fitted to a normalized exponential function:

$$\frac{1}{N_{J\gamma}}\frac{\mathrm{dN_{J\gamma}}}{\mathrm{d}\Delta\phi_{\mathrm{J}\gamma}} = \frac{\mathrm{e}^{(\Delta\phi-\pi)/\sigma}}{(1-\mathrm{e}^{-\pi/\sigma})\,\sigma}. \quad (1)$$

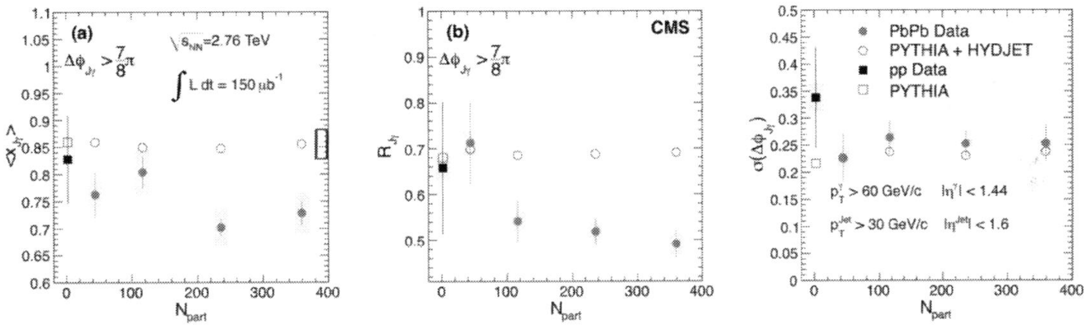

Figure 1. (Left) Average ratio of jet transverse momentum to photon transverse momentum as a function of N_{part}. The empty box at the far right indicates the correlated systematic uncertainty. (Middle) Average fraction of isolated photons with an associated jet above 30 GeV/c as a function of N_{part}. (Right) Azimuthal angle correlation is consistent with MC for all centrality bins. In all panels, the yellow boxes indicate point-to-point systematic uncertainties and the error bars denote the statistical uncertainty.

The resulting $\sigma(\Delta\phi_{J\gamma})$ values in *PbPb* do not show a significant centrality dependence within the present statistical and systematic uncertainties. For central *PbPb* collisions, $\sigma(\Delta\phi_{J\gamma})$ is similar to the PYTHIA reference based on the Z2 tune, and comparison with other PYTHIA tunes shows a theoretical uncertainty that is larger than the difference between the data and MC. Comparing the PYTHIA tune Z2 with tune D6T [4, 5] shows an 8% difference in $\sigma(\Delta\phi_{J\gamma})$, which is expected because these two tunes differ in their parton shower ordering resulting in a different $\Delta\phi$ correlation. The result that $\sigma(\Delta\phi_{J\gamma})$ is not found to be significantly modified by the medium is consistent with the earlier observation of an unmodified$\Delta\ \phi$ correlation in dijet events [6].

2.2. Photon-jet momentum imbalance

The momentum asymmetry ratio $x_{J\gamma} = p_{\mathrm{T}}^{Jet}/p_{\mathrm{T}}^{\gamma}$ is used to quantify the photon-jet momentum imbalance. In addition to the jet and photon selections used in the$\Delta\ \phi_{J\gamma}$ study, we further impose a strict$\Delta\ \phi_{J\gamma} > \frac{7}{8}\pi$ cut to suppress contributions from background jets. Note that photon-jet pairs for which the associated jet falls below the 30 GeV/c threshold are not included in the $x_{J\gamma}$ calculation. This limits the bulk of the $x_{J\gamma}$ distribution to $r_{J\gamma} \gtrsim 0.5$. The $\langle x_{J\gamma}\rangle$

Journal of Physics: Conference Series **446** (2013) 012007 doi:10.1088/1742-6596/446/1/012007

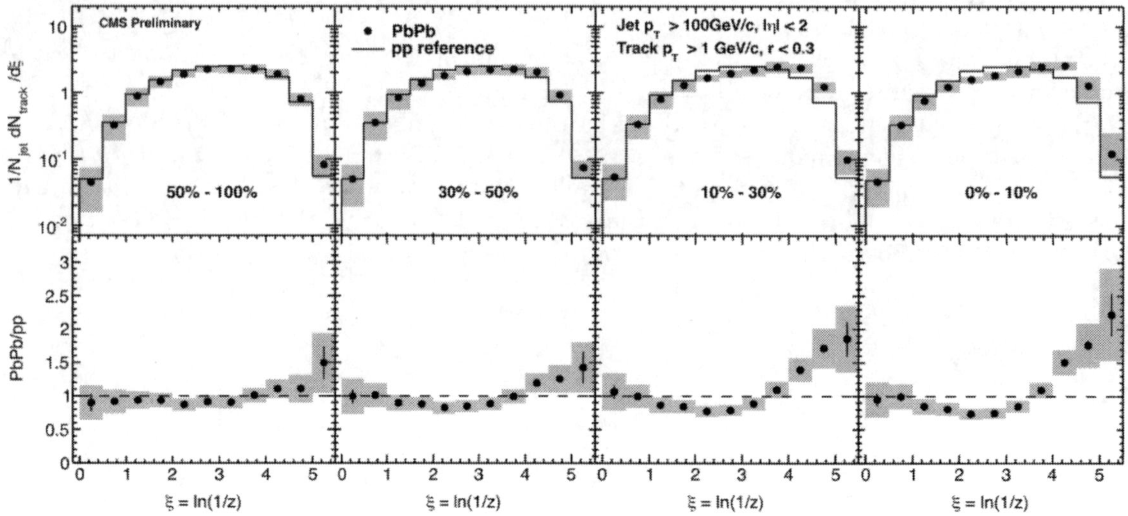

Figure 2. Top row shows fragmentation function in *PbPb* in bins of increasing centrality overlaid with *pp* reference data. The reference was made from the fragmentation function in *pp* collision with the same jet selection, but reweighted by the *PbPb* jet spectra and then smeared by the momentum resolution. The *PbPb* data is shown in the top row in four increasing centrality bins from left to right. The bottom row shows the ratio of each *PbPb* fragmentation function to its *pp* reference.

obtained from PYTHIA tunes Z2 and D6T agree to better than 1%. Overlaid in the peripheral bin is the $\langle x_{J\gamma}\rangle$ for 2.76 TeV *pp* data, showing consistency to the MC reference. However the poor statistics of the *pp* data does not allow a significant comparison. Further studies using the 7 TeV high statistics *pp* data showed a good agreement in $\langle x_{J\gamma}\rangle$ between data and PYTHIA justifying the use of PYTHIA + HYDJET as an un-modified reference.

While the photon-jet momentum ratio in the PYTHIA + HYDJET simulation shows almost no change in the peak location and only a modest broadening, even in the most central *PbPb* events, the *PbPb* collision data exhibit a change in shape, shifting the distribution towards lower $x_{J\gamma}$ as a function of centrality. It is important to note that, as discussed above, the limitation of $x_{J\gamma} \gtrsim 0.5$ limits the degree to which this distribution can shift.

It is important to keep in mind that the average energy loss of the selected photon-jet pairs does not constitute the full picture. There are genuine photon-jet events which do not contribute to the $\langle x_{J\gamma}\rangle$ distribution because the associated jet falls below the $p_T^{Jet} > 30$ GeV/c threshold.

As shown in Figure 1, the value of $r_{J\gamma}$ is found to decrease, from $r_{J\gamma} = 0.685 \pm 0.008$(stat.)–$0.698 \pm 0.006$(stat.) for the PYTHIA + HYDJET reference, as well as *pp* and peripheral *PbPb* data, to the significantly lower $r_{J\gamma} = 0.49 \pm 0.03$(stat.) ± 0.02(syst.)–0.54 ± 0.05(stat.) ± 0.02(syst.) for the three *PbPb* bins above 50% centrality.

3. Modification of Jet Fragmentation

Jet fragmentation functions are measured by correlating reconstructed charged-particle tracks falling within the jet cones, with the axis of the respective jet [7]. As done in previous measurements at hadron colliders [8], the fragmentation function is presented as a function of the variable

$$\xi = \ln\frac{1}{z}\ ;\quad z = \frac{p_{\parallel}^{\mathrm{track}}}{p^{\mathrm{jet}}}, \tag{2}$$

where $p_{\parallel}^{\mathrm{track}}$ is the momentum component of the track along the jet axis, and p^{jet} is the magnitude of the jet momentum. The tracks in a cone of$\Delta\ R = \sqrt{(\Delta\phi)^2 + (\ \Delta\eta)^2} < 0.3$ around the jet axis are selected for analysis. The fragmentation functions, defined as $1/\mathrm{N_{jet}}\ \mathrm{dN}/\mathrm{d}\xi$, are normalized to the total number jets (N_{jet}).

In order to quantify the medium-related effects, the results are compared to the references based on *pp* data. For constructing a proper reference distribution, the *pp* jet momenta are smeared and reweighted to match the *PbPb* jet distributions for a given analysis selection. As for the jet shape analysis, the background of tracks not correlated with the jet in *PbPb* collisions is estimated using the η-reflection method.

Figure 2 shows the comparison of fragmentation function between the *pp* reference and the *PbPb* data. The significant modification of the fragmentation function was found in central events and it grows with the collision centrality. In the 50-100% bin, the ratio of *PbPb* / *pp* is flat at unity which means no modification. However, an excess in high ξ is observed for more central events. In the most central 0–10% collisions and for the lowest charged particle momenta studied, the *PbPb* / *pp* fragmentation function ratio rises about a factor of 2. This implies that for central collisions the spectrum of particles in a jet has an enhanced contribution of soft particles compared to one from *pp* collisions.

4. Summary

The first study of isolated-photon-jet correlations and modification of the inclusive jet fragmentation function in *PbPb* collisions at $\sqrt{s_{NN}} = 2.76\mathrm{TeV}$ has been performed as a function of collision centrality using a dataset corresponding to an integrated luminosity of 150 $\mu\mathrm{b}^{-1}$. Isolated photons with $p_{\mathrm{T}}^{\gamma} > 60$ GeV/c were correlated with jets with $p_{\mathrm{T}}^{Jet} > 30$ GeV/c to determine the width of the angular correlation function, $\sigma(\Delta\phi_{J\gamma})$, the jet/photon transverse momentum ratio, $x_{J\gamma} = p_{\mathrm{T}}^{Jet}/p_{\mathrm{T}}^{\gamma}$, and the fraction of photons with an associated jet, $r_{J\gamma}$. Also, in order to study the jet fragmentation function to high ξ region with high statistics, high p_T jets were selected instead of isolated-photon-jet events.

In summary, no angular broadening of recoiled jets was observed beyond that seen in the *pp* data and MC reference at all centralities. Due to the hot and dense medium created in central *PbPb* collisions, there was a significant energy loss of jets and a remarkable modification of fragmentation function was found.

References

[1] S. Chatrchyan et al. The CMS experiment at the CERN LHC. *JINST*, 03:S08004, 2008.
[2] S. Chatrchyan *et al.* Studies of jet quenching using isolated-photon+jet correlations in PbPb and pp collisions at $\sqrt{s_{NN}} = 2.76$ TeV. arXiv:1205.0206. 2012.
[3] CMS Collaboration. Detailed Characterization of Jets in Heavy Ion Collisions Using Jet Shapes and Jet-Track Correlations. PAS-CMS-HIN-12-013. 2012.
[4] R. Field. Physics at the Tevatron. *Acta Phys. Polon. B*, 39:2611, 2008.
[5] P. Bartalini and L. Fanò. Multiple Parton Interactions Studies at CMS. CMS-CR-2011-048. 2011.
[6] S. Chatrchyan *et al.* Observation and studies of jet quenching in PbPb collisions at $\sqrt{s_{NN}} = 2.76\mathrm{TeV}$. *Phys. Rev. C*, 84:024906, 2011.
[7] T. Affolder et al. Measurement of the inclusive jet cross section in $p\bar{p}$ collisions at $\sqrt{s} = 1.8$ tev. *Phys. Rev. D*, 64:032001, 2001.
[8] F. Abe et al. Jet-fragmentation properties in $p\bar{p}$ collisions at $\sqrt{s} = 1.8$ tev. *Phys. Rev. Lett.*, 65:968, 1990.

Journal of Physics: Conference Series **446** (2013) 012008 doi:10.1088/1742-6596/446/1/012008

Systematic Monte-Carlo studies of dijets at RHIC using the VNI/BMS Parton Cascade

C.E.Coleman-Smith

E-mail: cec24@phy.duke.edu
Department of Physics, Duke University, Durham, NC 27701

Abstract. We present a study of the dijet suppression at RHIC using the parton cascade model. We examine the modification of the dijet asymmetry A_j and the fragmentation distributions z and j_T in terms of: $\hat{q}$; the path length of leading and sub-leading jets; cuts on the jet energy distributions; jet cone angle and the jet-medium interaction mechanism. We have introduced a string hadronization model and present hadronic jet fragmentation distributions. We find that A_j is most sensitive to $\hat{q}$ and less sensitive to the nature of the jet-medium interaction mechanism. The fragmentation distributions show jet modification and differentiate between elastic and radiative+elastic modes.

Measurements showing the modification of high energy $E_t \sim 100-200$ GeV dijets at the Large Hadron Collider (LHC) [1, 2] in Pb+Pb collisions have sparked an interest in using dijets to quantify jet modification in hot QCD matter. Proposed upgrades to the PHENIX experiment, and the continuing development of jet measurements at STAR promise access to dijets with $E_t \sim 15-65$ GeV at RHIC. The dijet asymmetry

$$A_j = \frac{E_{t,\ell} - E_{t,s}}{E_{t,\ell} + E_{t,s}}, \tag{1}$$

where $E_{t,\ell}$ is the transverse energy of the leading jet and $E_{t,s}$ is that of the sub-leading jet, has been successfully reproduced with a variety of models [3, 4, 5, 6, 7]. It is not obvious that the dijet asymmetry is the most suitable quantity for measuring the modification of dijets. Given that the dijet program at RHIC is still in its infancy it is worthwhile to consider the existence of other perhaps more differential observables.

We have used the VNI/BMS [8, 9] parton cascade to systematically explore dijet suppression at RHIC energy scales under controlled medium conditions. We examine the modification of dijets under variation of the medium radius and temperature, the strong coupling constant and the jet definition in terms of the Anti-Kt cone angle and the leading jet energy cut. The VNI/BMS parton cascade model provides access to the full jet/medium development at a fixed temperature and density, and thus fixed $\hat{q}$. We run the code in a static uniform-medium mode.

The full set of results from our systematic analysis can be found in [10]. In this proceeding we outline the model, introduce the hadronization process and show some selected results from the analysis.

The parton cascade model (PCM) is a Monte-Carlo implementation of the relativistic

Journal of Physics: Conference Series **446** (2013) 012008 doi:10.1088/1742-6596/446/1/012008

Boltzmann transport of quarks and gluons

$$p^\mu \frac{\partial}{\partial x^\mu} F_k(x,p) = \sum_i \mathcal{C}_i F_k(x,p). \tag{2}$$

The collision term $\mathcal{C}_i$ includes all possible $2 \to 2$ interactions and final-state radiation $1 \to n$

$$\mathcal{C}_i F_k(x,\vec{p}) = \frac{(2\pi)^4}{2S_i} \int \prod_j d\Gamma_j |\mathcal{M}_i|^2 \delta^4 (P_{in} - P_{out}) D(F_k(x,\vec{p})), \tag{3}$$

$d\Gamma_j$ is the Lorentz invariant phase space for the process j, D is the collision flux factor and S_i is a process dependent normalization factor. A geometric interpretation of the total cross-section is used to select pairs of partons for interaction. Between collisions, the partons propagate along straight line trajectories.

In the VNI/BMS PCM outgoing (from a hard scattering) off-shell partons are brought back on-shell through a medium modified time-like branching, this includes a Monte-Carlo Landau-Pomeranchuk-Migdal (LPM) effect [11, 5, 12, 13]. The radiated partons are assigned a formation time which is dynamically modified by further elastic interactions which generally serve to reduce the initial formation time. The shower partons are unable to radiate until their formation time has expired. This naturally builds a spatial dependence into the radiation process.

The strong coupling constant is fixed at $\alpha_s = 0.3$. A QGP medium is simulated as a box of thermal quarks and gluons generated at some fixed temperature. Periodic boundary conditions are imposed on the box, whose size is selected to be large enough that if a simulated jet wraps around it will not interact with its own wake.

The partonic contents of jets created by a suitable event generator and evolved down to $Q_0 = 1$ GeV, in this study PYTHIA 8 [14], are directly inserted into the box of thermal quarks and gluons. Each jet from the dijet pair is evolved separately. The jet partons propagate for a given distance and their evolution is recorded. Each particle in the initially inserted jet is marked as being "jetty", this jettiness tag is then iteratively applied to all partons that interact with an already jetty-labeled parton.

We have developed a simple scheme for hadronizing the contents of these jets which follows the Les Houches accord for particle color labeling. We extract the colors labels of the initial partons from the event generator. As these partons interact over the evolution of the jet the color labels are propagated appropriately. Each medium parton is given the next free color label. The partons interact through color averaged cross-sections. After including contributions from the medium the jet is in usually a non singlet state. We neutralize this color structure by generating a minimal set of thermal partons with appropriate color labels to reduce the jet to a color singlet state. The final set of partons is hadronized through the Lund string model [15].

In Fig. 1 we show a comparison of the partonic and hadronic transverse and longitudinal momentum distributions for a 100 GeV quark jet evolved in a box at $T = 0.35$ GeV for 4 fm/c. The dashed blue curve shows the distributions of the jet partons along with the associated thermal color-matching partons, these additional partons lead to an enhancement at small P_T and small z. Hadronization smooths and softens the longitudinal momentum distribution, conservation of momentum leads to an associated constriction of the transverse distribution.

We consider dijets at RHIC scales $\sqrt{s_{NN}} = 200$ GeV, all jets are reconstructed with the Anti-Kt algorithm [16]. The following kinematic cuts were always imposed: the minimum jet cut is set to $p_{t,min} = 5$ GeV, $\Delta|\eta_{12}| < 1.1$, $\Delta\phi_{12} > \pi/2$ where $\Delta\eta_{12}$ and $\Delta\phi_{12}$ are measured between the two reconstructed jets in the event. In this kinematic region the mean jet $\langle Et \rangle \sim 30$ GeV.

In Fig. 2 we show the influence of varying the medium temperature on the dijet asymmetry A_j. Recall that perturbative calculations for $\hat{q}$ scale like T^3. The higher temperature $T = 0.35$ GeV

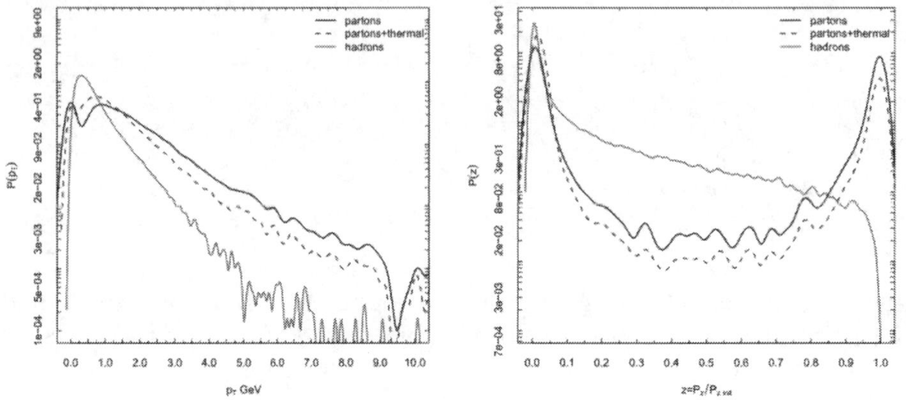

Figure 1. (Color Online) Transverse (left) and longitudinal (right) momentum distributions for the contents of 100 GeV quark jet evolved in a box at $T = 0.35$ GeV for 4 fm. Partonic distributions are shown as the blue curves, hadronic curves in red. The dashed blue curve represents the partonic distributions after the addition of the thermal partons required for hadronization

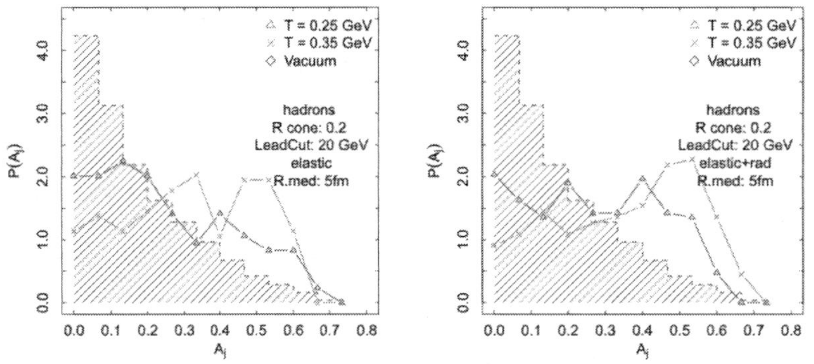

Figure 2. (Color Online) Elastic (left) and Elastic+Radiation (right) distributions of A_j as a function of the medium temperature, for hadrons. The shaded blue histogram shows the vacuum (PYTHIA) results.

(green curve, crosses) leads to a strong increase in the observed asymmetry for both interaction mechanisms. The full simulation (right panel), including elastic and radiative results, shows a somewhat stronger asymmetry than the elastic only result. This shift is observed to be strongest at small jet cone radii, here we have used $R = 0.2$. Increasing the cone radius "recaptures" the escaped energy and so tends to reduce the observed asymmetry, see Fig. 3 and [10].

We plot the longitudinal jet fragmentation $z = \frac{p_{t,h}}{p_{t,jet}} \cos \Delta r$ in Fig. 3, for leading (closed symbols) and sub-leading (open symbols) dijets. The plots show the vacuum distributions (blue, dashed) and the evolved distributions (red, solid). Let us first consider the left panel, which shows jets reconstructed at $R = 0.2$. There is a strong peak at $z = 1$. The limited phase space for the vacuum evolution of the jets at RHIC scales frequently gives rise to jets which have not undergone many splittings. The longitudinal momentum distribution for these jets is dominated by one or two very hard partons. This peak is diminished by evolution, with a concomitant enhancement at small z for the evolved jets. Since z is normalized to the final-state measured jet momentum, the medium induced energy loss is mostly scaled out.

The right hand panel of Fig. 3 shows the fragmentation distribution for the same events but calculated with a much wider jet cone angle $R = 1.0$. The peak at $z = 1$ is still just evident in the vacuum distributions. There is a clear separation between the evolved and vacuum distributions (blue,red). The small z peak is enhanced by medium induced radiation, forward scattering and the fragmentation process itself (as shown in Fig. 1). These two extremes paint a picture of jets as having a hard core surrounded by a more diffuse cone of soft partons from medium induced radiation and forward elastic scattering.

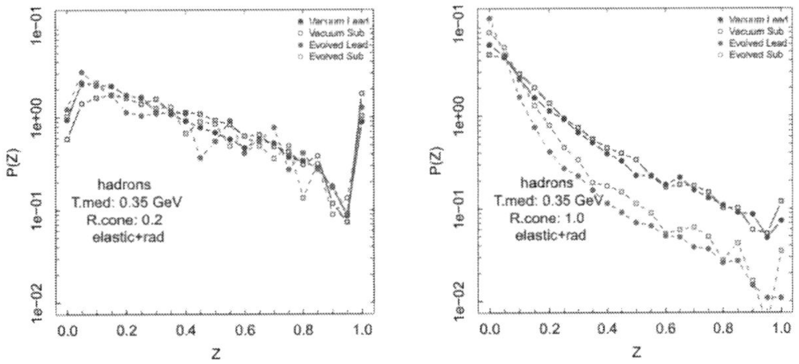

Figure 3. (Color Online) Longitudinal jet fragmentation z distributions for evolved (red, solid lines) and vacuum (blue, dashed lines) dijets reconstructed at $R = 0.2$ (left) and $R = 1.0$ (right). The (sub-leading, leading) dijet results are shown with (filled,open) symbols. Note that these curves are normalized to unit area.

Dijets at RHIC scales are likely to be strongly modified by the presence of the deconfined QGP medium. The observables we have discussed are sensitive to many aspects of this modification and suggest that further jet measurements at RHIC will provide valuable insights into the nature of the QGP.

We acknowledge support by DOE grants DE-FG02-05ER41367 and DE-SC0005396. This research was done using computing resources provided by OSG EngageVo funded by NSF award 075335.

References

[1] Aad G *et al.* (Atlas) 2010 *Phys. Rev. Lett.* **105** 252303
[2] Chatrchyan S *et al.* (CMS Collaboration) 2011 *Phys.Rev.* **C84** 024906
[3] Qin G Y and Müller B 2011 *Phys. Rev. Lett.* **106** 162302
[4] Young C, Schenke B, Jeon S and Gale C 2011 (*Preprint* 1103.5769)
[5] Coleman-Smith C E, Bass S A and Srivastava D K 2011 (*Preprint* 1101.4895)
[6] He Y, Vitev I and Zhang B W 2011 (*Preprint* 1105.2566)
[7] Renk T 2012 (*Preprint* 1202.4579)
[8] Geiger, Klaus and Müller, Berndt 1992 *Nucl. Phys.* **B369** 600–654
[9] Bass S A, Müller B and Srivastava D K 2003 *Phys. Lett.* **B551** 277–283
[10] Coleman-Smith C E and Muller B 2012 *Phys.Rev.* **C86** 054901 (*Preprint* 1205.6781)
[11] Zapp K, Stachel J and Wiedemann U A 2009 *Phys. Rev. Lett.* **103** 152302
[12] Landau L D and Pomeranchuk I J 1953 *Dolk. Akad. Nauk. SSSR* **92**
[13] Migdal A B 1956 *Phys. Rev.* **103** 1811–1820
[14] Sjöstrand T, Mrenna S and Skands P Z *JHEP* **0605** 026
[15] Andersson B 1997 *Camb.Monogr.Part.Phys.Nucl.Phys.Cosmol.* **7** 1–471
[16] Cacciari M, Salam G P and Soyez G 2008 *JHEP* **04** 063 (*Preprint* 0802.1189)

Journal of Physics: Conference Series **446** (2013) 012009 doi:10.1088/1742-6596/446/1/012009

Jet-Hadron Correlations in pp and Pb-Pb Collisions with ALICE

Megan Connors
Yale University, New Haven, CT 06520
E-mail: `Megan.Connors@Yale.edu`

Abstract. Jet quenching has been observed at both RHIC and LHC energies, indicating that partons lose energy as they traverse the medium. To probe the effects of this partonic energy loss, measurements of the angular correlations between fully reconstructed jets and charged tracks in Pb-Pb collisions are studied. Fully reconstructing a jet provides access to the kinematics of the initial hard scattering while allowing us to study the distribution of hadrons on the away side from the modified recoil jet. Here we present first measurements of jet-hadron correlations in pp collisions at $\sqrt{s}$=2.76 TeV and an outlook for Pb-Pb collisions. The jets in this analysis are reconstructed from the 2011 data set using both charged tracks and neutral energy measured in the ALICE tracking system and the electromagnetic calorimeter respectively.

1. Introduction

Jets are the result of a hard scattering process in the initial phase of the collision. In heavy ion collisions the partons from the hard scattering process are modified by the presence of the quark-gluon plasma (QGP). This modification has been observed as a suppression of high-momentum particles at both RHIC and LHC energies [1, 2, 3, 4]. This suppression has also been observed for di-jets via two-particle correlations. A high-momentum particle is used as a trigger particle and is paired with all other particles in the event, referred to as associated particles. In pp collisions, the distribution of associated particles shows a peak on the near side at$\Delta\ \varphi = 0$ and on the away side at$\Delta\ \varphi = \pi$, where $\Delta\varphi$ is the azimuthal angle between the two particles. The near-side peak results from the jet particles associated with the same jet as the trigger particle. The away-side jet peak results from the opposing jet created in the hard scattering. In heavy ion collsions the away-side jet peak is suppressed [5, 6]. In addition to the suppression on the away side, modifications to the shape have also been observed on both the near and away side. In this analysis we reconstruct the jet and use the reconstructed jet axis to measure$\Delta\ \varphi$ distributions with associated charged tracks, instead of using a single trigger particle as a proxy for the jet.

Fully reconstructed jets are very versatile triggers because there are several variables associated with jet reconstruction that can be adjusted. In principle, by varying the different cuts such as minimum constituent p_T, we can vary how surface biased the trigger jet is and thereby adjust the pathlength traversed by the opposing jet. This would allow us to map out energy loss as a function of path length and do true jet tomography. However, lowering the constituent cut to probe deeper into the medium also introduces more background from the underlying event which would need to be handled with care. By fully reconstructing the jet close to the QGP surface, we can better approximate the initial energy of the parton on the opposing side. While biasing the jet used as the trigger in our correlations to have a shorter pathlength through the medium, we also bias its fragmentation but the opposing jet is not biased by these cuts.

Jet-hadron correlations have been studied by the STAR collaboration at RHIC in Au+Au collisions at 200 GeV [7]. The results show that the suppression of pairs at high p_T is compensated by the production of additional low momentum particles. An apparent increase in the away-side width in the heavy ion environment compared to pp collsions was also observed but the significance of this difference is limited by the uncertainties due to the ambiguity in the jet v_n. Only very recently, ATLAS and STAR presented measurements of the jet v_2 for central collisions at 2.76 TeV and 200 GeV respectively [8, 9]. The impact of these measurements on the jet-hadron correlations is still being explored.

Making similar measurements at the LHC is important to help constrain theories which can teach us about how the partons lose energy in the QGP. These proceedings present a measure of the baseline jet-hadron correlations from pp data at 2.76 TeV and discuss the status of the jet-hadron measurement in PbPb collisons at ALICE.

2. The Data Sample

The data used in this analysis were collected by ALICE during the 2.76 TeV pp and 2.76 Pb-Pb LHC runs in 2011. The pp data include events that were triggered by using the ALICE electromagnetic calorimeter (EMCal). If a shower contained at least 3 GeV of energy, the event was accepted by the EMCal trigger. The statistics for the jet analysis is enhanced by using the triggered data sample. However, tracks from the minimum bias events are used for event mixing to exclude the correlations due to the EMCal trigger. The Pb-Pb analysis shown in these proceedings use a combination of the minimum bias, central and semi central triggered events. This results in a relatively flat centrality distribution for the 0-10% most central events.

3. Jet Reconstruction

Jets are reconstructed using both clusters in the EMCal and charged particles reconstructed in the ALICE tracking system. Contribution from the deposition of energy by charged tracks in the EMCal has been removed from the clusters by subtracting up to 100% of the momentum of the tracks matched to a given cluster. If the sum of the momentum of matched tracks is greater than the energy in the cluster, the cluster is simply excluded. We use the anti-$k_{\rm T}$ algorithm with a resolution parameter, $R = 0.4$, from the FastJet package [10]. Only jets which are fully contained within the area of the EMCal are considered. To avoid contributions from the underlying event in heavy ion collisions, we require the jets to have an area $A > 0.4$ [11] and use a minimum constituent cut of 3 GeV/c for both clusters and tracks. The constituent cut also helps bias the trigger jet closer to the surface [12]. Note that a 3 GeV/c cut for jets in the range 20-60 GeV has a more significant effect on the surface bias than a 1 GeV/c cut has on a 100 GeV jet as shown in [13]. To increase the surface bias and reject more background, we require that the leading particle in the jet satifies a $p_T > 6$ GeV/c cut.

4. Correlations Measurement

4.1. pp

A correlation function is defined as the number of same-event pairs over the number of mixed-event pairs where the mixed-event pair distribution is normalized such that the bin at $\Delta\eta = \Delta\phi = 0$ is 1. For two-particle correlations the yield is typically expressed as number of pairs per trigger, $N_{\rm trig}$, as in Eqn. 1, where ϵ is the efficiency of the associated particles.

$$\frac{1}{\mathrm{N_{trig}}}\frac{\mathrm{dN}}{\mathrm{d}\Delta\varphi}=\frac{1}{\epsilon\mathrm{N_{trig}}}\frac{\frac{\mathrm{dN^{same}_{pairs}}}{\mathrm{d}\Delta\varphi}}{\frac{\mathrm{dN^{mixed}_{pairs}}}{\mathrm{d}\Delta\varphi}} \tag{1}$$

Since the ALICE tracking detectors have full azimuthal coverage, the mixed-event pairs are approximately flat as a function of$\Delta\ \varphi$ while the finite η acceptance causes a drop towards large$\Delta\ \eta$. However, for jet-hadron correlations the$\Delta\ \eta$ dependence is rather flat in the range $|\Delta\eta| < 0.4$ since the range in η for the jets is more restricted than the η range of the associated tracks. To remove the contribution of the underlying event from the correlation function, a flat pedestal is subtracted. The resulting jet functions for three different associated momentum bins are shown in Fig. 1. The pedestal is determined from a fit which consists of two Gaussians plus the flat pedestal. The uncertainty on the pedestal subtraction is shown as a gray band around zero. There is an overall uncertainty of 7% on the normalization due to the uncertainty on the efficiency correction.

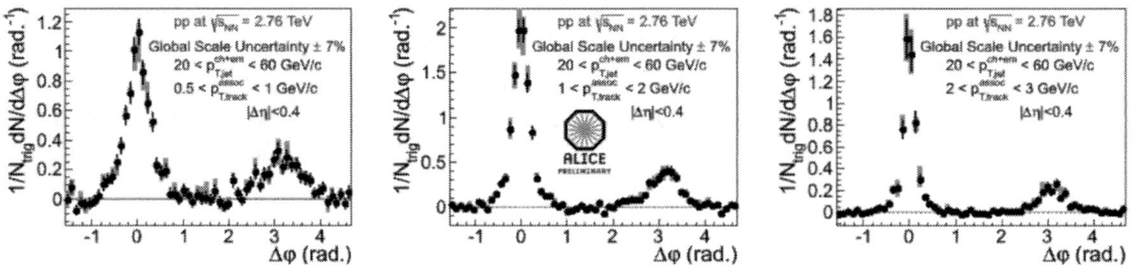

Figure 1. Jet-hadron$\Delta\ \varphi$ correlations from pp data. The underlying event has been subtracted.

The measured jet-hadron correlations qualitatively agree with expectations. The peaks of the near-side and away-side jets are clearly observed. The away-side peak appears broader and is lower than the near-side peak due to the effects of k_{T}, the intristic tranverse momentum [14, 15], which causes the two jets to not be directly back to back. Since we limit the correlations to be within $|\Delta\eta| < 0.4$, we do not capture all the jet particles.

4.2. Pb-Pb

In heavy ion collisions, the underlying event in the correlation functions is modulated by flow. To measure the effects of the QGP on the properties of the jet, we must remove the contribution to the correlation function from flow as in Eqn. 2. This requires a measure or an assumption of jet v_n.

$$\frac{1}{\mathrm{N_{trig}}}\frac{\mathrm{dN}}{\mathrm{d}\Delta\varphi}=\frac{1}{\epsilon\mathrm{N_{trig}}}\frac{\frac{\mathrm{dN^{same}_{pairs}}}{\mathrm{d}\Delta\varphi}}{\frac{\mathrm{dN^{mixed}_{pairs}}}{\mathrm{d}\Delta\varphi}}-b_0(1+\Sigma\ v^{\mathrm{trig}}_{\mathrm{n}}v^{\mathrm{assoc}}_{\mathrm{n}}\cos(n\Delta\varphi)) \tag{2}$$

A first look at the jet-hadron correlations from the Pb-Pb data is shown in Fig. 2 as a function of$\Delta\ \varphi$ and$\Delta\ \eta$. Correlations due to flow have not been removed and the normalization scale has not been corrected for efficiency. There is a clear near-side jet peaked observed near $\Delta\varphi = \Delta\eta = 0$ but nothing more about the jets can be concluded until the flow is removed. Once this is done, one can compare the shapes and yields of the distributions observed in Pb-Pb to pp to study any potential modication due to the QGP.

Quantifying the modification of the jet-hadron correlations at LHC energies, will provide additional constraints on energy loss models and will lead to a better understanding of the energy loss mechanisms and properties of the QGP.

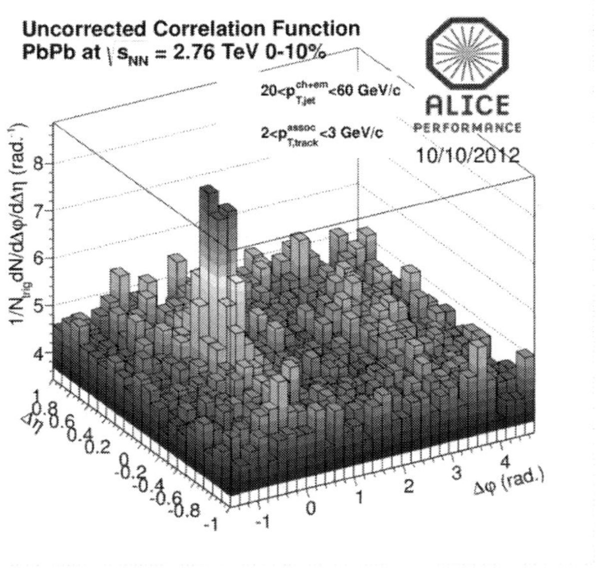

Figure 2. Jet-hadronΔ $\eta - \Delta\varphi$ correlations from 0-10% most central Pb-Pb collisions at 2.76 TeV. Contributions from flow are not removed and efficiency corrections have not been applied.

5. References

[1] J. Adams et al. [STAR Collaboration], Phys. Rev. Lett. 91, 172302 (2003)
[2] K. Adcox et al. [PHENIX Collaboration], Phys. Lett. B 561, 82 (2003)
[3] K. Aamodt et al. [ALICE Collaboration], Phys. Lett. B 696, 30 (2011)
[4] S. Chatrchyan et al. [CMS Collaboration], Eur. Phys. J. C 72, 1945 (2012)
[5] K. Aamodt et al. Phys. Rev. Lett. 108, 092301 (2012)
[6] C. Adler et al. [STAR Collaboration], Phys. Rev. Lett. 90, 082302 (2003)
[7] Alice Ohlson (for the STAR collaboration) Quark Matter 2011 Proceedings arXiv:1106.6243
[8] Alice Ohlson (for the STAR collaboration) Quark Matter 2011 Proceedings
[9] ATLAS Collaboration ATLAS-CONF-2012-116
[10] M. Cacciari, G. P. Salam and G. Soyez, JHEP 0804, 063 (2008)
[11] Peter Jacobs, Nucl. Phys. A855, 299 (2011).
[12] Thorsten Renk arXiv:1210.1330 [hep-ph]
[13] Thorsten Renk arXiv:1202.4579v2 [hep-ph]
[14] S. S. Adler et al. [PHENIX Collaboration], Phys. Rev. D 74, 072002 (2006)
[15] R. P. Feynman, R. D. Field, and G. C. Fox, Nucl. Phys. B128, 1 (1977).

Journal of Physics: Conference Series **446** (2013) 012010 doi:10.1088/1742-6596/446/1/012010

Interference effects between the initial and final state radiation in a QCD medium

Nestor Armesto,[1] Hao Ma, [1] Mauricio Martinez,[1] Yacine Mehtar-Tani,[2] and Carlos A. Salgado,[1,3]

[1] Departamento de Física de Partículas and IGFAE, Universidade de Santiago de Compostela, E-15782 Santiago de Compostela, Galicia-Spain.
[2] Institut de Physique Théorique, CEA Saclay, F-91191 Gif-sur-Yvette, France.
[3] Physics Department, Theory Unit, CERN, CH-1211 Genéve 23, Switzerland.

E-mail: mauricio.martinez@usc.es

Abstract. Coherence between multiple emitters is one of the most remarkable properties of QCD jet physics in vacuum. In the presence of a QCD medium one expects that the coherence pattern between multiple emitters is affected. In this work we calculate the gluon emission spectrum off an "asymptotic quark" traversing a dilute QCD medium. When interferences between initial and final quark are included, the medium induced gluon distribution gets modified. The coherent, incoherent and soft limit of the medium induced spectrum are studied. In the soft limit we find an elegant and intuitive probabilistic interpretation. We comment on possible phenomenological applications of this setup for studying coherence effects on observables studied in high energy nuclear collisions.

1. Introduction

Jets produced in high energetic hadronic collisions are an important tool to understand the theory of strong interactions, i.e., QCD. One of the most important properties of jets is the coherence pattern of soft gluon emissions which is related with the quantum interferences among different emitters in a parton cascade. One phenomenological consequence of color coherence phenomena in QCD is the suppression of radiation at large angles thanks to the angular ordering of subsequent soft gluon emissions. To illustrate this, let us consider soft gluon radiation in a deep inelastic process for the color singlet configuration (t channel). In this case, the probability to emit a gluon from any of the emitters is [1]

$$\langle dN_{in}\rangle_\phi = \frac{\alpha_s C_F}{\pi}\frac{d\omega}{\omega}\frac{d(\cos\theta)}{1-\cos\theta}\Theta(\cos\theta-\cos\theta_{qq})\,, \tag{1}$$

where θ is the angle of emitted gluon and θ_{qq} is the angle between the incoming and the outcoming parton due to hard scattering with the virtual photon. From last expression we observe the presence of soft and collinear divergences which must be resummed[1]. In addition, Eq.((1)) indicates us that gluon radiation is going to be confined inside the cone defined by the opening angle θ_{qq} along the emitter. Hence, there is no large angle emission.

In the presence of a QCD medium of finite size the efforts so far have been concentrated to understand the medium induced gluon radiation due to one emitter[2]. Nevertheless, not much is known about color coherence effects in the presence of a QCD medium. Recently, this important issue has been started to be studied by considering the radiation pattern of a $q\bar{q}$ antenna immersed in a QCD medium [3]. In this work, we briefly review an extension of color coherence studies inside a QCD medium to a space-like (t-channel) scattering process.

2. Scattering amplitude from Classical Yang Mills equations

A suitable method to describe the radiation of soft gluons is by using semiclassical methods of pQCD. In this approach the radiated gluon A^a_μ is treated as a fluctuation on top of a static background gauge field $A_{\rm med}$ (the medium). In the presence of the current generated by the incoming and outcoming quark, J^μ, the CYM equations read

$$[D_\mu, F^{\mu\nu}] = J^\mu \tag{2}$$

where $D_\mu = \partial_\mu - igA_\mu$, $F_{\mu\nu} = \partial_\mu A_\nu - \partial_\nu A_\mu - ig[A_\mu, A_\nu]$ and g is the gluon coupling constant. For the case of interest, we have two currents associated to the incoming quark[1] is given by $J^{\mu,a}_{inc} = g\, p^\mu/E\Theta(-t)\delta^{(3)}(\vec{x} - \vec{p}/Et)\Theta(-t)Q^a_{inc}$, where $p \equiv (E, \vec{p})$ is the energy and 3-momentum and Q^q_{inc} the color charge which are found through the continuity equation, $[D_\mu, J^\mu] = 0$ with $J = J_{inc} + J_{out}$. Finally, the amplitude of emitting a gluon with momentum $k \equiv (\omega, \vec{k})$ is given by

$$\mathcal{M}^a_\lambda(\vec{k}) = \lim_{k^2\to 0} -k^2 A^a_\mu(k)\epsilon^\mu_\lambda(\vec{k})\,, \tag{3}$$

where $\epsilon^\mu_\lambda(\vec{k})$ is the gluon polarization vector. In the following, we will work in light-cone gauge $A^+ = 0$. We consider in addition that the medium is created at $t = 0$. The amplitude for soft gluon emission reads[4]

$$\begin{aligned}\mathcal{M}^a_\lambda \;&=\; 2ig^2 \int \frac{d^2\boldsymbol{q}}{(2\pi)^2} \int_0^{L^+} dx^+\, [T\cdot A^-_{\rm med}(x^+, \boldsymbol{q})]^{ab} \Bigg\{ Q^b_{in}\, \frac{\boldsymbol{\kappa} - \boldsymbol{q}}{(\boldsymbol{\kappa} - \boldsymbol{q})^2} \\ &-Q^b_{out}\left[\frac{\bar{\boldsymbol{\kappa}} - \boldsymbol{q}}{(\bar{\boldsymbol{\kappa}} - \boldsymbol{q})^2}\left[1 - \exp\left(i\frac{(\bar{\boldsymbol{\kappa}} - \boldsymbol{q})^2}{2k^+}x^+\right)\right] + \frac{\bar{\boldsymbol{\kappa}}}{\bar{\boldsymbol{\kappa}}^2}\exp\left(i\frac{(\bar{\boldsymbol{\kappa}} - \boldsymbol{q})^2}{2k^+}x^+\right)\right]\Bigg\}\,,\end{aligned} \tag{4}$$

where $Q^b_{in(out)}$ is the color charge of the in(out)coming parton, $\boldsymbol{\kappa} = \mathbf{k} - x\mathbf{p}$ is the transverse momentum of the gluon relative to the incoming parton (similar definition for $\bar{\boldsymbol{\kappa}}$) and L^+ is the length of the medium.

3. Gluon Spectrum and its asymptotic limits

After taking the square of the scattering amplitude ((4)), the medium induced gluon spectrum is given by [4]

$$\begin{aligned}\omega\frac{dN^{\rm med}}{d^3\vec{k}} &= \frac{4\alpha_s C_F\hat{q}}{\pi}\int_0^{L^+} dx^+ \int \frac{d^2\boldsymbol{q}}{(2\pi)^2}\mathcal{V}^2(\boldsymbol{q})\left[\frac{1}{(\boldsymbol{\kappa} - \boldsymbol{q})^2} - \frac{1}{\boldsymbol{\kappa}^2} + 2\frac{\bar{\boldsymbol{\kappa}}\cdot\boldsymbol{q}}{\bar{\boldsymbol{\kappa}}^2(\bar{\boldsymbol{\kappa}} - \boldsymbol{q})^2}\left(1 - \cos\left[\frac{(\bar{\boldsymbol{\kappa}} - \boldsymbol{q})^2}{2k^+}x^+\right]\right)\right. \\ &\left.+2\left\{\frac{\boldsymbol{\kappa}\cdot\bar{\boldsymbol{\kappa}}}{\boldsymbol{\kappa}^2\bar{\boldsymbol{\kappa}}^2} - \frac{\bar{\boldsymbol{\kappa}}\cdot(\boldsymbol{\kappa} - \boldsymbol{q})}{\bar{\boldsymbol{\kappa}}^2(\boldsymbol{\kappa} - \boldsymbol{q})^2}\right\} + 2\left\{\frac{\bar{\boldsymbol{\kappa}}\cdot(\boldsymbol{\kappa} - \boldsymbol{q})}{\bar{\boldsymbol{\kappa}}^2(\boldsymbol{\kappa} - \boldsymbol{q})^2} - \frac{(\bar{\boldsymbol{\kappa}} - \boldsymbol{q})\cdot(\boldsymbol{\kappa} - \boldsymbol{q})}{(\bar{\boldsymbol{\kappa}} - \boldsymbol{q})^2(\boldsymbol{\kappa} - \boldsymbol{q})^2}\right\}\left(1 - \cos\left[\frac{(\bar{\boldsymbol{\kappa}} - \boldsymbol{q})^2}{2k^+}x^+\right]\right)\right],\end{aligned} \tag{5}$$

The gluon spectrum is composed by three parts: (i) the gluon emission off the incoming parton which is the bremstrahlung of an accelerated color charge which undergoes rescattering, (ii)

[1] For the outoming quark one must change the argument of the Heaviside step function $\Theta(-t) \to \Theta(t)$ and the 4-momentum $p^\mu \to \bar{p}^\mu$.

Journal of Physics: Conference Series **446** (2013) 012010 doi:10.1088/1742-6596/446/1/012010

gluon emission off outcoming quark is identified with the medium induced radiation of the N=1 opacity expansion (GLV spectrum) [2] and (iii) the interferences between the incoming and the outcoming parton. In addition, the spectrum contains soft and collinear divergences.

3.1. Incoherent limit

For short formation times $\tau_f << L^+$, the phases cancel and the gluon spectrum ((5)) simplifies to [4]

$$\omega \frac{dN^{\rm med}}{d^3\vec{k}}\bigg|_{\tau_f \ll L^+} = \frac{4\,\alpha_s C_F\,\hat{q}}{\pi} \int_0^{L^+} dx^+ \int \frac{d^2\boldsymbol{q}}{(2\pi)^2} \mathcal{V}^2(\boldsymbol{q}) \Big\{ \bar{\boldsymbol{L}}^2 + \boldsymbol{C}^2(\boldsymbol{\kappa}-\boldsymbol{q}) - \boldsymbol{C}^2(\boldsymbol{\kappa}) \Big\}, \tag{6}$$

where we use the definition of the Lipatov vertex $\bar{\boldsymbol{L}}$ in the light cone gauge and the transverse emission current $\boldsymbol{C}(\boldsymbol{\kappa})$ respectively[2]

$$\bar{\boldsymbol{L}} = \frac{\bar{\boldsymbol{\kappa}}-\boldsymbol{q}}{(\bar{\boldsymbol{\kappa}}-\boldsymbol{q})^2} - \frac{\bar{\boldsymbol{\kappa}}}{\bar{\boldsymbol{\kappa}}^2}, \qquad \boldsymbol{C}(\boldsymbol{\kappa}) = \frac{\boldsymbol{\kappa}}{\boldsymbol{\kappa}^2} - \frac{\bar{\boldsymbol{\kappa}}}{\bar{\boldsymbol{\kappa}}^2}. \tag{7}$$

This limit indicates the two main mechanisms of gluon radiation: the first term of Eq. ((6)) is the radiation of an asymptotic parton that suffers a scattering with the medium. The remaining two terms correspond to bremsstrahlung associated with the hard scattering followed by the radiated gluon suffering a classical sequential process of rescattering with the medium.This result is a generalization of the probabilistic interpretation of the incoherent limit of the GLV spectrum by including interferences [2].

3.2. Coherent limit

In the case of large formation times $\tau_f \gg L^+$, the gluon spectrum is reduced to [4]

$$\omega \frac{dN^{\rm med}}{d^3\vec{k}}\bigg|_{\tau_f \gg L^+} = \frac{4\,\alpha_s C_F\,\hat{q}}{\pi} \int_0^{L^+} dx^+ \int \frac{d^2\boldsymbol{q}}{(2\pi)^2} \mathcal{V}^2(\boldsymbol{q}) \Big\{ \frac{1}{(\boldsymbol{\kappa}-\boldsymbol{q})^2} - \frac{1}{\boldsymbol{\kappa}^2} + 2\frac{\bar{\boldsymbol{\kappa}}\cdot\boldsymbol{\kappa}}{\bar{\boldsymbol{\kappa}}^2\boldsymbol{\kappa}^2} - 2\frac{\bar{\boldsymbol{\kappa}}\cdot(\boldsymbol{\kappa}-\boldsymbol{q})}{\bar{\boldsymbol{\kappa}}^2(\boldsymbol{\kappa}-\boldsymbol{q})^2} \Big\} \tag{8}$$

The first two terms in the last expression are the classical broadening of the gluon emission off the incoming quark. The next two terms are interferences. In the absence of interferences, the independent GLV gluon spectrum gets suppressed due to the LPM effect while in the case studied here some of the interferences among the initial and final parton remain.

3.3. Soft limit

In the soft limit $\omega \to 0$ it is possibe to separate the contributions associated with the initial and final state of the total gluon spectrum (vacuum + medium induced) which reads as [4]

$$\omega \frac{dN^{\rm tot}}{d^3\vec{k}}\bigg|_{\omega\to 0} = \frac{\alpha_s C_F}{(2\pi)^2} \left(\mathcal{P}^{tot}_{in} + \mathcal{P}^{tot}_{out}\right)\big|_{\omega\to 0}, \tag{9}$$

where

$$\mathcal{P}^{tot}_{in} = (1-\Delta_{med})\left(\frac{1}{\boldsymbol{\kappa}^2} - \frac{\bar{\boldsymbol{\kappa}}\cdot\boldsymbol{\kappa}}{\bar{\boldsymbol{\kappa}}^2\boldsymbol{\kappa}^2}\right), \qquad \mathcal{P}^{tot}_{out} = \left(\frac{1}{\bar{\boldsymbol{\kappa}}^2} - (1-\Delta_{med})\frac{\bar{\boldsymbol{\kappa}}\cdot\boldsymbol{\kappa}}{\bar{\boldsymbol{\kappa}}^2\boldsymbol{\kappa}^2}\right). \tag{10}$$

[2] An identical definition follows for $\boldsymbol{C}(\boldsymbol{\kappa}-\boldsymbol{q})$ by changing $\boldsymbol{\kappa} \to \boldsymbol{\kappa}-\boldsymbol{q}$.

The fact that we can split the contribution of each emitter to the gluon spectrum allow us to have a probabilistic interpretation: The first term $\mathcal{P}_{in}^{tot}$ is the coherent gluon emission off the incoming quark reduced by the probability Δ_{med} that the emitted gluon interacts with the QCD medium. In addition, the angular ordering property observed in the vaccuum case will be preserved of the case of the initial state radiation. The second term $\mathcal{P}_{out}^{tot}$ accounts for the partial decoherence of the emitted gluon due to the scatterings with the QCD medium which share some similarities observed in the $q\bar{q}$ antenna case [3]. Such decoherence is measured by the probability of an interaction with the medium Δ_{med}. As a consequence, the medium opens the phase space for large angle emissions, a property called antiangular ordering [3].

4. Summary

In this work we consider the impact of the interference pattern between the initial and final state radiation in the presence of a QCD medium. By using semiclassical methods of pQCD we determine the medium induced gluon spectrum of an asymptotic parton which suffers a hard collision and crosses a QCD medium afterwards. The gluon spectrum has three contributions: the independent gluon emissions associated with the incoming and outgoing parton as well as interference terms between both emitters. To shed light on the structure of the spectrum, we consider three asymptotic limits of the gluon spectrum: the incoherent, coherent and soft sector. In the incoherent limit, we find a probabilistic interpretation for the gluon spectrum is found which is a generalization of the GLV spectrum by accounting interferences among the two emitters. For the coherent limit the medium induced gluon spectrum reduces to the classical broadening contribution associated exclusively to the initial state and some of the interferences remain. In the soft limit, we provide a probabilistic interpretation for the radiation pattern of the full gluon spectrum which has two main important features: (i) gluon emissions from the initial state remain coherent as in vacuum but now these are reduced by the probability to interact with the medium measured by $\Delta_{med} \sim \hat{q}/m_D^2$ and (ii) gluon emission off the outcoming parton loses its vacuum coherence, i.e., large angle emissions (antiangular ordering) arise from the medium induced coherence between both emitters.

Our work is a first principles approach for the proper inclusion of coherence effects on observables sensitive to initial state radiation in high energy nuclear collisions.

Acknowledgements

This work is supported by European Research Council grant HotLHC ERC-2001-StG-279579; by Ministerio de Ciencia e Innovación of Spain under projects FPA2008-01177, FPA2009-06867-E and FPA2011-22776; by Xunta de Galicia (Consellería de Educación and Consellería de Innovación e Industria – Programa Incite); by the Spanish Consolider-Ingenio 2010 Programme CPAN and by FEDER. C.A.S. is a Ramón y Cajal researcher. The work of Y.M.T. is supported by the European Research Council under the Advanced Investigator Grant ERC-AD-267258.

References

[1] A. Bassetto, M. Ciafaloni, G. Marchesini, Phys. Rept. **100** (1983) 201.

[2] R. Baier, Y. L.Dokshitzer, A. H. Mueller, S. Peigne, D. Schiff, Nucl. Phys. **B 483** (1997) 291, Nucl. Phys. **B 484** (1997) 265; B. G. Zakharov, JETP Lett. **63** (1996) 952, JETP Lett. **65** (1997) 615; U. A. Wiedemann, Nucl. Phys. **B 588** (2000) 303; M. Gyulassy, P. Levai, I. Vitev, Phys. Rev. Lett. **85** (2000) 5535, Nucl. Phys. **B 594** (2001) 371.

[3] Y. Mehtar-Tani, C. A. Salgado, K. Tywoniuk, Phys.Rev.Lett. **106** (2011) 122002, JHEP **1204** (2012) 064, Phys.Lett. **B 707** (2012) 156, arXiv:1205.5739; N. Armesto, H. Ma, Y. Mehtar-Tani, C. A. Salgado, K. Tywoniuk, JHEP **1201** (2012) 109; Y. Mehtar-Tani, K. Tywoniuk, arXiv:1105.1346; J. Casalderrey-Solana, E. Iancu, JHEP **1108** (2011) 015.

[4] N. Armesto, H. Ma, M. Martinez, Y. Mehtar-Tani, C. A. Salgado, Phys.Lett. **B717** (2012) 280-286.

Journal of Physics: Conference Series **446** (2013) 012011 doi:10.1088/1742-6596/446/1/012011

An overview of (selected) recent results in finite-temperature lattice QCD

Alexei Bazavov

Physics Department, Brookhaven National Laboratory, Upton, NY 11973, USA

E-mail: obazavov@quark.phy.bnl.gov

Abstract. I discuss recent results on the lattice with the main emphasis on the thermodynamics of the crossover region, restoration of the chiral symmetry and fluctuations of conserved charges as an indicator of deconfinement, that may also be used to determine the chemical freeze-out conditions in heavy-ion collision experiments.

1. Introduction

It has been established by lattice QCD calculations that at the physical values of light quark masses and at vanishing chemical potential there is no genuine phase transition in QCD, but rather an analytic crossover [1, 2, 3]. While the low-temperature phase exhibits confinement and breaking of the chiral symmetry, at high temperatures the behavior of the theory is qualitatively different – the interaction between quarks and gluons decreases due to the asymptotic freedom, leading to deconfinement, and the chiral symmetry is restored (see [4] for a recent review). At the crossover the QCD partition function does not exhibit a singularity, so it is not a surprise that different physical observables show a change in the their temperature dependence at somewhat different temperatures.

2. Restoration of chiral symmetry and T_{pc} in QCD

Let us start the discussion with the phenomena of restoration of chiral symmetry. The QCD Lagrangian at zero light quark mass allows for independent left and right rotations, *i.e.* possesses $SU(2)_L \times SU(2)_R$ symmetry. This symmetry is however spontaneously broken by the vacuum, as indicated by condensate of quark anti-quark pairs. It is expected on general grounds [5] that at zero light quark mass there is a second-order phase transition in QCD (in the $O(4)$ universality class) for which the chiral condensate is the order parameter. The renormalized quark condensate Δ_{ls} (I reserve the name "chiral condensate" for the light quark condensate at zero mass) is shown in Fig. 1 and defined as:

$$\Delta_{ls} = \frac{\langle\bar{\psi}_l\psi_l\rangle_T - \frac{m_l}{m_s}\langle\bar{\psi}_s\psi_s\rangle_T}{\langle\bar{\psi}_l\psi_l\rangle_0 - \frac{m_l}{m_s}\langle\bar{\psi}_s\psi_s\rangle_0}, \quad \langle\bar{\psi}_q\psi_q\rangle = \frac{T}{V}\frac{\partial \ln Z}{\partial m_q} = \frac{1}{N_s^3 N_\tau}\left\langle \mathrm{Tr} M_q^{-1}\right\rangle, \quad q = l, s, \tag{1}$$

where M_q is the fermion matrix, *i.e.* the discretized version of the Dirac operator, $N_s^3 N_\tau$ is the four-dimensional lattice volume. The disconnected susceptibility which is proportional to fluctuations of the quark condensate

$$\chi_l^{disc} = \frac{1}{N_s^3 N_\tau} \left\{ \left\langle \left(\mathrm{Tr} M_l^{-1} \right)^2 \right\rangle - \left\langle \mathrm{Tr} M_l^{-1} \right\rangle^2 \right\} \tag{2}$$

is shown in Fig. 2. The results are with the asqtad and HISQ/tree action at the light quark mass $m_l = m_s/20$ (somewhat heavier but close to the physical $m_l = m_s/27$).

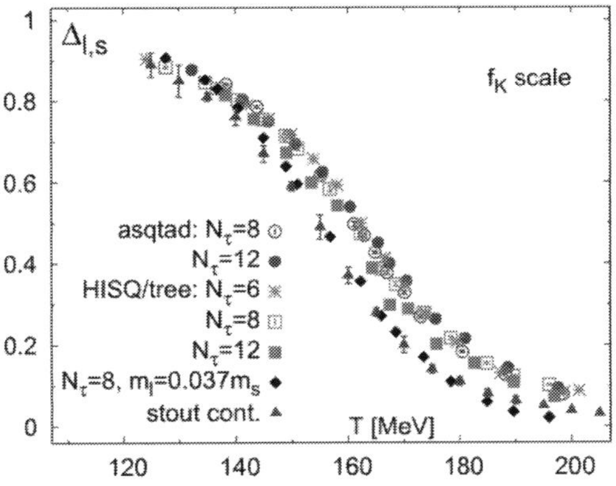

Figure 1. The renormalized quark condensate with the asqtad and HISQ/tree action.

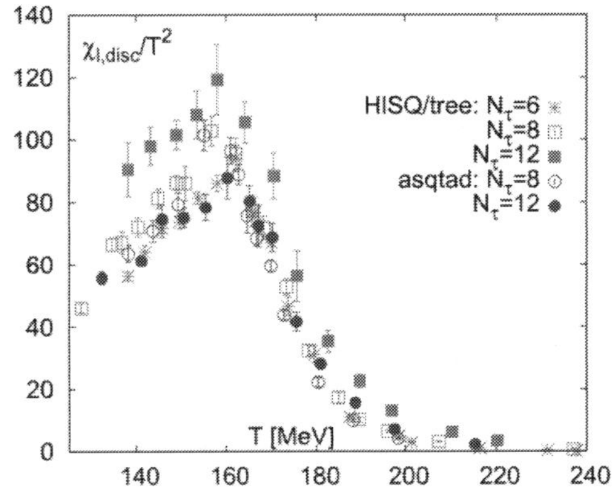

Figure 2. The disconnected susceptibility with the asqtad and HISQ/tree action.

If the physical light quark mass is small enough one may hope that remnants of criticality (plus subleading corrections) still govern the crossover region. Applicability of the critical scaling was extensively studied in Ref. [6] with staggered fermions and the scaling was indeed observed. (An additional complication for staggered fermions is that at non-zero lattice spacing the relevant universality class is $O(2)$ rather than $O(4)$. Luckily, the numerical difference between $O(2)$ and $O(4)$ fits is not significant for our discussion, comparable with other uncertainties, and we refer to this analysis as $O(N)$ scaling.) This approach has been taken by the HotQCD collaboration, who calculated the light quark condensate and its susceptibility with the asqtad and HISQ/tree actions on lattices with the temporal extent $N_\tau = 6$, 8 and 12 at several values of the light quark mass and then performed fits to $O(N)$ scaling functions complemented by non-singular terms. The pseudo-critical temperature, T_{pc}, defined this way as a location of the peak of the chiral susceptibility, reduces to the true critical temperature in the chiral limit. The result in the continuum limit at the physical light quark mass is $T_{pc} = 154(9)$ MeV [7]. This is compatible with the earlier results by the Budapest-Wuppertal collaboration that are in a range $T_{pc} = 147 - 157$ MeV, depending on what chiral observable is picked to determine T_{pc} [8].

The studies discussed so far rely on staggered fermions (one of the numerically cheapest lattice fermion formulations). The fermion doubling problem leads to 16 species of lattice fermions per one in continuum and staggered fermions reduce them to 4 species. The latter are however non-degenerate (and are labeled by "taste" to distinguish from flavor). In other words, taste symmetry is broken, for particular actions we use at $O(a^2)$, a being the lattice spacing. This leads to distorted hadron spectrum, and taste-breaking effects have been identified as the largest source of systematic errors in staggered simulations (a discussion in relation to thermodynamics is presented in Ref. [9]). It is desirable to have independent tests of the results also with other types of lattice fermions. The latter require more computational resources so calculations directly at the physical mass may not be feasible. However, some conclusions can be drawn from simulations at heavier masses. The Budapest-Wuppertal collaboration showed agreement between the continuum-extrapolated light quark condensate for staggered and Wilson fermions at the pion mass $m_\pi = 545$ MeV [10], Fig. 3 (left). Another study by the same group compared the condensate with overlap fermions (that provide exact chiral symmetry at non-zero lattice spacing) again to staggered at $m_\pi = 350$ MeV [11]. Although the continuum limit for overlap has not been taken, it seems, Fig. 3 (middle), that disagreement would be unlikely. The HotQCD collaboration pursued simulations with domain-wall fermions at $m_\pi = 200$ MeV [12]. A comparison of the staggered and domain-wall disconnected chiral susceptibility is shown in Fig. 3 (right). Note similar location of the peak in the domain-wall (DWF) and HISQ/tree $N_\tau = 12$ data. (Strictly speaking, results for different lattice actions should be compared in the continuum limit, however, Fig. 3 (right) provides a very encouraging check.)

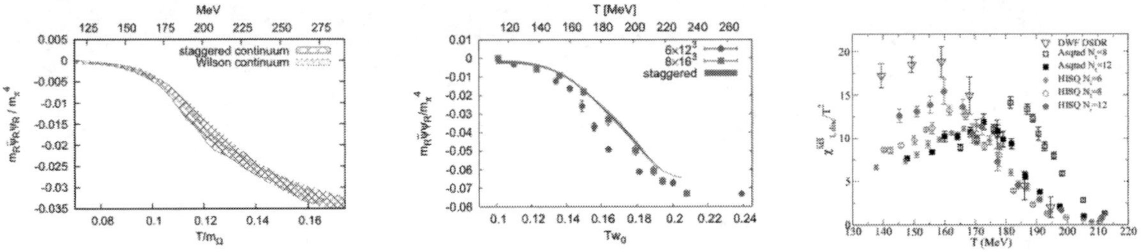

Figure 3. Comparison of the observables indicating restoration of the chiral symmetry between staggered and three other fermion formulations (Wilson, overlap and domain-wall). See text for details.

3. Deconfinement and fluctuations of conserved charges

QCD with infinitely heavy quarks, *i.e.*, $SU(3)$ pure gauge theory, exhibits first-order phase transition at $T_c \simeq 260$ MeV [13]. The order parameter is the renormalized Polyakov loop:

$$L_{ren}(T) = \exp(-cN_\tau/2) \left\langle \frac{1}{3} \mathrm{Tr} \prod_{x_0=1}^{N_\tau} U_0(x_0, \vec{x}) \right\rangle, \tag{3}$$

where $U_0(x_0, \vec{x})$ is the gauge field in the time direction. The exponential prefactor takes care of the multiplicative renormalization. L_{ren} can be related to the free energy of a static test quark $F_Q(T)$ as $L_{ren}(T) = \exp(-F_Q/T)$. In pure gauge theory the Svetitsky-Yaffe argument [14] allows to interpret deconfinement as a transition from the symmetric phase, $L_{ren} = 0$, into the broken phase, $L_{ren} \neq 0$, where the Polyakov assumes one of the three equally possible values, related by Z_3 symmetry. So, historically, based on the experience from pure gauge theory, the Polyakov loop served as an indicator of deconfinement in lattice QCD. Dialing quark masses to smaller values breaks the Z_3 symmetry (making one of the values of L_{ren} preferable over others, just like a magnetic field in spin systems). Due to string breaking by light dynamical quarks in QCD the value of L_{ren} is always non-zero in the confined phase. Thus, in QCD the Polyakov loop loses its meaning of the order parameter and is not related to any singularity of the partition function (in the chiral limit). As one can see in Fig. 4 (left) the temperature dependence of the Polyakov loop in $SU(2)$ and $SU(3)$ pure gauge theory and in QCD is quite different. Sharp behavior in the former case gives way to smoother change over wide temperature range. The behavior of the Polyakov loop in QCD at low temperatures can be understood in terms of the hadronic degrees of freedom [15, 16]. However, this description breaks down in the vicinity of the transition region.

Another possibility to probe deconfinement is through fluctuations of conserved charges. For instance, light and strange quark number fluctuations, that are suppressed at low temperature,

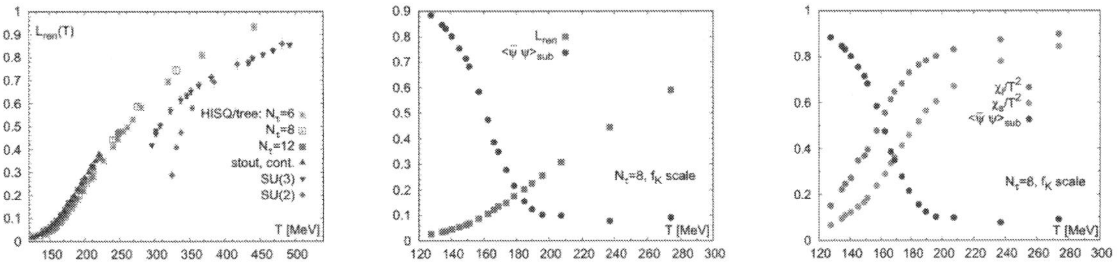

Figure 4. Temperature dependence of the Polyakov loop in $SU(2)$ and $SU(3)$ pure gauge theory and QCD (left). Comparison of the behavior of the renormalized light quark condensate and the Polyakov loop (middle) and light and strange quark number fluctuations (right).

signal liberation of the degrees of freedom with quantum numbers of quarks. In Fig. 4 (middle, right) we compare the light quark condensate to the Polyakov loop, light and strange quark number susceptibility. As one can see, the quantities associated with deconfinement rise somewhat slowly and over substantial temperature range. Thus, deconfinement is a gradual phenomenon in full QCD, and associating a particular temperature with it (by searching for inflection points or alike) does not appear meaningful. Rather, full temperature dependence of various observables should be studied. In particular, one can study fluctuations and correlations of baryon number B, electric charge Q and strangeness S that are given by the derivatives of the pressure with respect to the corresponding chemical potentials

$$\chi_{ijk}^{BQS} = \frac{\partial^{i+j+k}(p/T^4)}{\partial(\mu_B/T)^i \partial(\mu_Q/T)^j \partial(\mu_S/T)^k}. \tag{4}$$

The fluctuations of strangeness, baryon number and electric charge, calculated with the HISQ/tree action and extrapolated to the continuum, are shown in Fig. 5 (HotQCD collaboration [17]). Solid curves represent the Hadron Resonance Gas model, which is in remarkable agreement with the lattice data up to $T \sim 150 - 160$ MeV. (Electric charge fluctuations are the most sensitive to the cutoff effects in the pion sector and obtaining the continuum limit is, thus, a more subtle issue.) The results from the Budapest-Wuppertal collaboration [18] for the same quantities are in complete agreement.

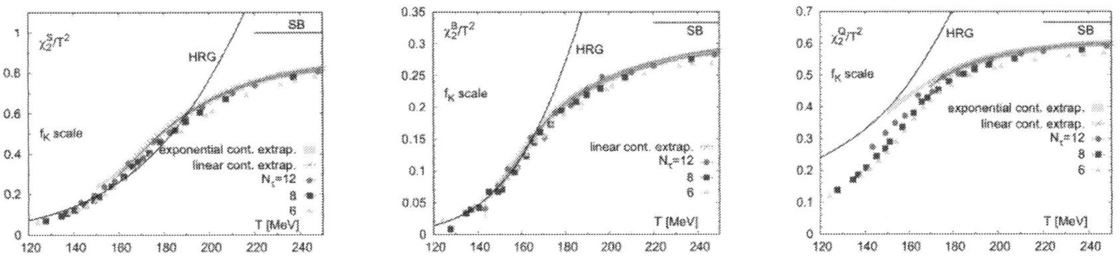

Figure 5. Fluctuations of strangeness, baryon number and net electric charge with the HISQ/tree action extrapolated to the continuum and compared to the Hadron Resonance Gas model.

To probe the critical behavior of QCD further one can consider higher-order cumulants of conserved charges. Comparing ratios of cumulants calculated on the lattice with experimental data, under assumption that at the time of chemical freeze-out the system created in heavy-ion

Journal of Physics: Conference Series **446** (2013) 012011 doi:10.1088/1742-6596/446/1/012011

collisions can be described by equilibrium thermodynamics with temperature T_f and baryon chemical potential μ_B^f, one can extract these freeze-out parameters. A recent attempt to use electric charge fluctuations for this purpose has been presented in Ref. [19]. By constraining the mean net strangeness and electric charge to those of the incident nuclei:

$$M_S = 0, \;\; M_Q = rM_B, \;\; (r \simeq 0.4 \text{ in gold-gold and lead-lead collisions}), \tag{5}$$

one can determine the strange and electric charge chemical potential of the system. To the next-to-leading order (NLO) in the baryon chemical potential:

$$\frac{\mu_Q}{T} = q_1 \frac{\mu_B}{T} + q_3 \left(\frac{\mu_B}{T}\right)^3, \quad \frac{\mu_S}{T} = s_1 \frac{\mu_B}{T} + s_3 \left(\frac{\mu_B}{T}\right)^3. \tag{6}$$

The q_i, s_i coefficients are related to ratios of generalized susceptibilities, *i.e.* various combinations of derivatives of the partition function with respect to chemical potentials, Eq. (4). It has been shown in Ref. [19] that NLO corrections are small and at NLO the dependence of μ_Q/μ_B and μ_S/μ_B on μ_B at several temperatures is shown in Fig. 6. The bands give total uncertainty of the lattice calculation and dashed lines represent the HRG model. Once μ_Q and μ_S satisfying the constraint (5) are fixed, ratios of cumulants can be evaluated, and Ref. [19] focused on

$$R_{12}^X \equiv \frac{M_X}{\sigma_X^2} = \frac{\mu_B}{T}\left(R_{12}^{X,1} + R_{12}^{X,3}\left(\frac{\mu_B}{T}\right)^2 + \mathcal{O}((\mu_B/T)^4)\right), \tag{7}$$

$$R_{31}^X \equiv \frac{S_X\sigma_X^3}{M_X} = R_{31}^{X,0} + R_{31}^{X,2}\left(\frac{\mu_B}{T}\right)^2 + \mathcal{O}((\mu_B/T)^4), \tag{8}$$

where $X = B$ or Q, M_X is mean, σ_X^2 is variance and S_X skewness for corresponding conserved charges. The results are shown in Fig. 7 and 8. If R_{31}^Q is determined from experiment, one can use the band in Fig. 7 to convert it to the freeze-out temperature T_f. Then from an experimentally determined ratio R_{12}^Q, T_f and the corresponding band in Fig. 8 one can find the freeze-out chemical potential μ_B^f. Once the full experimental analysis of R_{31}^Q and R_{12}^Q is available, the results of Ref. [19] can determine T_f and μ_B^f for various center-of-mass energies and thus map a part of the freeze-out curve on the QCD phase diagram.

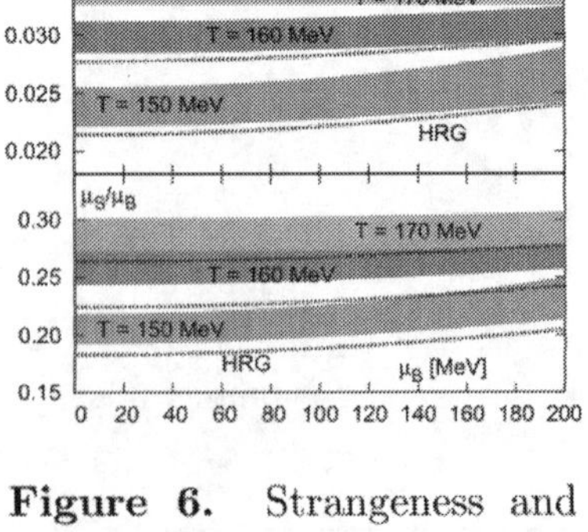

Figure 6. Strangeness and electric charge chemical potential at several temperatures with the constraints (5).

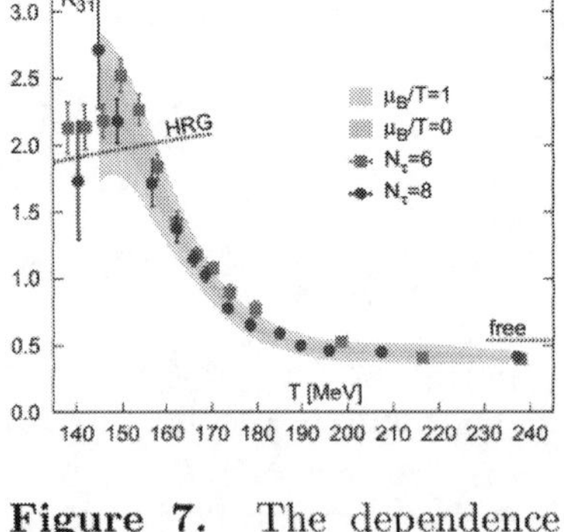

Figure 7. The dependence $R_{31}^Q(T)$ allows for determination of the freeze-out temperature T_f.

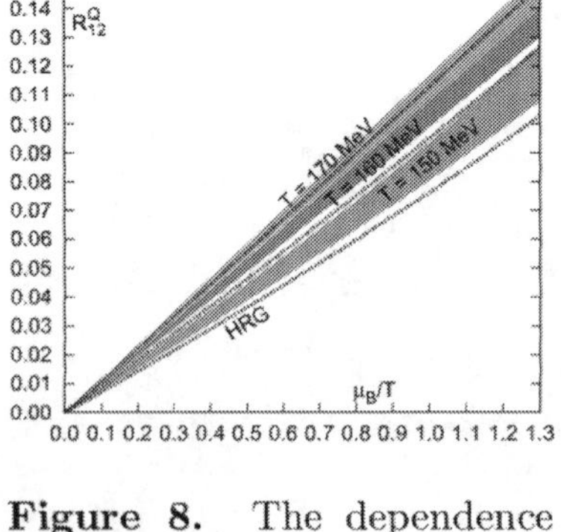

Figure 8. The dependence $R_{12}^Q(T)$ at given T_f determines the freeze-out chemical potential μ_B^f.

4. A list of other results on the lattice

Due to the space constraints many other recent interesting calculations on the lattice could not be included. Among those I would like to mention:

- Progress on the equation of state with 2+1 flavors of staggered fermions: an update of the previous results from the Budapest-Wuppertal [20] and HotQCD [21, 22] Collaboration. Preliminary results still indicate a discrepancy in 150 – 300 MeV range, which may be resolved by ongoing calculations.
- 2+1 flavor equation of state with Wilson fermions from the WHOT-QCD collaboration [23].
- The equation of state with dynamical charm quark, see Refs. [20] and [24] for preliminary results.
- Calculation of the charmonium screening masses from spatial correlation functions in 2+1 flavor QCD [25] and Maximum Entropy Method (MEM) reconstruction of the charmonium spectral functions in quenched QCD [26].
- Thermodynamics of the crossover region with high magnetic fields [27].

Acknowledgments

I would like to thank the organizers of Hot Quarks 2012 for a very productive and interesting meeting and Frithjof Karsch and Peter Petreczky for careful reading and comments on the manuscript. This work was in part supported by U.S. Department of Energy under Contract No. DE-AC02-98CH10886.

References

[1] Bernard C *et al.* [MILC Collaboration] 2005 *Phys. Rev.* D **71** 034504 (*Preprint* hep-lat/0405029)
[2] Cheng M, Christ N H, Datta S, Van der Heide J *et al.* 2006 *Phys. Rev.* D **74** 054507 (*Preprint* hep-lat/0608013)
[3] Aoki Y, Endrodi G, Fodor Z, Katz S D and Szabo K K 2006 *Nature* **443** 675 (*Preprint* hep-lat/0611014)
[4] Petreczky P 2012 *J. Phys.* G **39** 093002 (*Preprint* 1203.5320)
[5] Pisarski R D and Wilczek F 1984 *Phys. Rev.* D **29** 338
[6] Ejiri S, Karsch F, Laermann E, Miao C *et al.* 2009 *Phys. Rev.* D **80** 094505 (*Preprint* 0909.5122)
[7] Bazavov A, Bhattacharya T, Cheng M, DeTar C *et al.* [HotQCD Collaboration] 2012 *Phys. Rev.* D **85** 054503 (*Preprint* 1111.1710)
[8] Borsanyi S *et al.* [Wuppertal-Budapest Collaboration] 2010 *JHEP* **1009** 073 (*Preprint* 1005.3508)
[9] Bazavov A and Petreczky P 2010 *PoS* **LATTICE2010** 169 (*Preprint* 1012.1257)
[10] Borsanyi S, Durr S, Fodor Z, Hoelbling C *et al.* 2012 *JHEP* **1208** 126 (*Preprint* 1205.0440)
[11] Borsanyi S, Delgado Y, Durr S, Fodor Z *et al.* 2012 *Phys. Lett.* B **713** 342 (*Preprint* 1204.4089)
[12] Bazavov A, Bhattacharya T, Buchoff M I, Cheng M *et al.* [HotQCD Collaboration] 2012 *Phys. Rev.* D **86** 094503 (*Preprint* 1205.3535)
[13] Boyd G, Engels J, Karsch F, Laermann E *et al* 1996 *Nucl. Phys.* B **469** 419 (*Preprint* hep-lat/9602007)
[14] Svetitsky B and Yaffe L G 1982 *Nucl. Phys.* B **210** 423
[15] Megias E, Ruiz Arriola E and Salcedo L L 2012 *Phys. Rev. Lett.* **109** 151601 (*Preprint* 1204.2424)
[16] Bazavov A and Petreczky P 2013 (*Preprint* 1301.3943)
[17] Bazavov A, Bhattacharya T, DeTar C E, Ding H-T *et al.* [HotQCD Collaboration] 2012 *Phys. Rev.* D **86** 034509 (*Preprint* 1203.0784)
[18] Borsanyi S, Fodor Z, Katz S D, Krieg S, Ratti C and Szabo K 2012 *JHEP* **1201**, 138 (*Preprint* 1112.4416)
[19] Bazavov A, Ding H T, Hegde P, Kaczmarek O *et al.* 2012 *Phys. Rev. Lett.* **109** 192302 (*Preprint* 1208.1220)
[20] Borsanyi S, Endrodi G, Fodor Z, Katz S D *et al.* 2011 *PoS* **LATTICE2011** 201 (*Preprint* 1204.0995)
[21] Petreczky P [HotQCD Collaboration] 2012 *PoS* **LATTICE2012** 069 (*Preprint* 1211.1678)
[22] Bazavov A [HotQCD Collaboration] 2012 (*Preprint* 1210.6312)
[23] Umeda T *et al.* [WHOT-QCD Collaboration] 2012 *PoS* **LATTICE2012** 074 (*Preprint* 1212.1215)
[24] Bazavov A *et al.* [MILC Collaboration] 2012 *PoS* **LATTICE2012** 071
[25] Karsch F, Laermann E, Mukherjee S and Petreczky P 2012 *Phys. Rev.* D **85** 114501 (*Preprint* 1203.3770)
[26] Ding H T, Francis A, Kaczmarek O, Karsch F, Satz H and Soeldner W 2012 *Phys. Rev.* D **86** 014509 (*Preprint* 1204.4945)
[27] Bali G S, Bruckmann F, Endrodi G, Fodor Z, Katz S D and Schafer A *Phys. Rev.* D **86** 071502 (*Preprint* 1206.4205)

Dipole asymmetry at $\sqrt{s_{NN}} = 200$ GeV Au+Au Collisions from STAR experiment at RHIC

Yadav Pandit (for the STAR Collaboration)
Department of Physics, UIC, Chicago, USA
E-mail: ypandit@uic.edu

Abstract. We report the first measurements of azimuthal anisotropy at midrapidity originated from the dipole asymmetry due to fluctuations in the initial geometry in Au+Au collisions at $\sqrt{s_{NN}} = 200$ GeV, based on data from the STAR experiment at the Relativistic Heavy Ion Collider. The signal is almost symmetric in pseudorapidity and we report it as a function of pseudorapidity and transverse momentum for different centrality. Results are compared with available model predictions.

1. Introduction

Recent developments in understanding the initial geometry in heavy ion collisions points to a lumpy initial state [1, 2]. Event-by-event fluctuations break the symmetry of the initial density profile, and as a result there is, in general, one direction where the profile is steepest. This effect can be quantified as a dipole asymmetry in the initial density [3],

$$\epsilon_1 e^{i\Phi_1} = -\frac{\langle r^3 e^{i\phi} \rangle}{\langle r^3 \rangle} \quad (1)$$

where Φ_1 corresponds to the steepest direction of the density profile in polar coordinate system (r, ϕ) and ϵ_1 is the magnitude of the dipole asymmetry. This can be measured in heavy ion collision experiment as a first harmonic coefficient in the azimuthal distribution of the particles with respect to the Φ_1 plane [4]. This signal associated with dipole asymmetry is symmetric with respect to pseudorapidity (η), where as the conventional directed flow, also the first flow harmonic (v_1), originated from the bounce off motion of the nucleons is an odd function of pseudorapidity [5, 8]. To measure this signal we need to correct for the effect of momentum conservation and suppress rapidity odd directed flow.

2. The STAR experiment and Analysis Details

About ten million Au + Au collisions at $\sqrt{s_{NN}} = 200$ GeV have been used in this study, all acquired in the year 2004 using the STAR detector with a minimum-bias trigger. The Time Projection Chamber (TPC) of STAR covers covers pseudorapidity $|\eta| < 1$ and has been used as a main tracking detector for this experiment. The centrality definition of an event is based on the number of charged tracks in the TPC within $|\eta| < 0.5$. In this analysis, events are required to have vertex z coordinate (along the beam direction) within 30 cm from the center of the TPC.

As proposed by Luzum and Ollitrault [4], we used the modified event vector method to measure this signal. The event plane vector is reconstructed from tracks with transverse momentum (p_T) up to 2 GeV/c and within $|\eta| < 1.0$. The event plane vectors are defined by [9],

$$Q_x = \sum_i w_i \cos\phi_i, \tag{2}$$

$$Q_y = \sum_i w_i \sin\phi_i \tag{3}$$

where weight is taken as

$$w_i = p_T - \frac{< p_T^2 >}{< p_T >}. \tag{4}$$

The $< p_T >$ and $< p_T^2 >$ represent event averaged quantities. The choice of this weight corrects the effect of the momentum conservation and also it cancels out the conventional pseudorapidity asymmetric directed flow [4]. We used scalar product method as well as the event plane method to measure the signal. In the scalar product method, the particles with $0.5 < \eta < 1.0$ were assigned to one subevent and particles with $-1.0 < \eta < -0.5$ to the other subevent, separated by η gap of 1.0 units between two subevents and at least 0.5 unit with the particle of interest and the subevent to which it is correlated. In order to remove the acceptance effects we applied recentering correction [9] to the flow vectors. With unit vector defined as $u_i = e^{i\phi}$, the first harmonic coefficient v_1 is evaluated as,

$$v_1\{\eta > 0\} = \frac{\langle \mathrm{Q_a}(\eta < 0)\mathrm{u_i^*})\rangle}{\sqrt{\langle \mathrm{Q_a Q_b}\rangle}} \tag{5}$$

$$v_1\{\eta < 0\} = \frac{\langle \mathrm{Q_b}(\eta > 0)\mathrm{u_i^*})\rangle}{\sqrt{\langle \mathrm{Q_a Q_b}\rangle}} \tag{6}$$

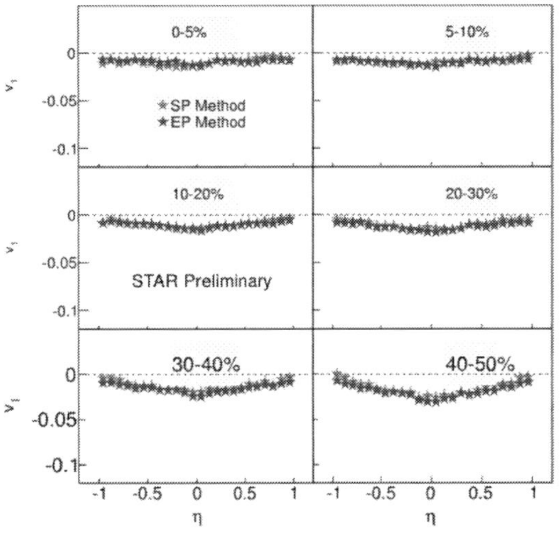

Figure 1. The first flow harmonic associated with dipole asymmetry at $\sqrt{s_{NN}} = 200$ GeV Au+Au collisions from central to peripheral collisions as a function of pseudorapidity.

In the event plane method, the particles with $0.1 < \eta < 1.0$ were assigned to one subevent and particles with $-1.0 < \eta < -0.1$ to the other subevent, so that subevents were separated by η gap of 0.2 units. Larger pseudorapidity separation was desired but we are limited by the event plane resolution. In this method, we first evaluate the event plane angle Ψ_1 from the event plane vector (Q_1) as,

$$\Psi_1 = \left(\tan^{-1}\frac{Q_y}{Q_x}\right), \tag{7}$$

The event plane angle was flattened using a shifting correction method to correct for the detector acceptance effects. The particle correlation with event plane angle Ψ_1 is given by,

$$v_1\{\eta > 0\} = \frac{\langle \cos(\phi - \Psi_\mathrm{a}(\eta < 0))\rangle}{\sqrt{\langle \cos(\Psi_\mathrm{a}(\eta < 0) - \Psi_\mathrm{b}(\eta > 0))\rangle}}. \tag{8}$$

Journal of Physics: Conference Series **446** (2013) 012012 doi:10.1088/1742-6596/446/1/012012

$$v_1\{\eta < 0\} = \frac{\langle\cos(\phi - \Psi_b(\eta > 0))\rangle}{\sqrt{\langle\cos(\Psi_a(\eta < 0) - \Psi_b(\eta > 0))\rangle}}. \quad (9)$$

3. Result and discussion

Figure 1 presents the first flow harmonic associated with dipole asymmetry for Au+Au collisions as a function of pseudorapidity(η) in 0-5% through 40-50% central collisions in $0.15 < p_T < 2.0$ GeV/c . Results are presented from event plane method and scalar product method. Results from both methods are consistent with each other within statistical errors. It was expected that very little η dependence would be found [4], which is true for central collisions. However in more peripheral collisions it is observed that the signal has some η dependence. This small η dependence may be a viscous effect and/or may be coming from non flow correlation which increases from central to peripheral collisions. This effect may also come from imperfect cancelation of odd component of v_1. We are not aware of any model results studying this effect, and future theoretical studies may shed light on it.

In Fig. 2, the v_1 signal along with v_2 and v_3 as a function of transverse momentum up to $p_T \sim 5$ GeV/c is shown for central (0-10%) and mid central (30-40%) collisions. The magnitude of v_1 signal at intermediate transverse momentum ($3 < p_T < 5$ GeV/c) is comparable to the elliptic flow signal and triangular flow in central collisions (0-10%) and smaller than v_2 but larger than v_3 in mid central collisions (30-40%) collisions. Results form the ATLAS Collaboration at $\sqrt{s_{NN}} = 2.76$ TeV Pb+Pb collisions also show similar trend [10], however the measurement technique is slightly different. They used two-particle correlation data and employed a two-component fit ansatz to separate the pseudorapidity-even first harmonic coefficient from the signal due to momentum conservation.

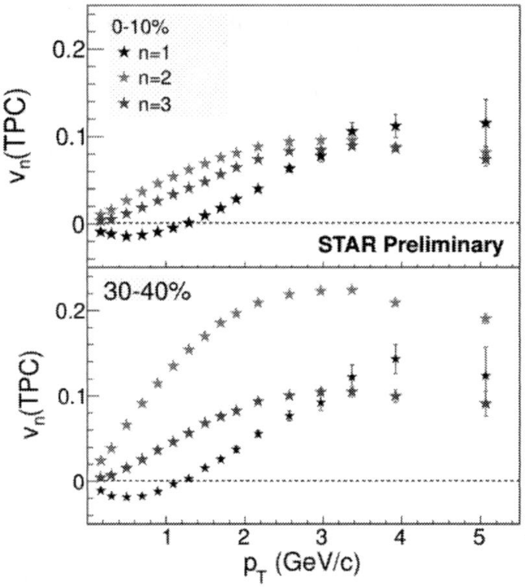

Figure 2. First,second and third flow harmonics near midrapidity($|\eta| < 1.0$) as a function of transverse momentum for 0-10% and 30-40% central Au+Au collisions

In Fig. 3, the first harmonic coefficient as a function of transverse momentum in 0-10%, 10-20%, 20-30%, 30-40% and 40-50% central collisions are compared to the viscous hydrodynamic model prediction with fluctuating initial conditions from Monte Carlo Glauber [11] and small viscosity setting($\eta/S = 0.16$) of reference [12]. These model predictions do a very good job in describing the data at low transverse momentum and in central collisions and they deviate at higher p_T in more peripheral collisions.

Results in the figures above are presented with only statistical errors. In these studies, contribution from short range correlation such as Bose-Einstein correlations, coulomb interactions are studied using a pseudorapidity gap of at least 0.5 units in pseudorapidity between the event vector and the particle of interest using scalar product method. Comparison to event plane method with smaller pseudorapidity gap suggests that non-flow contribution from short range correlations are small. To reduce the effects from high p_T particles in the estimation of the event plane, we used p_T weight only up to 2 GeV/c. The non-flow contribution from the jets/minijets is not known and might be a significant contributor to the systematic uncertainties especially at peripheral collisions.

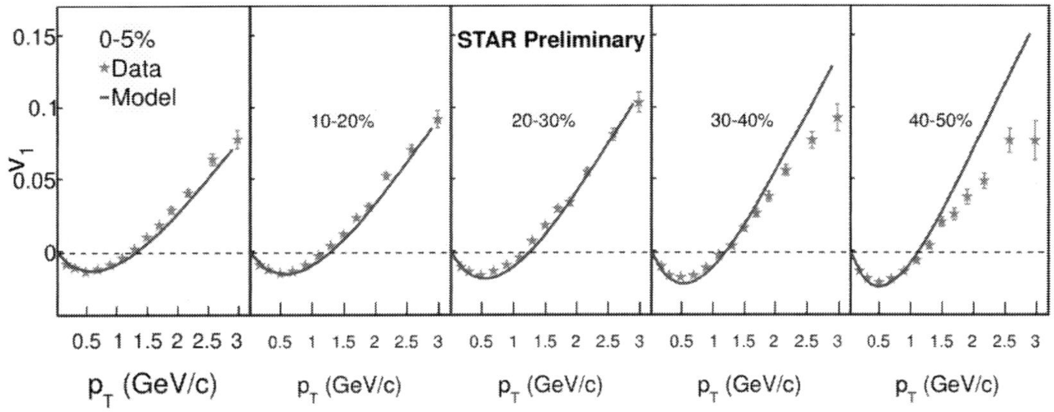

Figure 3. First flow harmonics as a function of transverse momentum compared with hydro model prediction.

4. Summary

In summary, we present first measurements of directed flow associated with dipole asymmetry from Au + Au collisions at $\sqrt{s_{NN}}$ = 200 GeV from STAR collaboration at RHIC. We report result from the scalar product method as well as standard event plane method. The results from both methods are consistent with each other. The signal is symmetric with respect to pseudorapidity unlike conventional directed flow which is anti-symmetric with respect to pseudorapidity. At intermediate p_T range 3-5 GeV/c, this signal is comparable to the elliptic flow signal in central collisions and triangular flow at central and mid central collisions. We also compared our results with hydrodynamic model calculation from E. Retinskaya et al [12] which is consistent with the data at low transverse momentum and in central collisions.

References

[1] P. Sorensen, B. Bolliet, A. Mocsy, Y. Pandit and N. Pruthi, Phys. Lett. B **705** 71 (2011).
[2] B. Alver and G. Roland, Phys. Rev. C **81**, 054905 (2010).
[3] D. Teaney and L. Yan Phys. Rev. C**83**, 064904 (2011).
[4] Matthew Luzum and J. Y. Ollitrault Phys. Rev. Lett. **106**, 102301 (2011).
[5] G. Agakishiev et al. (STAR Collaboration) Phys. Rev. C**85**, 014901 (2012)
[6] B. H. Alver, C. Gombeaud, M. Luzum and J. Y. Ollitrault, Phys. Rev. C **82**, 034913 (2010).
[7] B. Schenke, S. Y. Jeon, and C. Gale, Phys. Rev. Lett. **106**, 042301 (2011).
[8] Y. Pandit, Azimuthal anisotropy in Heavy ion collisions, Ph.D Dissertation, Kent State University, 2012.
[9] A. M. Poskanzer and S. A. Voloshin, Phys. Rev. C **58**, 1671 (1998).
[10] G. Aad et. al.(ATLAS Collaboration) Phys. Rev. C **86**, 014907(2012)
[11] M. L. Miller, K. Reygers, S. J. Sanders and P. Steinberg, Ann. Rev. Nucl. Part. Sci. **57**, 205 (2007); B. Alver et al., Phys. Rev. C **77** , 014906 (2008).
[12] E Retinskaya et al Phys. Rev. Lett. **108**, 252302 (2012)

Journal of Physics: Conference Series **446** (2013) 012013 doi:10.1088/1742-6596/446/1/012013

Femtoscopy at High m_T in Heavy-Ion Collisions with ALICE

Hans Beck for the ALICE collaboration

Institut für Kernphysik, Johann Wolfgang Goethe-Universität, Frankfurt, Germany

E-mail: Hans.Beck@cern.ch

Abstract. Femtoscopic two-particle correlations carry important information about the particle emitting source. Of particular interest is the m_T dependence of extracted source radii, which is introduced by the dynamics of the system created in heavy-ion collisions. The reach in m_T of the traditional pion-pion correlation measurements is limited by the small rest mass of the examined particles. Thus, it can be extended with the analysis of pairs of heavier particles: kaons and protons, and proton-lambda pairs, being the heaviest system to extract source sizes so far. The excellent particle identification capabilities and high statistics datasets of ALICE allow to extract source radii differentially in transverse mass and provide new constraints on hydrodynamic models.

1. Introduction

Femtoscopy allows to extract radii that describe the regions of homogeneity of the particle emitting source in a heavy-ion collision, i.e., the fireball at kinetic freeze-out. The m_T dependence of these radii is sensitive to a variety of properties of the produced matter, e.g., the order of the phase transition from the QGP to hadrons, the existence of pre-flow, the viscosity of the QGP, the time scales of the evolution, and the temperature [1, 2]. The standard way of carrying out a femtoscopic analysis of heavy-ion collisions is to investigate pairs of identically charged pions. Their reach in pair transverse mass m_T in ALICE is limited due to their small pair transverse mass for vanishing pair transverse momentum $m_T(k_T \to 0)$ of only 0.14 GeV/c^2. Thus, the reach in m_T can easily be extended by investigating pairs of heavier particles: charged kaons, neutral kaons, protons, and pairs of protons and lambdas, with a $m_T(k_T \to 0)$ of $0.49, 0.50, 0.94$, and 1.03 GeV/c^2, respectively. These systems are affected differently by several sources of correlations. Pairs of identical particles are subject to quantum statistical effects; charged particles will be affected by the Coulomb interaction and all particles will interact strongly, although this interaction can be neglected for the charged pions and charged kaons.

The effect of quantum statistics introduces for bosons a tendency to occupy the same state. The way to experimentally access this enhancement of particles with the same quantum numbers is to divide the two-particle distribution by the product of single-particle distributions. This is realized by dividing pairs in real events by pairs in mixed events. The result is the two-particle correlation function $C_2(q_{\rm inv})$, here measured as a function of the generalized invariant relative momentum of the pair: $q_{\rm inv} = |q - P(qP)/P^2|$, with $q = p_1 - p_2, P = p_1 + p_2$, and p being the momentum of the particles [3]. The correlation function carries information about the spatial extent of the source. In the case of ideal bosons, the correlation function will be a constant

at unity for high relative momenta plus a Gaussian-shaped enhancement with the maximum at $q_{\rm inv} = 0$ and the width being inversely proportional to the size of the source. Impurities in the boson sample will lower the height of the correlation function, but will not affect the width, leaving the size measurement undisturbed. The impact of quantum statistics goes into the opposite direction for fermions, leading to a depletion for small momentum differences. To extract a source size from particle pairs that interact strongly, a sufficient knowledge of the interaction is neccessary. Precise measurements for all discussed pairs here are existing. A model [4] can be utilized to translate the source size via the known interaction to an observable correlation function and vice versa. The correlation functions with a large contribution from the final state strong interaction, i.e., these of pairs of protons and protons and lambdas, are sensitive in height and shape to the size of the source.
While for low $m_{\rm T}$ pairs the region of homogeneity, i.e. the region from which correlated pairs get emitted, still spans over the full source, the expansion of the medium introduces a reduction of this homogeneity region for high $m_{\rm T}$ pairs. With the strong flow developed in Pb-Pb collisions at the LHC the apparent source size decreases as a function of $m_{\rm T}$. The flow velocity is competing with the thermal velocity $v_{\rm Therm} = \sqrt{T/m_{\rm T}}$, thus the dependence is expected to be most pronounced as a function of $m_{\rm T}$ [5].

2. Data analysis

ALICE is the dedicated heavy-ion experiment at the LHC. The analysis presented here was done with the central barrel; i.e. the Inner Tracking System (ITS), Time Projection Chamber (TPC), and Transition Radiation Detector (TRD) were used for tracking. Particle identification (PID) was provided by the TPC and the Time-Of-Flight (TOF) detector. In the following, as an example, a selection of primary protons will be discussed.
The measurement of the specific energy loss $\mathrm{d}E/\mathrm{d}x$ in the TPC and ITS allows to identify single particles in the region where the $\mathrm{d}E/\mathrm{d}x$ is proportional to $1/\beta^2$, with $\beta = v/c$ being the velocity of the particles. This region extends to roughly a momentum of 1 GeV/c for protons. TOF provides PID by measuring the time it takes for a particle to travel from the primary vertex to the TOF detector and is thus sensitive to the mass of the particle when the momentum measurement of the TPC gets included. It provides PID in the momentum region where the specific energy loss becomes the same for different particle species.
Depending on the momentum of the particle, different PID techniques were chosen. Namely the TPC for momenta $p < 0.75$ GeV/c, TPC and TOF for $0.75 \leq p < 1.0$ GeV/c and TOF alone for $1.0 \leq p < 3.25$ GeV/c. The proton selection for $p \leq 0.75$ GeV/c was done by applying an upper and lower cut on the measured $\mathrm{d}E/\mathrm{d}x$, which is based on a parametrization of the Bethe-Bloch curve. A clear separation of the protons is achieved for these momenta with the TPC alone. The purity of the proton sample was determined by fitting the $\mathrm{d}E/\mathrm{d}x$ distributions in slices of total momentum. The data are very well described by a combination of Gaussians for the different particle species. Even at momenta of 0.75 GeV/c the excellent TPC performance allows to select a proton sample of a purity higher than 99% with essentially no loss of protons by applying the selection criteria. A momentum of $0.75 \leq p < 1.0$ GeV/c, where the hybrid approach of using the TPC and TOF was chosen, is quite low for identifying protons with TOF alone. By applying a selection criterion based solely on the TOF information, a purity higher than 90% only is achieved. Looking at the time-of-flight spectrum, one sees that the main background in this momentum region comes from mismatched particles. This mismatch mainly originates from pions, the most abundant particles, which get assigned a wrong TOF cluster. Thus, one can easily apply a mild $\mathrm{d}E/\mathrm{d}x$ pre-selection which rejects pions and safely keeps all protons. Applying the $\mathrm{d}E/\mathrm{d}x$ pre-selection lifts the purity above 99%. For higher momenta the matching improves and no TPC pre-selection is neccessary. Out to $p < 3.25$ GeV/c, where the proton selection currently stops, purities over 99% are achieved.

As we are interested in studying the correlations of primary protons, another selection criterion on the transverse distance of closest approach of the track to the primary vertex (DCA_{xy}) was imposed. Templates for the shapes of the different distributions in DCA_{xy} for primary protons, protons from weak decays and protons produced in the detector material were obtained from Monte-Carlo simulations. It is important that these templates have a different shape in order to reliably discriminate between the contributions. This is the case: primary protons are peaked at $DCA_{xy} = 0$, protons from weak decays show broad shoulders and protons from material are flat in DCA_{xy}. Fitting the data with the templates, it was found that selecting only tracks with $DCA_{xy} \leq 0.1$ cm maximizes the significance of the primary protons. Furthermore, one can determine the residual contamination in the sample. In order to take into account the evolution of the amount of the different contributions and the slightly varying shapes in phase space, the fits were performed in bins of rapidity and transverse momentum.

2.1. Kaon analysis

We also report on femtoscopic radii obtained from correlations of charged and neutral kaons. K^0_s were reconstructed via their charged decay $K^0_s \to \pi^+\pi^-$. Oppositely charged tracks are paired and potential decay vertices are calculated by a V0 finder. Several analysis cuts were applied, e.g., PID and a maximum distance of closest approach of the daughters was required. K^0_s were finally selected by a window in the invariant mass around the value of the particle data group. A purity of $\approx 95\%$ could be achieved for the 10% most central events with higher purities for more peripheral events. $K^\pm$ could be identified with only the dE/dx measurement of the TPC up to transverse momenta of 0.6 GeV/c with a maximum contamination from $e^\pm$ in the region $p_T \approx 0.4$ GeV/c, where the specific energy loss of electrons and $K^\pm$ are equal, and an amount of $\pi^\pm$ in the region $0.5 \leq p_T < 0.6$ GeV/c rising with p_T. The growing contamination from $\pi^\pm$ could be reduced with the information from TOF and lead to the requirement of a TOF measurement for $p_T \geq 0.6$ GeV/c. Using the information from TOF, $K^\pm$ could be identified up to $p_T = 1.5$ GeV/c. The contamination in the final $K^\pm$ sample is significant only in the region $0.4 \lesssim p_T < 0.6$ GeV/c and is $\lesssim 20\%$.

3. Results

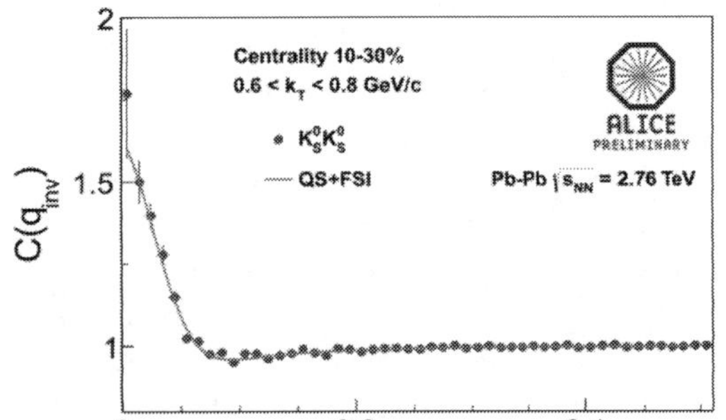

Figure 1: Examplary $K^0_sK^0_s$ correlation function

All correlation functions were obtained in common centrality bins: 0 – 10%, 10 – 30%, and 30 – 50%. The negatively and positively charged kaon pair distributions were combined after validating that the correlations functions coincide within errors. The $K^\pm$ correlation functions were obtained for seven bins in k_T ranging from 0.2 to 1.0 GeV/c and fitted with the Bowler-Sinyukov function:

$$C(q_{inv}) = 1 - \lambda + \lambda K(q_{inv})(1 + \exp(-R^2_{inv} q^2_{inv})).$$

$K(q_{inv})$ takes into account the Coulomb repulsion of the particles assuming factorization and is determined by the overlap of the square of the Coulomb wave function with the finite source [6, 7]. The data show the expected enhancement at low q_{inv} and are nicely described by the fit function. A clear k_T dependence is visible. The neutral kaons were fitted in four k_T bins from 0.2 to 1.5 GeV/c taking into account the attractive final state strong interaction and the quantum statistics. The extracted radii are in agreement with the charged kaon

analysis. Also the proton-proton correlation functions were obtained differentially in k_T. As for the other particles, a strong centrality dependence is observed. (See also [8].)
The extracted radii from the measurements discussed are shown together with the ones from pions in Fig. 2. The radii for all species and all m_T decrease for more peripheral events. The reach in m_T is doubled by the kaons and protons with respect to the pions, exhibiting the decrease of the radii with m_T out to $m_T = 1.6$ GeV/c^2. Comparing the radii for the different species, one observes only an approximate m_T scaling. To further study this slight tension, three-dimensional radii in the longitudinally co-moving system are under investigation in ALICE. The excellent PID and the high statistics datasets of ALICE allows for precise measurements over a wide range of m_T with overlapping regions of measurement in m_T for different particle species.

4. Outlook

At the SPS and RHIC, proton-Λ correlations provided femtoscopic radii at highest m_T [9, 10]. Fig. 3 shows the purity of $\bar{\Lambda}$ maximized for significance obtained with ALICE for the 10% most central collisions. Within $|\eta| < 0.9$ typical purities are on the order of 95%, increasing for more peripheral events. A radius parameter obtained from pairs of protons and Λ should be available soon.

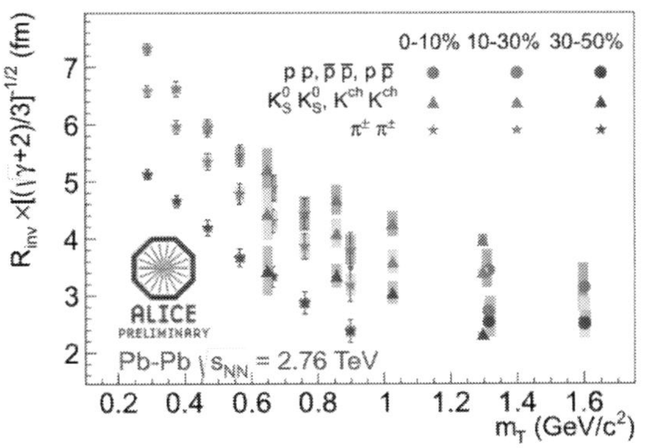

Figure 2: m_T dependence of radii from pions, kaons, and protons

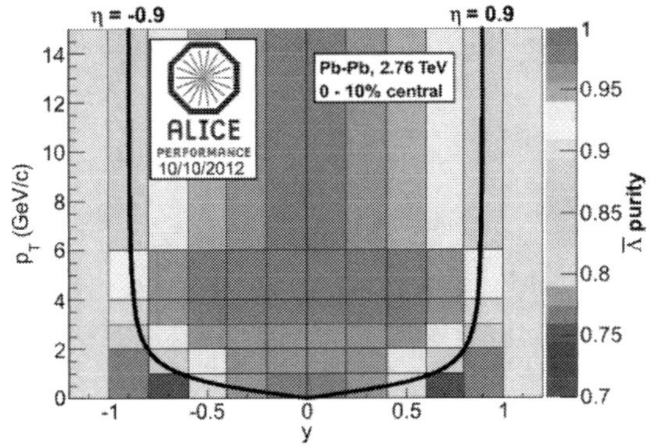

Figure 3: Purity of $\bar{\Lambda}$ as a function of their transverse momentum and rapidity for the 10% most central Pb-Pb collisions at $\sqrt{s_{NN}} = 2.76$ TeV.

5. Summary

ALICE provides an excellent environment for high purity and high statistics femtoscopy. The doubly differential centrality and m_T dependence of femtoscopic radii for kaons and protons was presented and compared to radii for pions obtained by ALICE. The obtained radii decrease with m_T, exhibiting features of strong collectivity in Pb-Pb collisions at the LHC.

References

[1] Pratt S 2009 *Phys. Rev. Lett.* **102** 232301
[2] Garishvili I *these proceedings*
[3] Lednicky R 2001 *Preprint* arXiv:nucl-th/0112011
[4] Lednicky R and Lyuboshits V L 1982 *Sov. J. Nucl. Phys.* **35** 770
[5] Lisa M A, Pratt S, Soltz R and Wiedemann U 2005 *Ann. Rev. Nucl. Part. Sci* **55** 357
[6] Bowler M G 1991 *Phys. Lett. B* **270** 69
[7] Sinyukov Y M, Lednicky R, Akkelin S V, Pluta J, Erazmus B 1998 *Phys. Lett. B* **432** 248
[8] Szymanski M P for the ALICE Collaboration 2012 *Preprint* arXiv:1211.3288 [nucl-ex]
[9] Anticic T *et al* (NA49 Collaboration) 2011 *Phys. Rev. C* **83** 054906
[10] Gos H P for the STAR collaboration 2007 *Eur. Phys. J. C* **49** 75

Journal of Physics: Conference Series **446** (2013) 012014 doi:10.1088/1742-6596/446/1/012014

Can baryon stopping explain the breakdown of constituent quark scaling and proposed signals of chiral magnetic waves at RHIC?

J M Campbell and M A Lisa

Department of Physics, Ohio State University, Columbus, Ohio 43210, USA

E-mail: campbell@physics.osu.edu

Abstract. Azimuthal emission spectra of various hadron species in ultra-relativistic heavy ion collisions at $\sqrt{s_{NN}} \approx 200$ GeV exhibit a curious hierarchy at intermediate p_t ($\approx 2-3$ GeV). Rather than being ordered by mass, the spectra seem to be ordered by whether the species is a baryon or meson. It is seen that when the elliptic flow ν_2 and transverse momentum p_T are both scaled by the number of quarks in each hadron, the spectra fall in line with each other. This number of constituent quark (NCQ) scaling suggests a system where the relevant degrees of freedom are colored partons as opposed to hadrons: the quark-gluon plasma (QGP). Thus, a break down of this scaling as beam energy is reduced could be indicative of the QGP threshold. However, at lower energies, there is also an increase in the number of entrance-channel partons transported to mid-rapidity due to baryon stopping, which can violate NCQ scaling even above the QGP threshold. We describe a specific pattern for the break down of the scaling that includes the observed difference in elliptic flow for positive and negative pions. We also contrast baryon stopping with the Chiral Magnetic Effect (CME)–an alternative model for π_+/π_- flow difference–and discuss results from tests that can distinguish between them.

1. Introduction

In non-central heavy ion collisions the azimuthal momentum distribution is anisotropic. This anisotropy is typically characterized in terms of Fourier components:

$$\frac{dN}{d\phi} \propto 1 + 2\sum_{n=1}^{\infty} \nu_n \cos(n\phi) \tag{1}$$

The second component ν_2–the elliptic flow parameter–is perhaps one of the most extensively analyzed quantities in the field. An important observation is the species dependence of ν_2 at top RHIC energies. At low p_T, ν_2 displays a simple mass ordering, and agrees well with hydrodynamic predictions, with lighter particles exhibiting a larger ν_2 at a given p_T. But at $p_T \geq 2$ GeV this ordering is broken, and mesons all begin to follow one common curve and baryons another, with baryons exhibiting a larger ν_2 than mesons. Intriguingly, scaling both ν_2 and p_T by the number of constituent quarks (NCQ) for each hadron–2 for mesons, 3 for baryons–causes $\nu_2(p_T)$ for all particle species to follow a common curve.

This result suggests that the relevant degrees of freedom in the collision system are quarks, not hadrons [1]. Once the flow field has been established, the system cools and the constituent quarks "coalesce" into the observed hadrons. This simple interpretation lends credence to claims that a new state of QCD matter–the quark-gluon plasma (QGP)–has been produced at RHIC.

If NCQ scaling is to be taken as a hallmark of QGP, it is compelling to search for the conditions under which it appears. A major thrust of the heavy ion community is an extensive beam energy scan program at RHIC [2], in which observables sensitive to the state of QCD matter are precisely measured as the collision energy $\sqrt{s_{NN}}$ is measured in small steps. The *breakdown* of NCQ scaling as $\sqrt{s_{NN}}$ is reduced may indicate the QGP threshold energy.

However, it is important to investigate less exciting phenomena that could bring about NCQ scaling violations. Two proposed mechanisms are baryon stopping [3] and chiral magnetic waves (CMW) [4].

2. Baryon stopping and Multi-component coalescence

In a recent publication [3], Dunlop *et al.* show that baryon stopping alone–with no change in the phase of matter–may lead to a characteristic pattern of NCQ-scaling breakdown. This is based on the speculation that entrance channel quarks must suffer many collisions to be *transported* to mid-rapidity. These collisions should give them higher elliptic flow profiles than the *produced* quarks ($\nu_2^T > \nu_2^P$) which are created pairwise *at* mid-rapidity. The observed hadrons, however, will coalesce out of the combined quark population. We shall call this the multi-component coalescence (MCC) scenario.

If the data from the RHIC energy scan program show a breakdown of NCQ scaling at low energy, then comparing the measured species-dependent pattern of the breakdown with theoretical expectations may help distinguish the mechanism behind the breaking. Presumably, if the mechanism is a change of phase from partonic to hadronic matter, the observed ν_2 should be ordered according to hadron cross-section, rather than the pattern predicted [3] in the MCC scenario.

Among other consequences of multi-component coalescence, the elliptic flow of negative pions should be larger than that of positive pions: $\Delta\nu_2^{\pi} \equiv \nu_2^{\pi^-} - \nu_2^{\pi^+} > 0$. This is due to the relative number of entrance channel quarks that will coalesce into exit channel negative pions versus positive pions. More specifically, in the MCC scenario, the resulting baryon (meson) ν_2 is given by adding the flow profiles of the three (two) quark species that make it up. In pions, for example:

$$\begin{aligned}\nu_2^{\pi^-}[d\bar{u}] &= X_{d^T}\nu_2^{d^T} + (1 - X_{d^T})\nu_2^{d^P} + \nu_2^{\bar{u}^P}\\ \nu_2^{\pi^+}[u\bar{d}] &= X_{u^T}\nu_2^{u^T} + (1 - X_{u^T})\nu_2^{u^P} + \nu_2^{\bar{d}^P}\end{aligned} \tag{2}$$

where the ν_2 values on the right side of the equations represent quark (or anti-quark) flow strengths. Superscripts T and P denote transported and produced quarks, respectively. The quantity X_{d^T} (X_{u^T}) represents the fraction of all d (u) quarks at mid-rapidity that arise from transport from the entrance channel:

$$X_{d^T} \equiv \frac{N_{d^T}}{N_{d^T} + N_{d^P}}, \qquad X_{u^T} \equiv \frac{N_{u^T}}{N_{u^T} + N_{u^P}} \tag{3}$$

Because coalescence and MCC only consider valence quarks, the incoming nuclei have no anti-quarks or strange quarks to transport: $X_{s^T} = X_{\bar{u}^T} = X_{\bar{d}^T} = X_{\bar{s}^T} = 0$.

Let us consider the simple case where all produced quarks have the same flow strength, and all transported quarks have the same flow strength. Then,

$$\begin{aligned}\nu_2^P &\equiv \nu_2^{u^P} = \nu_2^{d^P} = \nu_2^{s^P} = \nu_2^{\bar{u}^P} = \nu_2^{\bar{d}^P} = \nu_2^{\bar{s}^P}\\ \nu_2^T &\equiv \nu_2^{d^T} = \nu_2^{u^T}\end{aligned} \tag{4}$$

Combining equations (2) and (4) yields:

$$\Delta \nu_2^{\pi} = \Delta X_T(\nu_2^T - \nu_2^P) \tag{5}$$

where $\Delta X_T \equiv (X_{dT} - X_{uT})$. Since the quantity $\nu_2^T - \nu_2^P$ is assumed to be positive, equation (5) implies a *positive correlation* between $\Delta\nu_2^{\pi}$ and ΔX_T.

3. Pion elliptic flow and chiral magnetic waves

Burnier *et al.* [4] have proposed an explanation for non-zero $\Delta\nu_2^{\pi}$ that arises from chiral magnetic waves (CMW). The CMW induce an electric quadrupole moment in the QGP, which is predicted to lead to more elliptic flow for negative pions than for positive pions. The magnitude of this difference is predicted to be proportional to the charge asymmetry:

$$\Delta\nu_2^{\pi} \approx rA_{\pm} \tag{6}$$

where $A_{\pm} \equiv (N_+ - N_-)/(N_+ + N_-)$ is the charge asymmetry of the plasma, N_+ (N_-) is the number of positively (negatively) charged particles in the plasma, and r is the (positive) normalized electric quadrupole moment of the system. This relation is derived in the small $A_{\pm}$ limit.

Direct measurement of the electric quadrupole moment is impossible, but one may, for a given centrality selection, measure $\Delta\nu_2^{\pi}$ as a function of $A_{\pm}$. A direct proportionality would be consistent with the chiral magnetic wave prediction [4].

We would like to compare the predictions of the CMW and baryon stopping hypotheses by finding some meaningful way to compare equations (5) and (6). That is, what is the relation between ΔX_T and $A_{\pm}$?

To see how these quantities *might* be correlated, consider event by event fluctuations of both the number of stopped baryons and of isospin. Increasing the number of stopped baryons will inject net charge to mid-rapidity (by stopping more protons), thereby increasing $A_{\pm}$. It will also enrich the down quark population preferentially more than up quarks (since there are more down quarks in the entrance channel), thereby increasing ΔX_T: a *positive* correlation.

A positive fluctuation of isospin will also increase $A_{\pm}$ by stopping more protons (increasing $N_+ - N_-$) or fewer neutrons (decreasing $N_+ + N_-$). However, this will also enrich the up quark population preferentially more than down quarks, decreasing ΔX_T. Thus isospin fluctuations lead to a *negative* correlation.

Surely both types of fluctuations occur, but which one dominates?

4. UrQMD analysis

To determine the dominant source of fluctuations we seek to relate ΔX_T and $A_{\pm}$. Here we present results from UrQMD simulations of Pb+Pb collisions at $\sqrt{s_{NN}} = 17$ GeV. The distribution shown in figure 1 shows a strong linear relationship between ΔX_T and $A_{\pm}$. We are therefore justified in writing:

$$\Delta X_T = mA_{\pm} + b \tag{7}$$

where m is the slope of the distribution and b is some intercept. The data shown in Figure 1 yield $m = -1.64$ and $b = 0.15$. Combining this with equations (5) and (7) gives:

$$\Delta\nu_2^{\pi} = -1.64(\nu_T - \nu_P)A_{\pm} + 0.15(\nu_T - \nu_P) \tag{8}$$

We now have a prediction from the MCC scenario that is qualitatively different from the CMW prediction. From equation (6), CMW predicts a directly proportional, positive correlation between $\Delta\nu_2^{\pi}$ and $A_{\pm}$. MCC predicts a *negative* correlation, with some non-zero offset.

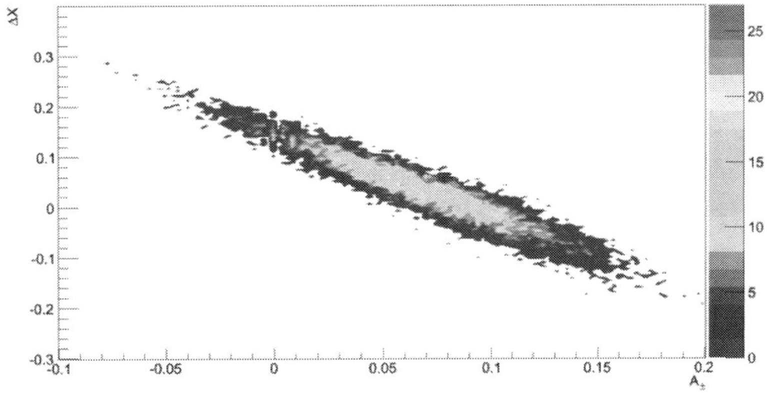

Figure 1. 10,000 events of Pb+Pb collisions at $\sqrt{s_{NN}} = 17$ GeV . $|\eta| \leq 0.5$.

5. Summary

Measured ν_2 parameters of hadrons at top RHIC energies follow a simple pattern that strongly suggests that it is the constituent quarks, rather than the hadrons themselves, that are experiencing collective flow. Searches for a break-down of this pattern are under way in the beam energy scan program at RHIC. Mechanisms behind such a break down include a change in the fundamental dynamical degrees of freedom in hot QCD matter and chiral magnetic wave effects.

Rather than evoking such interesting mechanisms, the multi-component coalescence approach asks what would be the consequence of changing nothing about the system except what we already *know* changes as the collision energy is lowered: increased baryon number transport. This simplistic approach yields a characteristic pattern of hadron ν_2 that should be compared to data arising from the energy scan. Indeed, the measured pattern–for a wide range of hadron species–should be compared to predictions of any proposed mechanism such as chiral magnetic waves.

Here we have compared MCC- and CMW-based predictions of the difference between the elliptic flow strengths of negative and positive pions. UrQMD simulations suggest that the two hypotheses have qualitatively different implications. Namely, CMW predicts a directly proportional, positively correlated, relationship between $\Delta\nu_2^\pi$ and $A_\pm$, whereas MCC predicts a negative correlation, with some positive $\Delta\nu_2^\pi$ at zero charge asymmetry.

6. Acknowledgments

This work is supported by the U.S. National Science Foundation under Grant PHY-0970048.

References

[1] Fries R, Greco V and Sorensen P 2008 *Ann. Rev. Nucl. Part. Sci.* **58** 177-205
[2] Aggarwal M *et al* 2010 *arXiv*:1007.2613
[3] Dunlop J, Lisa M and Sorensen P 2011 *Phys. Rev. C* **84** 044914
[4] Burnier Y, Kharzeev D, Liao J and Yee H 2011 *Phys. Rev. Lett.* **107** 052303

Hot Quarks 2012 IOP Publishing
Journal of Physics: Conference Series **446** (2013) 012015 doi:10.1088/1742-6596/446/1/012015

Beam Energy Scan Results from RHIC

P. Sorensen

Brookhaven National Laboratory, Physics Department, Bldg. 510a, Upton, NY 11973

E-mail: prsorensen@bnl.gov

Abstract. In 2010 and 2011, RHIC ran the first phase of a planned beam energy scan program to probe, among other things, the nature of the phase transition between hadrons and Quark Gluon Plasma as the matter vs anti-matter excess increases. Many experimental findings are now available from that scan. In this talk, I discuss the meaning of those results and the future plans and motivation for the second phase of the RHIC beam energy scan.

1. Introduction

The Relativistic Heavy Ion Collider (RHIC) was built to create and study Quark Gluon Plasma (QGP) [1, 2]. The matter created in collisions at RHIC is so hot and dense that quarks and gluons become the relevant degrees of freedom, not hadrons [3]. Even before the discovery of quarks and the advent of Quantum Chromodynamics, Hagedorn realized that the exponential increase in the hadronic mass spectrum implies that there is a maximum temperature (T) in a hadronic gas [4]. As we put more energy into the hadron gas, the energy can go into creating ever more higher mass states instead of increasing T (leading to a maximum T_{H}). With Quantum Chromodynamics, however, we realize that once the energy density becomes high enough, quarks and gluons become the relevant degrees of freedom. The left panel of Fig. 1 shows finite temperature Lattice QCD calculations of the energy density of nuclear matter scaled by T^4 which is proportional to the number of degrees of freedom in the system. As the temperature of the system increases, the degrees of freedom increase but then start to saturate above T=200 MeV. This saturation reflects the transition from a system where the energy pumped in to the system to heat it goes into creating heavier hadronic resonant states, to a system where hadronic states are irrelevant, quarks and gluons are the degrees of freedom, and pumping energy into the system goes into creating a hotter QGP. Lattice calculations have shown that when matter and anti-matter exist in equal amounts (zero baryon chemical potential μ_{B}) that transition between QGP and hadrons is a smooth cross-over [5].

Although Lattice calculations show that the QGP to hadron phase transition is a smooth crossover at zero baryon density, it is thought to be a first-order phase transition at higher baryon densities [7]. Lattice calculations suffer from the sign problem at non-zero baryon chemical potential so it is not easy to infer from QCD where the transition from a smooth crossover to a first order phase transition occurs. The point where the smooth crossover and first order phase transition meet is called a critical point and represents an important landmark on the phase diagram of nuclear matter. Figure 1 (right) shows a schematic diagram of the phases of QCD. Full energy heavy-ion collisions at RHIC and those that at the LHC, are nearly net baryon free and probe along the left axis of the diagram. The flexibility of RHIC makes it possible to probe

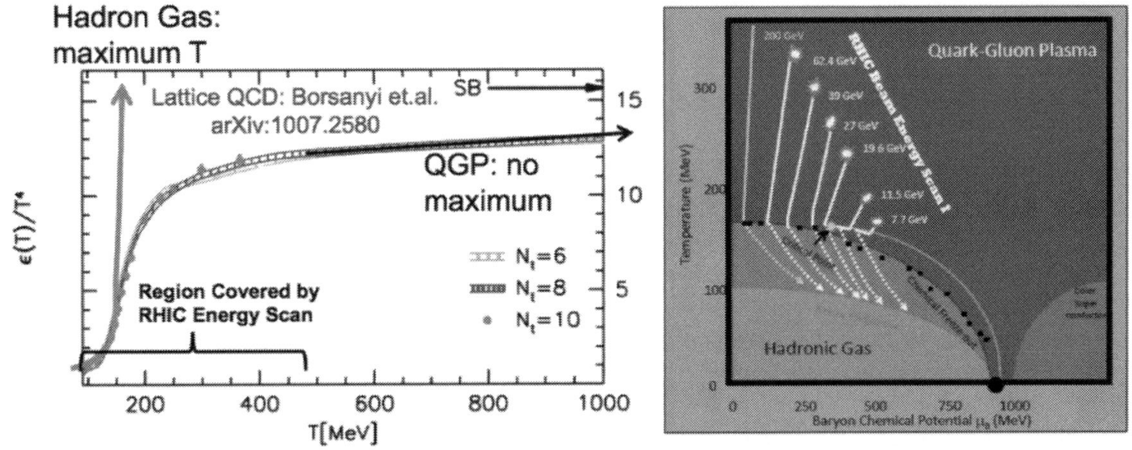

Figure 1. (Left) Lattice QCD calculations showing the transition from hadronic degrees of freedom to a QGP. (Right) the QCD phase diagram with trajectories showing the regions that can be probed by RHIC [6].

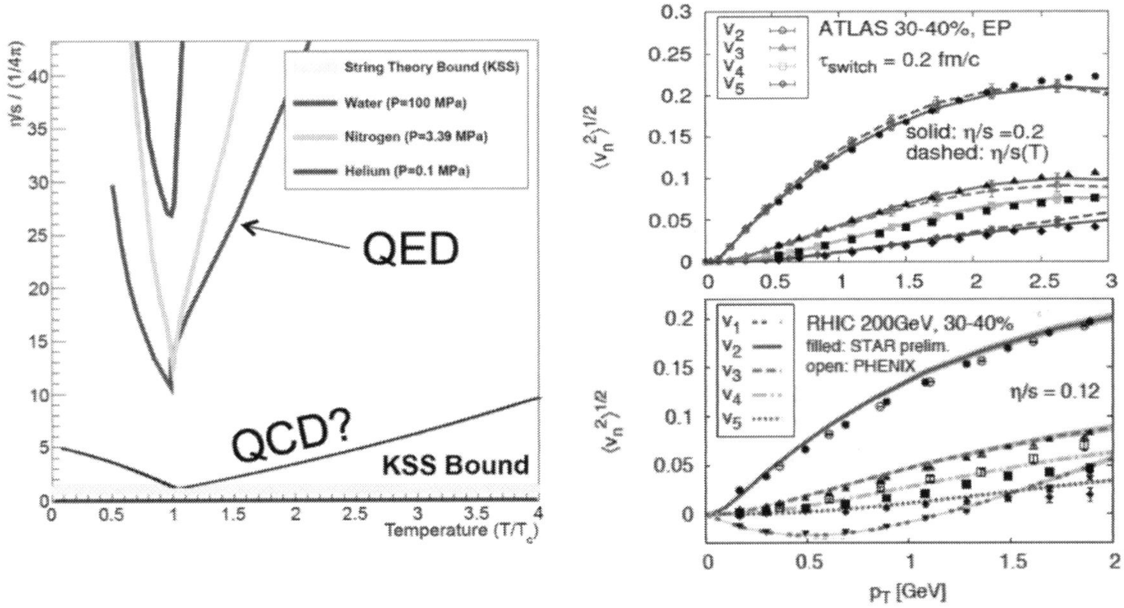

Figure 2. (Left) The temperature dependence of η/s for a variety of materials and a curve for the QCD equivalent which is not yet determined. (Right) Hydro calculations compared to data that can be used to better constrain the temperature dependence of η/s.

deeply into the non-zero baryon chemical potential axis. By lowering the energy of the colliding beams, one initiates a system with larger net baryon density. By scanning down to lower and lower energies, the RHIC experiments can search for evidence of a critical point or of a first order phase transition. The first phase of this program has been completed at RHIC [8] with many results already available.

When reducing the beam energy, one also reduces the initial temperature of the QGP phase until at low enough energies, one presumably does not heat the colliding nuclei enough to reach the QGP phase. One of the most important conclusions from the first five years of RHIC operations was that the QGP created at RHIC was a nearly perfect fluid [9, 10]. This conclusion was based on the observation that the QGP phase of the collisions appears to be well described by nearly perfect hydrodynamic models. Only a relatively small amount of viscosity can be introduced before the agreement between the models and the data are spoiled. Most calculations have relied on a constant viscosity but as shown in Fig. 2 (left) in most other systems, the viscosity to entropy ratio depends strongly on the temperature and has a minimum where a phase transition occurs [11]. It is possible therefore that the average viscosity of the QGP will be smaller at lower energies than it is at higher energies. Varying the beam energy and the initial temperature of the system will allow us to better extract the temperature dependence of the viscosity to entropy ratio η/s.

Figure 2 (right) shows data on the flow harmonics for collisions at 2.76 TeV and at 200 GeV with hydrodynamic model calculations adjusted to fit the data [12]. For the higher energy collisions, a value of $\eta/s = 0.2$ is preferred while for 200 GeV, a value of 0.12 is preferred. This is an early indication of a temperature dependence for η/s. It's also important to note that when looking only at the 2.76 TeV data, it is difficult to tell the difference between a constant η/s and a temperature dependent η/s; note the solid and dashed curves in the figure are almost indistinguishable. The study of the temperature dependence of η/s likely will require large data sets at a variety of beam energies such as have been generated with the first phase of the beam energy scan, or will be generated in the second phase.

2. Results from Beam Energy Scan Phase-One

Having reviewed the motivation for the beam energy scans, I will briefly discuss some of the results of the scan. In the first phase of the scan, small data-sets were collected at 7.7 and 11.5 GeV center of mass energy (several million events) with larger data sets (20 million events or larger) being collected at 19.6, 27, 39, and 62.4 GeV. The second phase of the beam energy scan will focus on increasing the precision of key results at the lower center of mass energies.

2.1. Energy Density and Flow

Simply by estimating the number of particles created in the collisions one can get a rough estimate of the initial energy density of the system. This requires one to guess at how long after the initial collision, some form of matter has been created (the formation time of a QGP τ_0). Figure 3 (left) shows estimates of the initial energy density scaled by τ_0 [13]. Lattice calculations show that the QGP should form when the energy density gets much higher than 0.6 GeV/fm^3. For $\tau_0 = 1$ fm, even the lowest energy collisions can be expected to create a QGP according to this criteria and estimate. One should not therefore jump to the conclusion that no QGP will be formed in the lowest energy collisions. Even for $\tau_0 = 2$ fm, the central 7.7 GeV collisions would reach energy densities high enough for QGP formation.

One of the signatures of QGP formation is a large azimuthal anisotropy in particle production thought to be generated by elliptic flow as the QGP responds to azimuthal variations of pressure gradients in the system. Figure 3 (right) shows the second harmonic variation v_2 vs p_T calculated from the four-particle cumulant method measured for several centralities and for energies ranging from 7.7 GeV up to 2.76 TeV [14]. It is striking how little this observable changes over such a wide range of energies. Note that the $\langle p_T \rangle$ of the particles does increase with energy so the average v_2 increases even if $v_2(p_T)$ does not. Also note that by looking at the ratio of different energies, one can see that the low-p_T data where most of the particles are does increase by 30-40% from 7.7 to 2.76 TeV. The relatively small change in $v_2(p_T)$ for energies all the way down to 7.7 GeV provides some indication that a QGP may still exist even for the lowest energies.

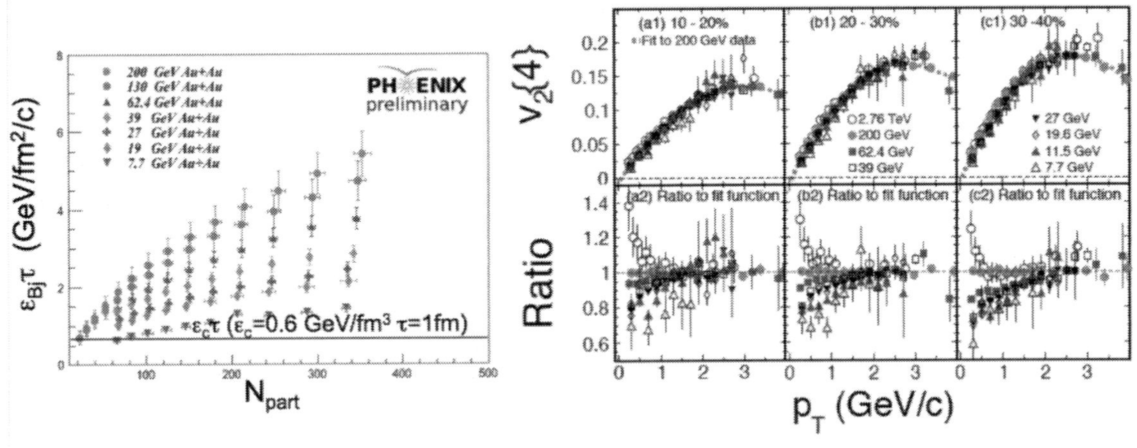

Figure 3. (Left) Estimates of the initial energy density for a variety of energies time the formation time. (Right) The variation of $v_2(p_T)$ from 7.7 GeV up to 2.76 TeV.

2.2. NCQ Scaling of Positively and Negatively Charged Particles

While as noted before, the $v_2(p_{\rm T})$ for charged hadrons does not change very much from 7.7 GeV to 2.76 TeV, when looking at the particle type dependence of $v_2(p_T)$ we do see changes in behavior as the energy is decreased [15]. The six panels on the left in Fig. 4 show $v_2(m_{\rm T} - m_0)$ for positively charged particles at six different energies. The panels on the right show the same for negatively charged particles. In the left panels, the baryons at all energies show a larger v_2 value at any given $m_{\rm T} - m_0$. This enhancement of baryon v_2 is taken as evidence that some baryons and mesons at intermediate p_T are created by the coalescence of constituent quarks at the hadronization phase boundary after the creation of a QGP [16]. As such, it is a good indicator for the presence of a QGP phase. For the positively charged particles in the left panels, this pattern of larger baryon v_2 exists at all energies. For negative particles shown in the left panel however, the pattern breaks down with anti-baryons no longer exhibiting an enhanced v_2 relative to mesons for energies below 19.6 GeV. The break down of the signal for coalescence can be traced to the reduced v_2 values for anti-baryons relative to baryons. This pattern and a more detailed look at differences between the other particles types suggests that this difference between positive and negative particles can either be related to the effect of mean-fields on partons [18] or on hadrons [17] in a system with ever increasing μ_B, or that the quark number transported to mid-rapidity leads to quarks with a larger v_2 that then coalesce into hadrons [19]. This could be either due to a concentration of net-baryon number in the in-plane region of the overlap volume or a larger flow being built up to bring the quarks carrying the baryon number to midrapidity through greater interactions. In either case, the data do not necessarily or even likely indicate a break-down of coalescence.

Note that the results for the ϕ-meson are very limited in their statistical precision. Larger data-sets need to be accumulated at the lower energies to better study the multi-strange hadron flow which is expected to reflect the partonic phase of the collisions more directly.

2.3. High $p_{\rm T}$ Suppression

$R_{\rm CP}$ (central spectra over peripheral each scaled by the number of binary collisions) also has a strong energy dependence. This ratio reflects changes in the shape of the spectra. At higher energies the higher momentum part of the spectra is suppressed because hard scattered partons lose energy in the QGP before they hadronize. At lower energies, instead of suppression an

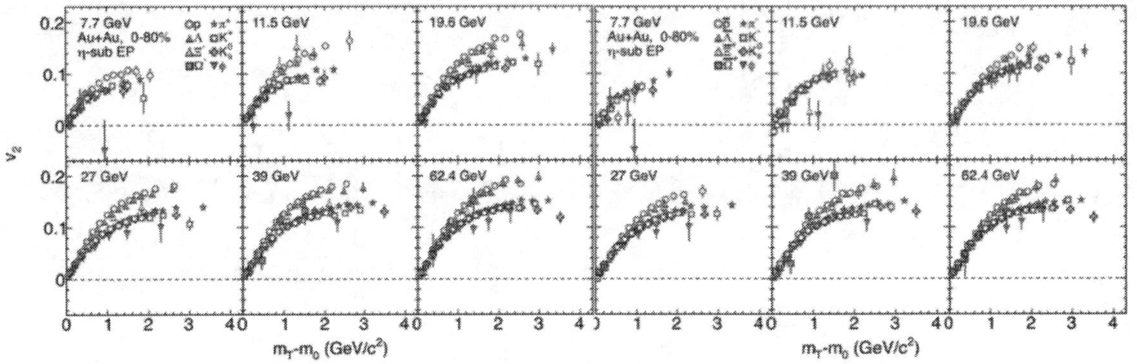

Figure 4. $v_2(m_{\mathrm{T}} - m_0)$ for 0-80% central Au+Au collisions at six different energies. Positively charged particles are shown in the left panels and negatively charged particles in the right.

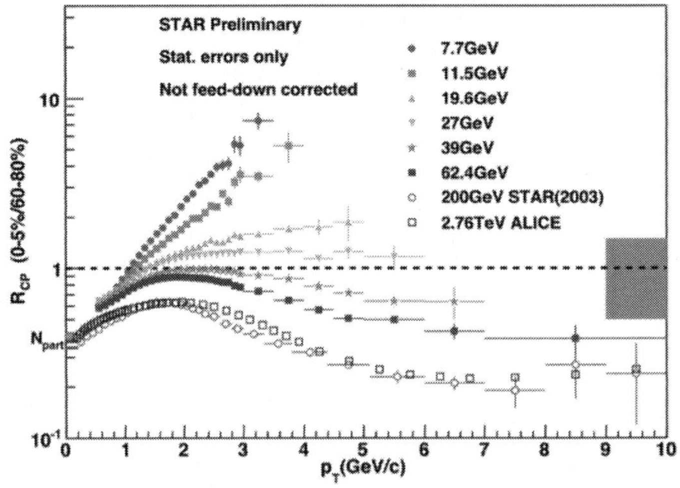

Figure 5. R_{CP} for charged hadrons at a variety of beam energies.

enhancement is seen [20]. As the energy is reduced, the number of hard partons is also reduced leading to a much softer spectra at lower energies. Flow can cause an enhancement in the spectra at higher p_{T} as abundant lower momentum particles are boosted into higher momentum where there are fewer particles to start with. This effect is stronger, for softer spectra. It's not clear to me therefore whether the lack of suppression in the low energy spectra are the result of flow and an already reduced number of hard partons that can be quenched or due to a disappearance of quenching in a QGP. In my judgement the former is more likely than the latter so the data are not evidence for the disappearance of the QGP.

2.4. Conclusions

RHIC provides experimental access to the most interesting regions needed to study QCD thermodynamics. At RHIC, one can reduce the initial temperature from several times the critical temperature to temperatures just above or below T_{C}. Comparisons to models suggests there is a temperature dependence to η/s with a smaller value needed to fit 200 GeV RHIC data than the 2.76 TeV LHC data. Data from the RHIC beam energy scan phase two can be used to map out the temperature dependence of η/s. The first phase of the beam energy scan led to several surprising discoveries including the μ_B dependence of v_2 and the non-monotonic variation of net proton v_1 trends (not shown in this proceedings). From the data at hand, it is plausible that a QGP may exist even at the lowest beam energy of 7.7 GeV but something interesting is happening below 20 GeV. A second phase of the scan is needed to accumulate sufficient events at lower energies to measure key observables.

Acknowledgements: I would like to thank the organizers for their kind invitation to present this talk at Hot Quarks 2012.

References

[1] W. Reisdorf and H. G. Ritter, Ann. Rev. Nucl. Part. Sci. **47**, 663 (1997).
[2] N. Herrmann, J. P. Wessels and T. Wienold, Ann. Rev. Nucl. Part. Sci. **49**, 581 (1999).
[3] S. Borsanyi *et al.*, JHEP **1011**, 077 (2010); F. Karsch, PoS C **POD07** (2007) 026; PoS **LAT2007** (2007) 015.
[4] R. Hagedorn, Statistical Thermodynamics of Strong Interactions at High Energies, Nuovo Cim. Suppl. 3, 147-186 (1965).
[5] Y. Aoki, G. Endrodi, Z. Fodor, S. D. Katz and K. K. Szabo, Nature **443**, 675 (2006).
[6] J. Cleymans, H. Oeschler, K. Redlich and S. Wheaton, Phys. Rev. C **73**, 034905 (2006).
[7] F. R. Brown *et al.*, Phys. Rev. Lett. **65**, 2491 (1990); A. Mocsy, F. Sannino and K. Tuominen, Phys. Rev. Lett. **92**, 182302 (2004); F. Karsch and M. Lutgemeier, Nucl. Phys. B **550**, 449 (1999); M. Asakawa and K. Yazaki, Nucl. Phys. A **504**, 668 (1989); A. Barducci, R. Casalbuoni, G. Pettini and R. Gatto, Phys. Rev. D **49**, 426 (1994); J. Berges and K. Rajagopal, Nucl. Phys. B **538**, 215 (1999); A. M. Halasz, A. D. Jackson, R. E. Shrock, M. A. Stephanov and J. J. M. Verbaarschot, Phys. Rev. D **58**, 096007 (1998); O. Scavenius, A. Mocsy, I. N. Mishustin and D. H. Rischke, Phys. Rev. C **64**, 045202 (2001); N. G. Antoniou and A. S. Kapoyannis, Phys. Lett. B **563**, 165 (2003); Y. Hatta and T. Ikeda, Phys. Rev. D **67**, 014028 (2003); M. A. Stephanov, Prog. Theor. Phys. Suppl. **153**, 139 (2004).
[8] P. Sorensen [STAR Collaboration], PoS **CPOD2006** (2006) 019; T. Satogata, *et.al.*, PoS **CPOD2007** (2007) 051.
[9] P. Huovinen, P. F. Kolb, U. W. Heinz, P. V. Ruuskanen and S. A. Voloshin, Phys. Lett. B **503**, 58 (2001); D. Teaney, Phys. Rev. C **68**, 034913 (2003).
[10] B. Muller, Acta Phys. Polon. B **38**, 3705 (2007); W. A. Zajc, Nucl. Phys. A **805**, 283 (2008).
[11] A. Majumder, B. Muller and X. -N. Wang, Phys. Rev. Lett. **99**, 192301 (2007).
[12] C. Gale, S. Jeon, B. Schenke, P. Tribedy and R. Venugopalan, Phys. Rev. Lett. **110**, 012302 (2013).
[13] J. D. Bjorken, Phys. Rev. D **27**, 140 (1983); J. T. Mitchell [PHENIX Collaboration], arXiv:1211.6139 [nucl-ex].
[14] L. Adamczyk *et al.* [STAR Collaboration], Phys. Rev. C **86**, 054908 (2012).
[15] L. Adamczyk, J. K. Adkins, G. Agakishiev, M. M. Aggarwal, Z. Ahammed, I. Alekseev, J. Alford and C. D. Anson *et al.*, arXiv:1301.2347 [nucl-ex]; L. Adamczyk *et al.* [STAR Collaboration], arXiv:1301.2348.
[16] R. J. Fries, V. Greco and P. Sorensen, Ann. Rev. Nucl. Part. Sci. **58**, 177 (2008).
[17] J. Xu, L. -W. Chen, C. M. Ko and Z. -W. Lin, Phys. Rev. C **85**, 041901 (2012).
[18] T. Song, S. Plumari, V. Greco, C. M. Ko and F. Li, arXiv:1211.5511 [nucl-th].
[19] J. C. Dunlop, M. A. Lisa and P. Sorensen, Phys. Rev. C **84**, 044914 (2011).
[20] M. Sumbera [for the STAR Collaboration], arXiv:1301.7224 [nucl-ex].

Influence of quark masses on the QCD phase diagram in the presence of a magnetic field

A. J. Mizher

Instituto de Ciencias Nucleares, Universidad Nacional Autónoma de México, Apartado Postal 70-543, México Distrito Federal 04510, México

E-mail: ana.silveira@correo.nucleares.unam.mx

Abstract.
The most recent lattice results on the QCD phase diagram in the presence of a strong magnetic field strongly disagree with results from previous lattice simulations and several model calculations. The most remarkable difference is the qualitative behavior of the critical temperature as a function of the magnetic field intensity, which in later results are shown to decrease in opposition to what was found previously. According to the authors, such a discrepancy in lattice simulations could be due to different lattice spacing, or different number of flavors and quark masses. We investigate the influence of quark masses on the Polyakov-Quark-Meson model and show that, although quantitatively the results are sensitive to those parameters, the qualitative behavior remains the same.

Introduction

The phase structure of the QCD phase diagram in the presence of magnetic field was studied in recent years by several model calculations and also by full lattice QCD simulations. Such an interest on the subject was motivated by the fact that it was proposed that a very intense magnetic field, of order $5 - 10m_\pi^2$ [1, 2], must be produced in non-central heavy-ion collisions and must have a considerable influence in the phase transitions suffered by the matter generated after the collision.

Most of the model calculations [3, 4, 5, 6] indicate an increasing behavior of the critical temperature as a function of the magnetic field and early lattice simulations [7] corroborated these results. Later lattice results [8], however, indicate a decrease of the critical temperature with the magnetic field intensity. In order to check their code and algorithm simulation, the authors of the latter have reproduced the results of the former group applying the same setup. Since in this context the previous results were recovered, the discrepancy was then associated to different lattice spacing or difference in number of flavors and quark masses. Approaches beyond mean field were considered [9], still obtaining an increase of the critical temperature with the magnetic field. A later approach [10] could reproduce qualitatively the lattice results in [8], considering the context of the MIT bag model.

According to [8] the difference between their result and the previous one can be traced back to the behavior of the chiral condensate as a function of the magnetic field. While in [7] it was found that the condensate has a monotonically increasing dependence on the field for all the range of temperature tested, [8] obtains a more complex behavior, in which such a dependence can increase or decrease depending on the temperature. In order to get some insight on the role of quark masses in model calculations we use the linear sigma model coupled to quarks and to the Polyakov loop to obtain a phase diagram as a function of the magnetic field for different values of constituent quark masses. Although we include the Polyakov loop, taking into account the interplay between the chiral and deconfining transition, we limit our analysis to the chiral condensate, following the indication in [8] that the chiral transition guides the general behavior of both transitions.

The Polyakov-Quark-Meson model

In this context, the deconfinement properties are encoded in the expectation value of the Polyakov loop, which works as an order parameter for this transition:

$$\text{Confinement}: \quad \begin{cases} \langle L \rangle = 0 \,, & \text{low } T \\ \langle L \rangle \neq 0 \,, & \text{high } T \end{cases} , \qquad L(x) = \frac{1}{3}\,\mathrm{Tr}\,\mathcal{P}\exp\Big[i\int_0^{1/T} \mathrm{d}\tau\, A_4(\vec{x},\tau)\Big] \,, \tag{1}$$

where A_4 is the matrix-valued temporal component of the Euclidean gauge field A_μ and the symbol $\mathcal{P}$ denotes path ordering. On the other hand, the expectation value of the chiral condensate works as the order parameter for the chiral transition, indicating if this symmetry is broken or restored for each value of external parameters:

$$\text{Chiral symmetry}: \quad \begin{cases} \langle \sigma \rangle \neq 0 \,, & \text{low } T \\ \langle \sigma \rangle = 0 \,, & \text{high } T \end{cases} , \qquad \begin{aligned} \phi &= (\sigma, \vec{\pi})\,, \\ \vec{\pi} &= (\pi^+, \pi^0, \pi^-)\,. \end{aligned} \tag{2}$$

The chiral sector coupled to quarks is given by the Lagrangian:

$$\mathcal{L} = \overline{\psi}_f \left[i\gamma^\mu \partial_\mu - g(\sigma + i\gamma_5 \vec{\tau}\cdot\vec{\pi}) \right] \psi_f + \frac{1}{2}(\partial_\mu \sigma \partial^\mu \sigma + \partial_\mu \vec{\pi} \partial^\mu \vec{\pi}) - V(\sigma, \vec{\pi}) \,, \tag{3}$$

which contains spontaneous and explicit symmetry breaking and the coupled of mesons and quarks via a Yukawa term. The quarks act as a heat bath in which the condensate evolves and quark loop corrections provides the thermal contribution to the potential. The implementation of the magnetic field is done replacing the simple derivative that describes the dynamics of the fermions by covariant derivative. Such a derivative contains an Abelian gauge field which can be seen as a vector potential for the magnetic field. Also the interaction with the pure gauge sector, through the Polyakov loop, can be implement in the same fashion, the Polyakov loop being parametrized in such a way that a non-Abelian gauge field interacting with the mesons can represent it. The 1-loop quark correction is then given by:

$$e^{iV_{3d}\Omega_q/T} = \left[\frac{\det(i\not{D}^{(q)} - m_q)}{\det(i\not{\partial} - m_q)}\right] \cdot \left[\frac{\det_T(i\not{D}^{(q)} - m_q)}{\det(i\not{D}^{(q)} - m_q)}\right] , \tag{4}$$

where the quark mass is given by $m_q = \sqrt{m_c^2 + (g\langle\sigma\rangle)^2}$, with m_c the constituent mass, and the covariant derivative contains both gauge fields, the Abelian and the non-Abelian.

To reproduce the nature of the phase transition predicted in [8], a crossover, we take into account the quark degrees of freedom in the vacuum. The non inclusion of this correction implies a first order phase transition and a discussion regarding this point is done in [4].

We consider several values for the constituent quark mass in a range between the chiral limit and the physical consituent quark mass and also an extrapolation of that, $m_q = 500$ MeV, in order to investigate the systematics. For a set of temperature values, we obtain the condensate as a function of the magnetic field. The results are plot in Fig.1 . In the chiral limit and for small values of the quark mass the condensate increases with the magnetic field and saturates at a constant value for all values of the temperature. This behavior is again in accordance with [7], which shows in their Fig.3 a similar saturation (regarding that the periodic behavior is an artefact from the fact that they have introduced the $U(1)$ field as a phase factor $Ae^{i\phi}$). As the quark mass is increased the value of magnetic field to reach saturation becomes larger, however the behavior is systematically the same.

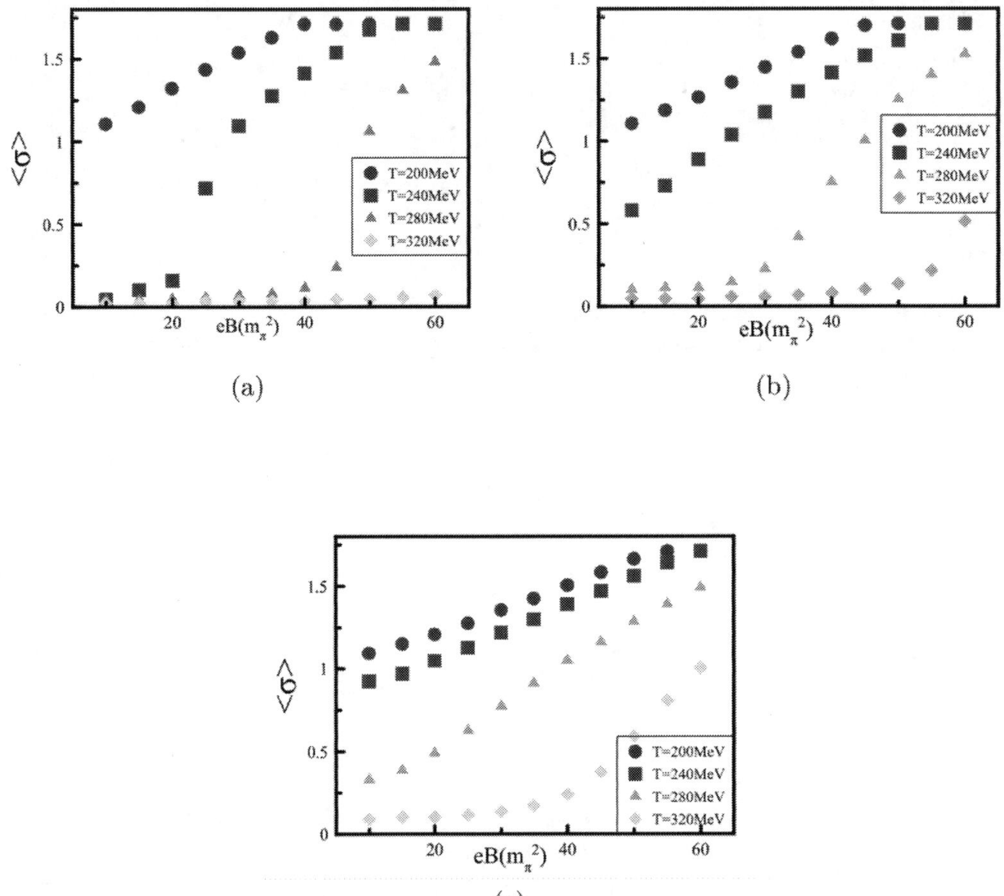

(c)

Figure 1. Expectation value of the chiral condensate for different values of magnetic fields. The different plots differ by the value of the quark masses considered and the different curves by temperature. We considered (a) $m_q = 0$ MeV, (b) $m_q = 300$ MeV and (c) $m_q = 500$ MeV.

Conclusions

In this work we have studied the role of quark masses for the chiral phase transition in the context of the Polyakov-Quark-Meson model in the presence of a magnetic field. Motivated by discrepant lattice results regarding the behavior of the chiral condensate as a function of the magnitude of the magnetic field we have studied the effect of going beyond the chiral limit, since this was presented as a potential cause for the different lattice results. We showed that for all values of quark masses considered, from chiral limit to physical constituent quark masses, the qualitative behavior doesn't change, in accordance with [7] and in disagreement with more refined results in [8]. A systematic saturation of the value of the chiral condensate for all values of temperature is found, also in agreement with [7]. As a consequence, the critical temperature increases with the magnetic field and it was seen no sign in favor of a decreasing for any value of the magnetic field.

Acknowledgments

A.J.M. thanks Cristian Villavicencio for discussions. Support for this work has been received in part from DGAPA-UNAM under Grant No. PAPIIT-IN103811 and CONACyT-México under Grant No. 128534.

Bibliography

[1] D. E. Kharzeev, L. D. McLerran and H. J. Warringa, Nucl. Phys. A **803**, 227 (2008).
[2] V. Skokov, A. Y. .Illarionov and V. Toneev, Int. J. Mod. Phys. A **24**, 5925 (2009).
[3] C. V. Johnson and A. Kundu, JHEP **0812**, 053 (2008).
[4] A. J. Mizher, M. N. Chernodub and E. S. Fraga, Phys. Rev. D **82**, 105016 (2010).
[5] R. Gatto and M. Ruggieri, Phys. Rev. D **82**, 054027 (2010).
[6] R. Gatto and M. Ruggieri, Phys. Rev. D **83**, 034016 (2011).
[7] M. D'Elia and F. Negro, Phys. Rev. D **83**, 114028 (2011).
[8] G. S. Bali, F. Bruckmann, G. Endrodi, Z. Fodor, S. D. Katz, S. Krieg, A. Schafer and K. K. Szabo, JHEP **1202**, 044 (2012).
[9] V. Skokov, Phys. Rev. D **85**, 034026 (2012).
[10] E. S. Fraga and L. F. Palhares, Phys. Rev. D **86**, 016008 (2012).

Journal of Physics: Conference Series **446** (2013) 012017 doi:10.1088/1742-6596/446/1/012017

Measurement of beam energy dependent nuclear modification factors at STAR

Stephen P. Horvat (for the STAR Collaboration)
Department of Physics, 217 Prospect Street, New Haven, CT 06511-8499
E-mail: stephen.horvat@yale.edu

Abstract. The nuclear modification factors R_{AA} and R_{CP} have been used to measure medium-induced suppression in heavy-ion collisions at $\sqrt{s_{NN}} = 200$ GeV which was among the earliest evidence for the existence of a strongly interacting medium called a quark-gluon plasma (QGP). Nuclear modification factors for asymmetric collisions (R_{dA}) have measured the Cronin Effect, an enhancement of high transverse momentum particle yields in deuteron-gold collisions relative to proton-proton collisions. A similar enhancement is observed in data presented in these proceedings and competes with the quenching caused by partonic energy loss in the QGP. In these proceedings we will present charged-hadron R_{CP} at mid-rapidity for $\sqrt{s_{NN}} = 7.7$ - 62.4 GeV as well as identified π^+, K^+, and p R_{CP}. Comparisons to HIJING motivate possible methods for disentangling competing modifications to nuclear transverse momentum spectra.

1. Introduction

The RHIC beam energy scan (BES) is a program to collide Au+Au ions at various collision energies in order to explore the QCD phase diagram; searching for a possible critical point and for a phase boundary marked by the disappearance of key signatures for the formation of a QGP [1]. The nuclear modification factor provides one of these signatures. The ratio of transverse momentum (p_T) differentiated spectra from central over peripheral collisions and scaled by the mean number of binary p+p-like collisions in each event is called the nuclear modification factor and is denoted by R_{CP}. If p+p collisions are used for the reference instead then the nuclear modification factor is denoted by R_{AA}. In the presence of a QGP, high-p_T particles are quenched, transferring energy to lower momentum particles, causing the nuclear modification factor to be less than unity at high p_T, or suppressed [2-6]. Quenching competes with any effects that would cause enhancement, such as radial boosts or the Cronin Effect [7]. The Cronin Effect has also been observed by STAR as the enhancement of the nuclear modification factor in asymmetric d+Au collisions at $\sqrt{s_{NN}} = 200$ GeV [8]. These spectra can be influenced by the spectators, non-interacting nucleons in the collision system, so that a peripheral collision does not equate directly to a p+p collision due to cold nuclear matter (CNM) effects. The goal of this analysis is to determine at what beam energy suppression turns off, and to begin disentangling the causes and relative effects of quenching and enhancement.

2. Charged-hadron R_{CP}

Au+Au collisions near the center of the detector were analyzed. A particle track was used if its distance of closest approach (dca) was less than one centimeter in order to reduce feed-down

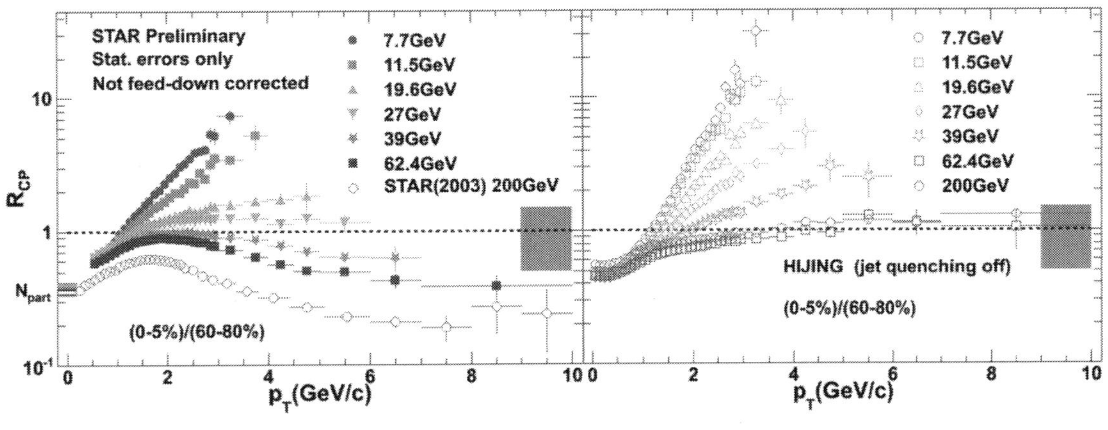

Figure 1. (Color online) Efficiency corrected charged hadron R_{CP} (left) for RHIC BES energies. p_{T} dependent errors are statistical only. The gray error band corresponds to the p_{T} independent uncertainty in N_{coll} scaling. Charged R_{CP} from HIJING simulated events (right) with jet quenching turned off.

contamination. Trigger efficiency, tracking efficiency, and acceptance corrected charged-hadron spectra from mid-rapidity $Au{+}Au$ collisions at $\sqrt{s_{\mathrm{NN}}} = 7.7, 11.5, 19.6, 27, 39$, and 62.4 GeV were produced for 0-5% central and 60-80% peripheral collisions in the STAR detector. The spectra were scaled by binary collisions with the scale factors obtained from a Monte Carlo Glauber model [9]. Taking the ratio of these scaled spectra for each energy gives the $R_{\mathrm{CP}}(\sqrt{s_{\mathrm{NN}}}, p_{\mathrm{T}})$ shown in Fig. 1 (left) along with STAR's published 200 GeV result [2]. These spectra were not feed-down corrected and were taken from $-0.5 < \eta < 0.5$. The global systematic uncertainty is dominated by the uncertainty in the centrality selection for the peripheral bins, which is used in the Glauber calculation and presents as an uncertainty on the binary collision scale factor. The same methods that produced the lower beam energy results were used to produce a 200 GeV result from 2010 data. This measurement disagreed with the feed-down corrected result from 2003 (Fig. 1 left) by 20% and so this was folded into the overall systematic uncertainty of the other results (Fig. 1 gray box). The cause of this discrepancy is under investigation. The efficiency correction is based on single particle embedding in the 39 GeV data set for $\pi^{\pm}$, $K^{\pm}$, and $p^{\pm}$ separately which were than combined and weighted by their relative yields for a charged hadron efficiency. The efficiency correction was extrapolated to the other data sets by making the assumption that the efficiency is the same for each particle at the same p_{T} and from the same multiplicity bin. This assumption was tested by producing the efficiencies from two data sets, 39 GeV from 2010 and 27 GeV from 2011, and ensuring that the predicted efficiency from the 39 GeV data set matched the 27 GeV efficiency. Then differences in acceptance due to detector performance between beam energies were accounted for by using stable portions of the detector as a reference. HIJING 1.35 [10] with jet quenching turned off was used to produce more than 10M collisions for each beam energy. The motivation for running the simulator with jet quenching turned off was the expectation that at sufficiently low beam energies, where medium-induced jet quenching has minimal effect, there would be a quantitative agreement between the charged hadron R_{CP} from simulation and data. If the simulation and the data agreed at low beam energies but deviated at higher beam energies then the beam energies where the deviation occurred could be considered candidates for the beam energies where a QGP is formed. Centrality selection was done using the same method as for the data; namely, counting the number of final state charged hadrons in $-0.5 < \eta < 0.5$ and then determining the 0-5% most

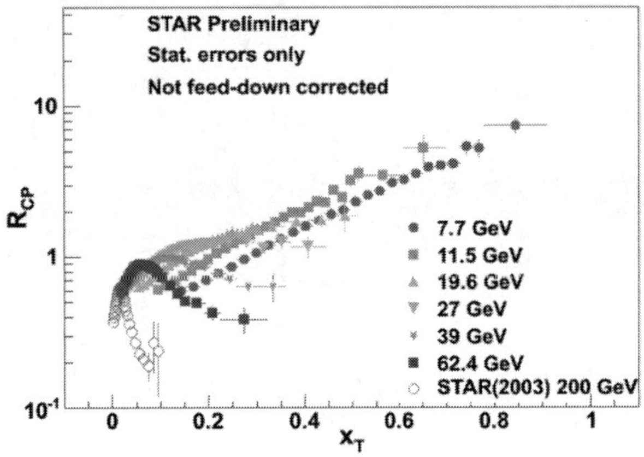

Figure 2. (Color online) The data from Fig. 1 (left) plotted vs. x_T rather than p_T.

central data as being the 5% of events with the highest multiplicity, and so forth for the other centralities. The result is shown in Fig. 1 on the right. Again, R_{CP} at lower beam energies is enhanced, although we do not see a quantitative agreement with charged hadron R_{CP} from data. We do not see suppression at higher collision energies, as expected since quenching was turned off. By running HIJING with jet quenching on and off and comparing with AMPT and other models we hope to disentangle the relative contributions of jet quenching, CNM effects, and possible contributions from radial flow or final state scattering. The results in Fig. 1 (left) are consistent with suppression for $\sqrt{s_{NN}} \geq 39$ GeV. This does not preclude medium induced energy losses at lower energies since other effects could be overwhelming this signature. The advantage for using charged hadron R_{CP} is that you can measure spectra to higher transverse momenta that you could not reach with particle identification. It was also considered that plotting R_{CP} vs. x_T rather than p_T might reveal trends in the data that were independent of collision energy. x_T is defined as $x_T = 2 * p_T/\sqrt{s_{NN}}$. This sort of scaling was applied to spectra previously [8] where it revealed $\sqrt{s_{NN}}$ independent trends at high p_T. Such a scaling is shown in Fig. 2, using the data from Fig. 1, and does not reveal any such trends.

3. Identified hadron R_{CP}

Particle yields were extracted from a simultaneous fit to dE/dx distributions measured in the STAR Time Projection Chamber and time of flight distributions measured in the STAR Time of Flight detector for each centrality and p_T bin at each beam energy. The functions used to extrapolate fit parameters for particle identification were varied in order to obtain the systematic errors for the high p_T bins. Efficiency corrections were obtained through track embedding and have a 5% systematic error associated with them. The differences between the published 62.4 GeV results [11] and those presented here were taken as a point by point systematic error and applied to all beam energies. The cause of this discrepancy is under investigation. The result (Fig. 3) is qualitatively consistent with published results [11] in that pions are less enhanced than protons, suggesting that pions may serve as a better gauge for jet quenching within the p_T range available through particle identification. Considering 2.5 GeV/c $< p_T <$ 4 GeV/c hadrons, the results from Fig. 3 show that protons are not suppressed at any beam energy and pions go from being suppressed at higher beam energies to being enhanced at lower beam energies with a transition near $\sqrt{s_{NN}} = 27$ GeV.

4. Summary

Charged hadron R_{CP} has been measured at mid-rapidity for a range of beam energies in order to determine at what beam energy the suppression of high p_T charged hadrons, a QGP signature,

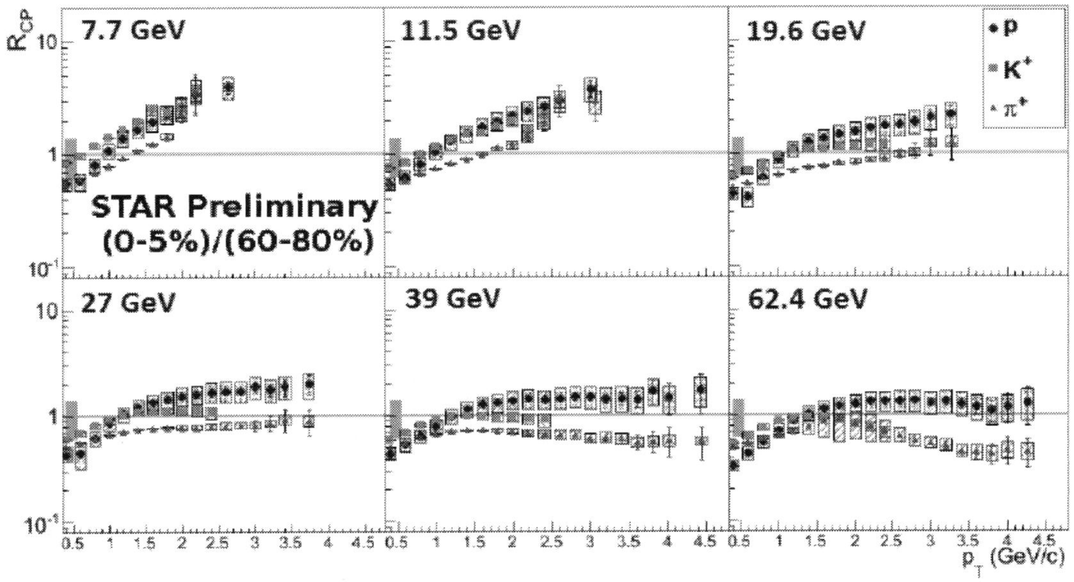

Figure 3. (Color online) $R_{\rm CP}$ for identified p, K^+, and π^+ for RHIC BES energies. The boxes are $p_{\rm T}$ dependent systematic uncertainties due to particle identification while the $p_{\rm T}$ independent uncertainty is from N_{coll} scaling.

turns off. Suppression is seen to turn off near 39 GeV for unidentified charged hadrons, but due to unquantified sources of enhancement it is currently unclear where medium-induced jet quenching turns off. Identified $R_{\rm CP}$ is qualitatively similar and promotes pions as a probe that is less effected by sources of enhancement. For pions, suppression at high $p_{\rm T}$ is seen to turn off near 27 GeV. A comparison to the HIJING event generator with jet quenching turned off did not reveal an energy at which the data and the model agree quantitatively, which precludes measuring the energy at which data and simulation deviate. This motivates the exploration of additional tunes of HIJING and other models in order to disentangle the competing effects which lead to suppression or enhancement.

References

[1] Mohanty B. (for the STAR Collaboration) 2011 *J. Phys. G: Nucl. Part. Phys.* **38** 124023
[2] Adams J. *et al.* (STAR Collaboration) 2003 *Phys. Rev. Lett.* **91** 172302
[3] Arsene I. *et al.* (BRAHMS Collaboration) 2005 *Nucl. Phys.* A **757** 1
[4] Back B. B. *et al.* (PHOBOS Collaboration) 2005 *Nucl. Phys.* A **757** 28
[5] Adcox K. *et al.*, (PHENIX Collaboration) 2005 *Nucl. Phys.* A **757** 184
[6] Aamodt K. *et al.* (ALICE Collaboration) 2011 *Phys. Lett.* B **696** 30
[7] Cronin J.W. *et al.* 1975 *Phys. Rev.* D **11** 3105
[8] Adams J. *et al.* (STAR Collaboration) 2006 *Phys. Lett.* B **637** 161
[9] Miller M. *et al.* 2007 *Ann. Rev. of Nucl. and Part. Sci.* **57** 205
[10] Gyulassy M. and Wang X. 1994 *Comput. Phys. Commun.* **83** 307
[11] Abelev B. *et al.* (STAR Collaboration) 2007 *Phys. Lett.* B **655** 104

"Chemical" composition of the Quark Gluon Plasma

Francesco Scardina[1,2], Maria Colonna[2], Salvatore Plumari[1,2] and Vincenzo Greco[1,2]

[1] Department of Physics and Astronomy, University of Catania, Via S. Sofia 64, I-95125 Catania (Italy)
[2] Laboratorio Nazionale del Sud, INFN-LNS, Via S. Sofia 63, I-95125 Catania (Italy)

E-mail: scardinaf@lns.infn.it

Abstract. In this article we discuss the issue of the quark to gluon ratio in the Quark Gluon Plasma (QGP). Our model to describe the QGP evolution is based on transport theory including the mean field dynamics described by a quasi-particle model.The last is able to take into account for the lattice QCD thermodynamics and implies a "chemical" equilibrium ratio between quarks and gluons strongly increasing as T approaches to the temperature of the phase transition T_c. We present first the tests performed in a fixed box to check that our code is able to reproduce the equilibrium ratio and then the results obtained for the simulations of ultra-Relativistic Heavy Ion Collisions (uRHIC's) at RHIC and LHC energies. We observe a rapid evolution from a gluon dominated initial state to a quark dominated plasma and we see that near T_c almost 80% of the particles composing the plasma are quarks. This has potentially a strong impact on several quantitative aspects of QGP probes and furnishes a justification to the coalescence hadronization model.

1. Introduction

The main purpose of this article is to study how the QGP created in uRHIC's at both RHIC and LHC energies approaches to chemical equilibrium, i.e the equilibrium value between quark and gluon abundancy. In fact the matter created just after the collision should be prevalently composed by gluons because most of the particles come from low x momentum fraction, where the nucleon parton distribution functions are gluon dominated and is sometimes called Glasma. During the evolution of the fireball the inelastic collisions should lead the system towards chemical equilibrium with a consequent enhancement of the number of quarks. A detailed analysis of the chemical equilibration of the plasma have not been performed yet and the several theoretical approaches describing the different QGP probes reveals that the different models assumes different chemical composition for the Quark Gluon Plasma. In some cases, as for examples in the most popular jet quenching models, the QGP is treated as a Gluon Plasma. In Hydrodynamics instead the assumption of a chemical equilibrated QGP is implicit in the employment of a lattice QCD equation of state $P(\epsilon)$. On the other hand the coalescence model [1, 2, 3] assumes a quark dominance in the region of phase transition.
The results obtained in our work clearly indicate that the inelastic collisions cause a quick change of the chemical composition of the plasma. More precisely during the evolution there is a substantial enhancement of the percentage of quarks and such enhancement is substantially larger if one considers that both quarks and gluons acquire a thermal mass.

2. Our model to describe the QGP

Usually the studies of the chemical composition of the QGP have been made considering the plasma as a massless gas. In this case the equilibrium value for $N_{q+\bar{q}}/N_g$ is simply given by the ratio of the degrees of freedom $N_{q+\bar{q}}/N_g = d_{q+\bar{q}}/d_g = 2N_c N_f/(N_c^2-1) = 9/4$ for a system with 3 flavors. On the other hand the data coming from lattice QCD for energy density, pressure

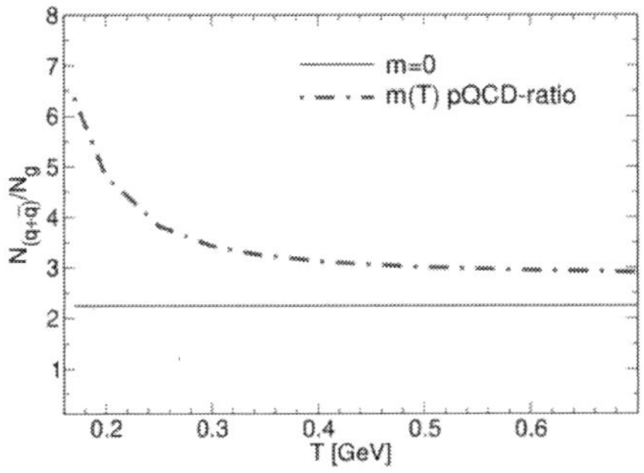

Figure 1. By dot-dashed line it is indicated the quark to gluon ratio at equilibrium as a function of the temperature, as predicted by QPM [4]; by thin solid line it is indicated instead the massless case.

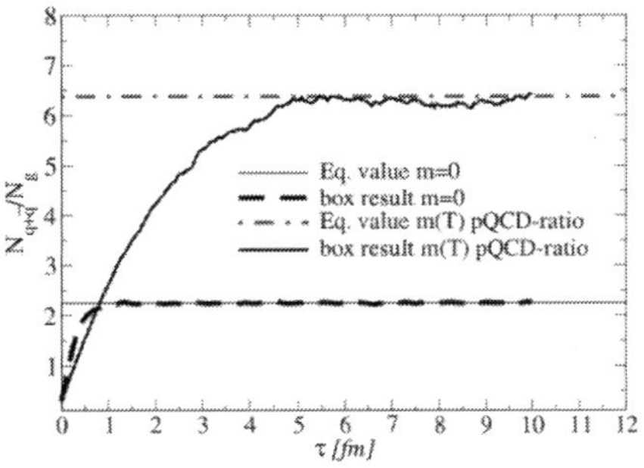

Figure 2. Time evolution of ratio $N_{q+\bar{q}}/N_g$ in box calculation. The solid line indicates the massive case while the dashed line indicates the massless case. The equilibria value are indicated as in Fig. 1.

and trace anomaly of the QGP are not compatible with those expected for a massless gas and strong deviations from the ideal gas behaviour have been observed. It has been shown [4] that the lattice data can be described by a Quasi-Particle Model (QPM) in which both quarks and gluons acquire thermal masses. We notice that such thermal masses has a strong impact on the ratio between quarks and gluons, as shown in Fig.1, and they introduce a temperature dependence on the ratio.

Our approach to describe the partonic stage of the matter created in ultra-relativistic heavy ion collisions is based on transport theory including the mean field dynamic described by a quasi-particle model. In such a picture the relativistic Boltzmann-Vlasov equation can be written as follows:

$$[p^\mu \partial_\mu + m^*(x)\partial_\mu m^*(x)\partial_p^\mu] f(x,p) = \mathcal{C}[f](x,p) \tag{1}$$

where $\mathcal{C}(x,p)$ is the Boltzmann-like collision integral, main ingredient of the several cascade codes:

$$\mathcal{C} = \iiint_{21'2'} (f_{1'}f_{2'} - f_1 f_2)|\mathcal{M}_{1'2'\to 12}|^2 \delta^4(p_1 + p_2 - p_1' - p_2') \tag{2}$$

where $\int_j = \int_j d^3p_j/(2\pi)^3\, 2E_j$, f_j are the particle distribution functions, while $\mathcal{M}_{f\to i}$ are the invariant transition matrix for elastic as well as inelastic processes; see article [5, 6] for more details. In order to include the QPM in the transport theory it is necessary to couple the equation (1) with the gap-like equation

$$\frac{\partial B}{\partial m_i} = -\int \frac{d^3\vec{p}}{(2\pi)^3} \frac{m_i(x)}{E_i(x)} f_i(x,p) \tag{3}$$

with $i = q, \bar{q}, g$. The previous equation assures the thermodynamical self consistency of the QPM and allows us to evaluate locally the masses appearing in Eq. (1), also in non-equilibrium conditions, guaranteeing the conservation of the energy momentum tensor of the fluid. The numerical solution of the Boltzmann equation is the same described in [7, 8, 9]. In order to check if our code is able to reproduce correctly the chemical equilibrium we have carried out simulations in which a particle ensemble, far from to the chemical equilibrium, is enclosed in a fixed box and evolves dynamically. In Fig.2 the results of such kind of simulations are shown for the massless case and for the massive case. We observe in both cases that the system reaches the expected equilibrium ratio.

3. Results for the simulations of heavy-ion collisions

We have carried out simulations for both RHIC (Au+Au at $\sqrt{s} = 200$ AGeV) and LHC energies $Pb + Pb$ at $\sqrt{s} = 5.5$ATeV. The initial conditions in the coordinate space are given by the standard Glauber model conditions. For the momenta space we use a Boltzmann-Juttner distribution function up to a transverse momentum equal to $p_T = 2$, while for larger momenta mini-jet distributions are implemented, as calculated by pQCD at NLO order [2]. Regarding the initial ratio between the number of quarks and the number of gluons we start from a situation in which 75 % of particles are gluons and 25 % are quarks. However our results are not significantly influenced by the initial quark to gluon ratio. We have considered in our simulations a running coupling $\alpha_s(Q^2)$ according to first order expansion and the energy scale given by $Q^2 = (\pi T)^2$. In Fig.3 the time evolution of the ratio $R_{qg} = N_{q+\bar{q}}/N_g$ for both the massless case (dashed lines) and the massive quasi-particle case (solid lines) is shown for both RHIC and LHC energies. We observe that in the massless case the system reaches very quickly the chemical equilibrium given by $R_{qg} \sim 2$. If the quasi-particle massive case is considered we can again see that one can

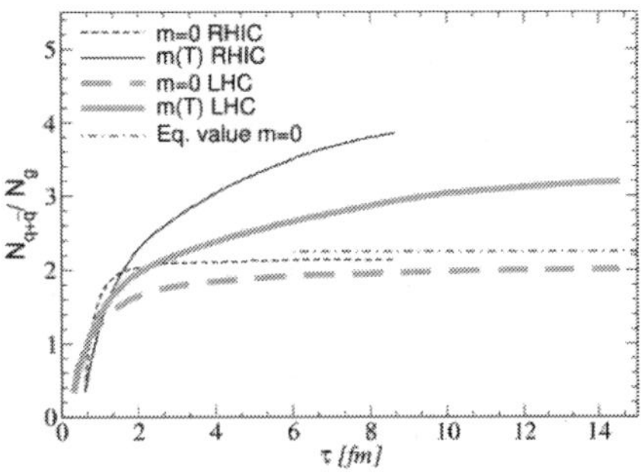

Figure 3. Quark plus anti-quark to gluon ratio as a function of τ at RHIC (black thin lines) and at LHC (light thick lines). Dashed lines are for the massless case and the solid for the massive case.

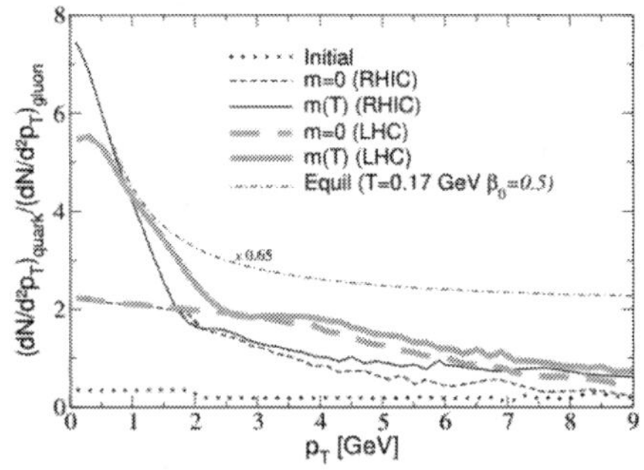

Figure 4. Quark to gluon ratio as a function of transverse momentum at the freez-out temperature. Labels as in Fig.3. The thin dot-dashed line represents the equilibrium ratio; see text for details.

reach also quickly a value of $R_{qg} \sim 2$, then there is a slower rise with a continue increase up to $R_{qg} \sim 4$ at RHIC . We have seen that in the massive case the equilibrium value, especially at low T, is higher compared to the massless case (see Fig.1), however as the temperature decreases the system becomes more dilute and it is not capable to reach the full chemical equilibration. Despite that in the region of phase transition ($T \sim T_c$) the fireball is composed by about 80% of quark plus anti-quarks. At LHC the trends are very similar but a longer part of the lifetime is spent in a T-region where the equilibrium R_{qg} is nearly constant and this implies a moderately smaller final ratio. Particularly interesting it is to analyze the momentum dependence of the

Journal of Physics: Conference Series **446** (2013) 012018 doi:10.1088/1742-6596/446/1/012018

quark to gluon ratio shown in Fig.4. We observe a strong decrease of the ratio with p_T that is due to the decreasing of the collision rate with the transverse momentum. We see a large difference between the massless and the massive case and such difference is much wider in the low p_T region. In particular in the massive case we note a quite strong dependence below $p_T = 2$ GeV that has not to be interpreted as a fast detachment from the chemical equilibrium. In fact the momentun dependence of the equilibrium ratio can be evaluated analytically [5, 6] at equilibrium and it is shown by the dot dashed line in Fig. 4. We observe that the results obtained in our simulations follow very closely the equilibrium behavior at least up to $p_T \sim 1.5$ GeV, nonetheless a significant gluon to quark conversion happens also at $p_T \geq 5$ GeV. In Fig.5, as example, we show how the ratios, $(p+\overline{p})/2\pi^0$, $(p+\overline{p})/(K+\overline{K})$, can be modified when the gluon conversion mechanism is included. We find that the effect of the inelastic collisions $q+\overline{q} \leftrightarrow g+g$ is to

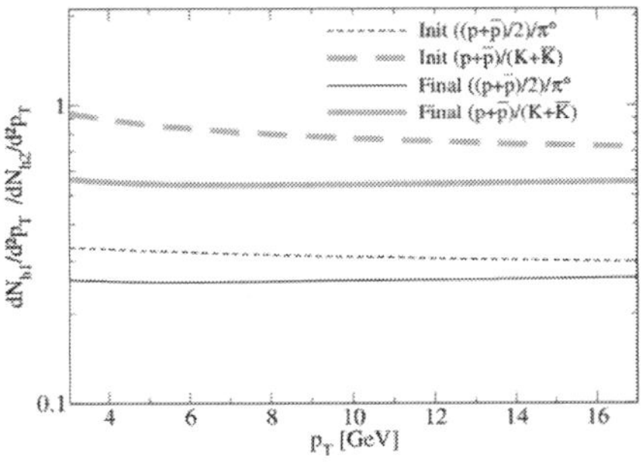

Figure 5. Ratio of hadrons obtained with the AKK fragmentation functions [10]. The dashed lines are the ratio according to the initial partons distributions, the solid lines are at the freeze-out considering the effect of the inelastic collisions.

reduce both the ratios. This disagrees with the results obtained in [11, 12, 13] in which it is shown that the inelastic collisions cause a decreasing of the proton/pion ratio able to explain the experimental observations. However this difference is due to the different choice of the initial parton spectra and in particular to the different initial q/g ratio.

4. Conclusions

Our studies show that the initially gluon dominated plasma created in uRHIC's evolves into a plasma dominated by quark plus anti-quarks close to the cross-over temperature T_c. This result supplies a theoretical justification of the *massive-quark* coalescence hadronization models able to explain the experimental observations of the quark-number scaling in the elliptic flow and the baryon over meson enhancement.

Moreover our results can have a wide impact on the main physical issues of the QGP physics as a quantitative estimate of the viscosity, the study of the identified high-p_T particles and the related jet quenching mechanism as well as on the physics of the quarkonia suppression.

References

[1] Greco V, Ko C and Levai P 2003 *Phys.Rev.Lett.* **90** 202302 (*Preprint* nucl-th/0301093)
[2] Greco V, Ko C and Levai P 2003 *Phys.Rev.* **C68** 034904 (*Preprint* nucl-th/0305024)
[3] Fries R J, Greco V and Sorensen P 2008 *Ann.Rev.Nucl.Part.Sci.* **58** 177–205 (*Preprint* 0807.4939)
[4] Plumari S, Alberico W M, Greco V and Ratti C 2011 *Phys.Rev.* **D84** 094004 (*Preprint* 1103.5611)
[5] Scardina F, Colonna M, Plumari S and Greco V 2012 (*Preprint* 1202.2262)
[6] Plumari S, Puglisi A, Colonna M, Scardina F and Greco V 2012 (*Preprint* 1209.0601)
[7] Xu Z and Greiner C 2005 *Phys.Rev.* **C71** 064901 (*Preprint* hep-ph/0406278)
[8] Ferini G, Colonna M, Di Toro M and Greco V 2009 *Phys.Lett.* **B670** 325–329 (*Preprint* 0805.4814)
[9] Plumari S, Baran V, Di Toro M, Ferini G and Greco V 2010 *Phys.Lett.* **B689** 18–22 (*Preprint* 1001.2736)
[10] Albino S, Kniehl B and Kramer G 2008 *Nucl.Phys.* **B803** 42–104 (*Preprint* 0803.2768)
[11] Liu W, Ko C and Zhang B 2007 *Phys.Rev.* **C75** 051901 (*Preprint* nucl-th/0607047)
[12] Liu W and Fries R 2008 *Phys.Rev.* **C77** 054902 (*Preprint* 0801.0453)
[13] Scardina F, Di Toro M and Greco V 2010 *Phys.Rev.* **C82** 054901

Meson Spectra and Thermodynamics in Soft-Wall AdS/QCD

Sean P. Bartz and Joseph I. Kapusta
Tate Laboratory of Physics, 116 Church St SE, Minneapolis, MN, 55455
E-mail: bartz@physics.umn.edu

Abstract. The AdS/CFT correspondence may offer new and useful insights into the non-perturbative regime of strongly coupled gauge theories such as Quantum Chromodynamics. Soft-wall AdS/QCD models have reproduced the linear trajectories of meson spectra by including background dilaton and chiral condensate fields. Efforts to derive these background fields from a scalar potential have so far been unsuccessful in satisfying the UV boundary conditions set by the AdS/CFT dictionary while reproducing the IR behavior needed to obtain the correct chiral symmetry breaking and meson spectra.

We present a three-field scalar parametrization that includes the dilaton field and the chiral and glueball condensates. This model is consistent with linear trajectories for the meson spectra and the correct mass-splitting between the vector and axial-vector mesons. We also present the resulting meson trajectories.

1. Introduction and Motivation

Quantum chromodynamics has been well tested for high-energy collisions, where perturbation theory is applicable. However, at hadronic scales, the interaction is non-perturbative, requiring a new theoretical model. The Anti-de Sitter Space/Conformal Field Theory (AdS/CFT) correspondence establishes a connection between n-dimensional Super-Yang Mills Theory and a weakly-coupled gravitational theory in $n+1$ dimensions [1]. Phenomenological models inspired by this correspondence are known as AdS/QCD, and have succeeded in capturing some features of QCD [2].

Quark confinement in QCD sets a scale that is encoded in a cut-off of the fifth dimension in the AdS theory. Soft-wall models use a dilaton as an effective cut-off to limit the penetration of the meson fields into the bulk. The simplest soft-wall models use a quadratic dilaton to recover the linear Regge trajectories [2], while models that modify the UV behavior of the dilaton more accurately model the ground state masses. We use the meson action from [3].

$$S_{meson} = \int d^5x\sqrt{-g}e^{-\Phi(z)}Tr\left[|DX|^2 + m_X^2|X|^2 + \kappa|X|^4 + \frac{1}{4g_5^2}(F_L^2 + F_R^2)\right]. \qquad (1)$$

Here, Φ is the dilaton, X is the scalar field, and $F_{L,R}$ contain the vector and axial vector gauge fields. The z-dependent vacuum expectation value of the scalar field encodes the chiral symmetry breaking,

$$\langle X\rangle = \frac{\chi(z)}{2}I, \qquad (2)$$

where I is the $N_F \times N_F$ identity matrix. Please see [3, 4] for details on the field content, the equations of motion, and the numerical methods for calculation of the meson spectra.

The models of [2, 3] use parametrizations for the background dilaton and chiral fields that are not derived as the solution to any equations of motion. A well-defined action provides a set of background equations from which these fields can be derived. In addition, this action provides access to the thermal properties of the model through perturbation of the geometry [5]. The full action contains the meson action (1) and the gravitational action, each of which has an intrinsic coupling to the dilaton,

$$\mathcal{S} = \int d^5x\sqrt{-g}e^{-\Phi}\mathcal{L}_{meson} + \int d^5x\sqrt{-g}e^{-2\Phi}\mathcal{L}_{grav}.$$

In the Einstein frame, the action for the background fields reads

$$S_{grav} = \frac{1}{16\pi G_5}\int d^5x\sqrt{-g_E}\left(R_E - \frac{1}{2}\partial_\mu\phi\partial^\mu\phi - \frac{1}{2}\partial_\mu\chi\partial^\mu\chi - V(\phi,\chi)\right). \tag{3}$$

The dilaton is scaled for a canonical action, $\phi = \sqrt{8/3}\Phi$.

One background equation does not depend on the potential,

$$6a\phi''(z) + [\phi'(z)]^2(6a^2-1) - [\chi'(z)]^2 + \frac{12a\phi'(z)}{z} = 0. \tag{4}$$

Examining (4), and requiring the IR behavior that $\phi = \lambda z^2$ and $\chi = \Gamma z$, determines $a = 1/\sqrt{6}$ [6]. This determines a relationship between λ, the parameter that sets the slope of the Regge trajectories, and Γ, which is related to the mass-splitting between the vector and axial vector mesons [3]

$$\Delta m^2 = \lim_{z\to\infty}\frac{g_5^2\chi^2}{z^2} = g_5^2\Gamma^2 = g_5^2 6^{3/2}\lambda. \tag{5}$$

Using the experimental value of λ gives a value of Δm^2 that is ~ 50 times larger than what is observed experimentally. Because this inconsistency arises from a background field that does not depend upon the choice of potential, this problem is pervasive in such models with two background fields.

2. Three-Field Model

We propose to solve this problem by adding another scalar field to the action for the background fields, G, dual to the glueball field. By this association, the UV boundary condition is $\lim_{z\to 0} G(z) = g_o z^4$, where g_o is the gluon condensate. To maintain linear confinement, $G \sim z$ in the IR. The background equations that result are

$$\chi'^2 + G'^2 = \frac{\sqrt{6}}{z^2}\frac{d}{dz}\left(z^2\phi'\right), \tag{6}$$

$$\tilde{V} + 12 = \frac{\sqrt{6}}{2}z^2\phi'' - \frac{3}{2}(z\phi')^2 - 3\sqrt{6}\phi', \tag{7}$$

$$\frac{\partial\tilde{V}}{\partial\phi} = 3z\phi', \tag{8}$$

$$\frac{\partial\tilde{V}}{\partial\chi} = z^2\chi'' - 3z\chi'\left(1 + \frac{z\phi'}{\sqrt{6}}\right), \tag{9}$$

$$\frac{\partial\tilde{V}}{\partial G} = z^2G'' - 3zG'\left(1 + \frac{z\phi'}{\sqrt{6}}\right), \tag{10}$$

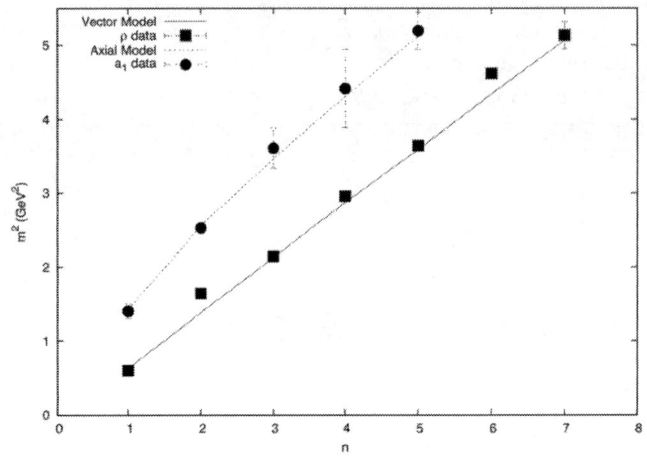

Figure 1. The ρ and a_1 meson masses [7] are fit well using the three-field parametrization.

where ($'$) indicates differentiation with respect to z. Here we have used a conformally transformed potential from the one in (3), $V(\phi,\chi) = e^{2\phi a}\tilde{V}(\phi,\chi)$. The potential depends on z only through the fields, so we may eliminate one of (8-10).

Examining (6) in the IR limit, we see that we can adjust the coefficients of the dilaton and chiral condensate fields independently. Thus, the addition of a third scalar background field resolves the phenomenological problem present in the two-field models of [6, 8].

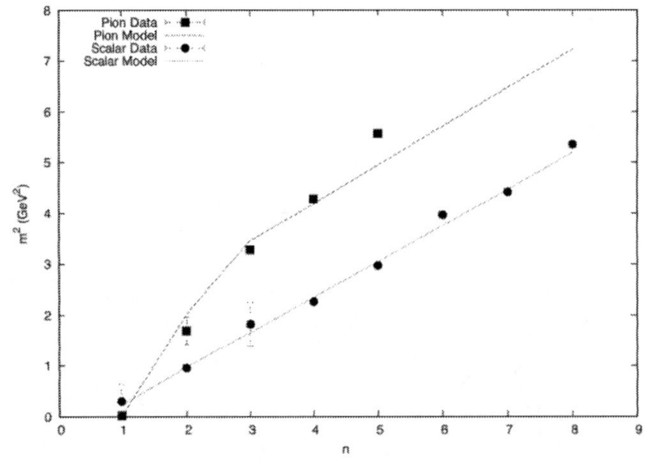

Figure 2. The f_0 meson and pion masses [7] are fit well using the three-field parametrization. The parameter κ is adjusted to avoid a virtual f_0 ground state. The pion ground state is massless, due to the zero quark mass in the model.

As a check on the three-field setup, we seek a parametrization for the chiral and glueball fields that yields an expression for the dilaton free of special functions. In addition, the parametrization must yield meson spectra that match well with experiment. The following expressions for the derivatives of the chiral and glueball fields match these criteria:

$$G'(z) = \frac{A}{B^3}\left(1-e^{-Bz}\right)^3, \qquad \chi'(z) = \frac{\alpha}{\beta^2}\left(1-e^{-\beta z}\right)^2. \tag{11}$$

The parametrization of the chiral field indicates a zero quark mass. The above parameters are defined: $\alpha = \sigma/3$, $\alpha/\beta^2 = \Gamma$, $A = g_o/4$. B is set by ensuring (6) is satisfied in the IR limit.

The values of the parameters, as determined by a least-squares fitting to the ρ and a_1 spectra, are: $\sigma = (0.375\,\text{GeV})^3$, $g_o = (1.5\,\text{GeV})^4$, $\Gamma = 0.25\,\text{GeV}$, $\lambda = (0.428\,\text{GeV})^2$. The vector and axial-vector meson spectra that result from this parametrization are shown in Figure 1.

2.1. Potentials for Power-Law Background Fields

We can learn about the asymptotic behavior of the potential by examining its form when the chiral and glueball fields are assumed to have generic power-law behavior: $\chi \sim z^n$, $G \sim z^m$. The dilaton is determined by (6). Adapting the potential ansatz from [6] to the three-field model,

$$\tilde{V} = -12 + 4\sqrt{6}\phi + c_2\phi^2 - \frac{3}{2}\chi^2 + c_3G^2 + c_4\chi^4 + c_5\phi\chi^2 + c_6G^4 + c_7\phi G^2 + c_8\chi^2G^2, \tag{12}$$

and inserting in equations (6-10), we solve for the coefficients c_i. There is a unique solution leaving c_2 free, which is identified with the dilaton mass as in [6].

$$c_3 = \frac{-3m - 6mn - 7m^2n + 2m^3n + 7mn^2 - 2mn^3}{2n(1+2m)} \tag{13}$$

$$c_4 = \frac{n^2(c_2 - 12n(1+n))}{48(1+2n)^2} \tag{14}$$

$$c_5 = \frac{6n^2 - c_2n}{2\sqrt{6}(1+2n)} \tag{15}$$

$$c_6 = \frac{m^2(c_2 - 12m(1+m))}{48(1+2m)^2} \tag{16}$$

$$c_7 = \frac{6m^2 - c_2m}{2\sqrt{6}(1+2m)} \tag{17}$$

$$c_8 = \frac{-m(6mn + 6n^2 + 12mn^2 - c_2n)}{24(1+2m)(1+2n)} \tag{18}$$

Thus, the ansatz (12) yields the asymptotic behavior of the potential. However, it remains to find a potential that is consistent with both the UV and IR limits of the background fields.

References

[1] Maldacena J M 1998 *Advances in Theoretical Mathematical Physics* **2** 231
[2] Erlich J, Katz E, Son D T and Stephanov M A 2005 *Physical Review Letters* **95** 261602
[3] Gherghetta T, Kapusta J I and Kelley T M 2009 *Physical Review D* **79** 076003
[4] Kelley T M, Bartz S P and Kapusta J I 2011 *Physical Review D* **83** 016002
[5] Kelley T M 2011 (*Preprint* `hep-ph/1108.0653`)
[6] Kapusta J I and Springer T 2010 *Physical Review D* **81** 086009
[7] Nakamura K and Group P D 2010 *Journal of Physics G: Nuclear and Particle Physics* **37** 075021+
[8] Batell B and Gherghetta T 2008 *Physical Review D* **78** 026002

Journal of Physics: Conference Series **446** (2013) 012020 doi:10.1088/1742-6596/446/1/012020

Overview of Flow and Hydrodynamic Modeling of Nuclear Collisions

Matthew Luzum[1,2]

[1] McGill University, 3600 University Street, Montreal QC H3A 2TS, Canada
[2] Lawrence Berkeley National Laboratory, Berkeley, CA 94720, USA

E-mail: `matthew.luzum@physics.mcgill.da`

Abstract. I give a brief overview of flow phenomena in relativistic heavy-ion collisions and the hydrodynamic models used to simulate these systems, aimed at new and young researchers, such as those attending the Hot Quarks 2012 conference.

1. Introduction

In an ultrarelativistic heavy-ion collision, experimenters detect the particles exiting the collision region after the initial collision and any further interactions have long ceased. This consists of anywhere from a few hundred to a few thousand charged hadrons, in addition to neutral hadrons, leptons, and electromagnetic radiation. With this information, we are tasked with inferring as much as we can about the dynamics of the collision and properties of any medium that may have been created during the short duration of the system evolution—hopefully a new, deconfined, phase of matter called the Quark-Gluon Plasma.

2. Measurements

After collecting this information for a large set of collision events, the events are sorted into centrality classes, and analyzed. Centrality selection is usually closely related to the total number of particles produced in a collision, which in a heavy-ion collision is closely related to impact parameter [1] (the latter relationship being much weaker for, e.g., proton-nucleus collisions [2], which were a topic of significant interest when brand new data were presented at this conference [3]). For the set of events in each centrality class, various observables are measured.

In general, the particles detected in a given collision event are governed by some underlying probability distribution, whose azimuthal dependence is often written as a Fourier series:

$$\frac{dN}{dp_T d\eta d\phi} \equiv \frac{1}{2\pi}\frac{dN}{dp_T d\eta}\left[1+\sum_{n=1}^{\infty} v_n \cos n(\phi-\Psi_n)\right]. \quad (1)$$

For a general distribution, the coefficients v_n and orientation angles Ψ_n can depend on transverse momentum p_T and pseudorapidity η, and there can be a different probability distribution for each type of particle and for each collision event (even within a narrow centrality bin).

These coefficients are of significant interest. However, the number of particles detected in a single event is typically insufficient to accurately map out this azimuthal anisotropy of the underlying probability distribution—a single-event measurement of v_n will typically have at least a 50% statistical uncertainty or worse. Therefore, experimenters look at correlations between detected particles, averaged over an ensemble of events.

The simplest example is a correlation between pairs of particles that are both detected coming from the same event. If each particle in a pair is restricted to be in a narrow bin in phase space, an event-averaged pair correlation can span 5 degrees of freedom: the transverse momentum and pseudorapidity of each particle, and the relative azimuthal angle (each collision has a random orientation in azimuth, so one can only measure rotationally invariant event-averaged quantities, and no meaningful information about the average azimuthal angle of the pair can be measured).

The probability distribution of pairs is determined by the underlying one-particle distribution, plus any intrinsic correlation between particle pairs δ. One can again write the azimuthal dependence as a Fourier series, this time with respect to the *relative* azimuthal angle between the pair, $\Delta\phi$ [4],

$$\frac{dN_{\text{pairs}}}{dp_T^a dp_T^b d\eta^a d\eta^b d\phi^a d\phi^b} = \frac{dN}{dp_T^a d\eta^a d\phi^a}\frac{dN}{dp_T^b d\eta^b d\phi^b} + \delta, \tag{2}$$

$$\frac{dN_{\text{pairs}}}{dp_T^a dp_T^b d\eta^a d\eta^b d\Delta\phi} \equiv \frac{dN_{\text{pairs}}}{dp_T^a dp_T^b d\eta^a d\eta^b}\left[1 + \sum_{n=1}^{\infty} V_{n\Delta}\cos n(\Delta\phi)\right], \tag{3}$$

where (a,b) represent bins in transverse momentum and pseudorapidity for the first and second particle, respectively, and any sine term is omitted since it vanishes when averaging over events. The resulting event-averaged Fourier coefficients thus in general measure the quantities:

$$\langle V_{n\Delta}\rangle = \langle v_n^a v_n^b \cos n(\Psi_n^a - \Psi_n^b)\rangle + \langle\delta_n\rangle. \tag{4}$$

If one can choose pairs such that δ_n is negligible, and if $\Psi_n^a \simeq \Psi_n^b$, this simplifies to

$$\langle V_{n\Delta}\rangle \simeq \langle v_n^a v_n^b\rangle. \tag{5}$$

It is believed that these are reasonably good approximations, and measurements of differential and integrated flow are derived from this observable. E.g., an integrated measurement $v_n\{2\}$ is the square root of this quantity when neither particle is restricted to a particular bin in phase space [5]:

$$v_n\{2\} = \sqrt{\langle V_{n\Delta}\rangle} \simeq \sqrt{\langle v_n^2\rangle} \tag{6}$$

It should be noted, however, that it is difficult to determine whether experimental procedures such as imposing a gap in pseudorapidity between the pair really accomplishes the former condition, and the latter is not exactly satisfied even in a purely hydrodynamic calculation [6, 7]. Nevertheless, they are well-defined observables in any case, that can easily be calculated in any theory.

Similarly, one can gain more information about, e.g., the event-by-event distribution of v_n coefficients and correlations between Ψ_n angles by looking at correlations between more than two particles [8, 9] or performing more complicated analyses [10, 11].

3. Hydrodynamic Models

If there is a sufficient separation between macroscopic and microscopic scales; that is, if a medium is created whose constituents interact strongly enough, the system is described by relativistic hydrodynamics. Models that include hydrodynamic evolution have been quite successful at describing a large number of correlation observables. Along with various other observables involving high momentum particles ("jet quenching") and heavy flavor bound states ("J/Ψ" suppression), this has lead to the standard interpretation of a strongly coupled quark gluon plasma having been produced in these collisions.

The equations of ideal relativistic fluid dynamics are given simply by conservation equations (energy, momentum, and any relevant conserved charges),

$$\partial_\mu T^{\mu\nu} = \partial_\mu j_i^\mu = 0, \tag{7}$$

under the assumption of isotropy in the zero-momentum frame:

$$T^{\mu\nu} = T^{\mu\nu}_{\text{ideal}} = (\epsilon + p)u^\mu u^\nu - pg^{\mu\nu}, \tag{8}$$

$$j_i^\mu = j^\mu_{\text{ideal}} = n_i u^\mu, \tag{9}$$

where ϵ is the energy density in the zero momentum frame ("local rest frame", $T^{0i} = 0$), u^μ is the velocity relating this frame to the lab frame, p is the isotropic pressure, and n_i the conserved charge density in the local rest frame. For example, a system in local thermal equilibrium will have this isotropic form, and is therefore described by these equations. Small deviations from local thermal equilibrium can be described by adding terms to these ideal forms, with each term having a corresponding transport coefficient (shear viscosity η, bulk viscosity ζ, etc.).

$$T^{\mu\nu} = T^{\mu\nu}_{\text{ideal}} + \Pi^{\mu\nu} \tag{10}$$

$$\Pi^{\mu\nu} \equiv \eta\nabla^{\langle\mu}u^{\nu\rangle} + \zeta\partial_\alpha u^\alpha(g^{\mu\nu} - u^\mu u^\nu) + ... \tag{11}$$

Here $\nabla^{\langle\mu}u^{\nu\rangle}$ is the gradient of the fluid velocity, $\partial^\mu u^\nu$, made traceless and orthogonal to the fluid velocity, while the second term is proportional to the trace. The transport coefficients are determined by the underlying dynamics of the medium in question, with the rule of thumb that stronger interactions result in smaller viscosity. The formalism is valid if the length scale set by these transport coefficients is much smaller than the macroscopic length scale set by the gradients.

Finally, the equations are closed by an equation of state $p = p(\epsilon)$.

Thus, a system in this regime has universal behavior, with the only connection to the microscopic physics of the medium in question coming through the equation of state and transport coefficients. These can be calculated from an underlying theory (e.g., the equation of state can be reliably calculated with lattice QCD) or taken as free parameters and constrained by experimental data.

To model a heavy-ion collision, one must also have a description of the earliest stages of the collision, in order to provide initial conditions that can be evolved in time with hydrodynamics. This early stage is not well understood, and a number of models have been used to generate initial conditions, using various physical pictures and assumptions. These include MC-Glauber [12, 13], NEXUS [14, 15], EPOS [16, 17], UrQMD [18, 19], AMPT [20, 21], DIPSY [22, 23, 24], MC-KLN [25, 26], MC-rcBK [27, 24], IP-Glasma [28, 29], etc., and there is much ongoing study [30].

Finally, at the end of the hydrodynamic evolution, one must convert from a fluid to a particle description. The formalism for this is the Cooper-Frye formalism [31], which is well-understood in the ideal case, but has significant uncertainties in the viscous case, since it depends on the microscopic dynamics of the system [32]. Essentially, particles are emitted independently from a fluid cell with a distribution in the fluid rest frame

$$f(p^\mu) = f_0(p \cdot u) + \delta f(p^\mu). \tag{12}$$

In ideal hydrodynamics, one has a thermal distribution f_0, while in general there is a viscous correction δf.

These particles are often simply treated as free particles (which can decay if they are unstable), or one can allow rescattering in a kinetic description (e.g., using models such as UrQMD [18, 33] or JAM [34, 26].

4. Results

Most models for the initial conditions are relatively smooth and uniform in the longitudinal direction—QCD matter tends to form flux tubes or string-like structures between the leading edge of colliding nuclei. This causes the distribution of particles (Eq. (1)) to vary slowly with (pseudo-)rapidity, and therefore causes the distribution of pairs to depend little on relative pseudorapidity $\Delta\eta$. On the other hand, the asymmetry v_n causes correlation of particles in relative azimuth $\Delta\phi$ according to Eq. (4). This agrees with the observed long-range structure of measured two-particle correlations, which in particular see a large number of particle pairs with the same azimuthal angle, but very different pseudorapidity (the "ridge"), which is difficult to explain in the absence of collective behavior in the collision system.

Various other generic features of correlations are also consistent with a hydrodynamic interpretation [35, 36], and actual hydrodynamic calculations have been very successful at describing a large number of 2-, 3-, 4-, 5-particle correlations and beyond [15, 37, 38].

Nevertheless, work is ongoing in understanding as much as we can about the correlations seen in these experiments, and what information we can extract from these data.

5. Summary

This was a brief overview of flow phenomena and hydrodynamic calculations in relativistic heavy-ion collisions, giving enough context and terminology for a young physicist to begin to read current literature of the field.

Acknowledgments

Funding was provieded by the Natural Sciences and Engineering Research Council of Canada

References

[1] Abelev B *et al.* (ALICE Collaboration) 2013 (*Preprint* 1301.4361)
[2] Bozek P 2012 *Phys.Rev.* **C85** 014911 (*Preprint* 1112.0915)
[3] Chatrchyan S *et al.* (CMS Collaboration) 2013 *Phys.Lett.* **B718** 795–814 (*Preprint* 1210.5482)
[4] Ollitrault J Y, Poskanzer A M and Voloshin S A 2009 *Phys.Rev.* **C80** 014904 (*Preprint* 0904.2315)
[5] Luzum M and Ollitrault J Y 2013 *Phys.Rev.* **C87** 044907 (*Preprint* 1209.2323)
[6] Gardim F G, Grassi F, Luzum M and Ollitrault J Y 2013 *Phys.Rev.* **C87** 031901 (R) (*Preprint* 1211.0989)
[7] Heinz U W, Qiu Z and Shen C 2013 (*Preprint* 1302.3535)
[8] Bhalerao R S, Luzum M and Ollitrault J Y 2011 *Phys.Rev.* **C84** 034910 (*Preprint* 1104.4740)
[9] Bilandzic A (ALICE Collaboration) 2012 (*Preprint* 1210.6222)
[10] Jia J (ATLAS Collaboration) 2012 (*Preprint* 1209.4232)
[11] Timmins A R (ALICE Collaboration) 2013 (*Preprint* 1301.6084)
[12] Miller M L, Reygers K, Sanders S J and Steinberg P 2007 *Ann.Rev.Nucl.Part.Sci.* **57** 205–243 (*Preprint* nucl-ex/0701025)
[13] Bozek P 2012 *Phys.Lett.* **B717** 287–290 (*Preprint* 1208.1887)
[14] Drescher H, Hladik M, Ostapchenko S, Pierog T and Werner K 2001 *Phys.Rept.* **350** 93–289 (*Preprint* hep-ph/0007198)
[15] Gardim F G, Grassi F, Luzum M and Ollitrault J Y 2012 *Phys.Rev.Lett.* **109** 202302 (*Preprint* 1203.2882)
[16] Werner K 2008 *Nucl.Phys.Proc.Suppl.* **175-176** 81–87
[17] Werner K 2012 (*Preprint* 1205.3379)
[18] Petersen H, Bleicher M, Bass S A and Stocker H 2008 (*Preprint* 0805.0567)
[19] Petersen H, La Placa R and Bass S A 2012 *AIP Conf.Proc.* **1441** 777–779
[20] Zhang B, Ko C, Li B A and Lin Z w 2000 *Phys.Rev.* **C61** 067901 (*Preprint* nucl-th/9907017)

[21] Pang L, Wang Q and Wang X N 2012 (*Preprint* 1211.1570)
[22] Arneodo M, Lamberti L and Ryskin M 1997 *Comput.Phys.Commun.* **100** 195–214 (*Preprint* hep-ph/9610286)
[23] Flensburg C 2012 *Prog.Theor.Phys.Suppl.* **193** 172–175
[24] Luzum M and Ollitrault J Y 2012 *Nucl.Phys.A* (*Preprint* 1210.6010)
[25] Drescher H J and Nara Y 2007 *Phys.Rev.* **C76** 041903 (*Preprint* 0707.0249)
[26] Hirano T, Huovinen P, Murase K and Nara Y 2013 *Prog.Part.Nucl.Phys.* **70** 108–158 (*Preprint* 1204.5814)
[27] Dumitru A and Nara Y 2012 *Phys.Rev.* **C85** 034907 (*Preprint* 1201.6382)
[28] Schenke B, Tribedy P and Venugopalan R 2012 *Phys.Rev.Lett.* **108** 252301 (*Preprint* 1202.6646)
[29] Gale C, Jeon S, Schenke B, Tribedy P and Venugopalan R 2013 *Phys.Rev.Lett.* **110** 012302 (*Preprint* 1209.6330)
[30] Chen G and Fries R J 2012 (*Preprint* 1212.4119)
[31] Cooper F and Frye G 1974 *Phys.Rev.* **D10** 186
[32] Dusling K, Moore G D and Teaney D 2010 *Phys.Rev.* **C81** 034907 (*Preprint* 0909.0754)
[33] Petersen H, Steinheimer J, Burau G, Bleicher M and Stocker H 2008 *Phys.Rev.* **C78** 044901 (*Preprint* 0806.1695)
[34] Nara Y, Otuka N, Ohnishi A, Niita K and Chiba S 2000 *Phys.Rev.* **C61** 024901 (*Preprint* nucl-th/9904059)
[35] Luzum M 2011 *Phys.Lett.* **B696** 499–504 (*Preprint* 1011.5773)
[36] Sorensen P, Bolliet B, Mocsy A, Pandit Y and Pruthi N 2011 *Phys.Lett.* **B705** 71–75 (*Preprint* 1102.1403)
[37] Schenke B, Jeon S and Gale C 2012 *Phys.Rev.* **C85** 024901 (*Preprint* 1109.6289)
[38] Qiu Z and Heinz U 2012 *Phys.Lett.* **B717** 261–265 (*Preprint* 1208.1200)

Initial Conditions with Flow from a McLerran Venugopalan model with Transverse Dynamics

Guangyao Chen and Rainer J. Fries

Cyclotron Institute and Department of Physics and Astronomy, Texas A&M University, College Station, TX 77843, USA

E-mail: glorychern@neo.tamu.edu

Abstract. Using a recursive solution of the Yang-Mills equation, we calculate analytic expressions for the gluon fields created in ultra-relativistic heavy ion collisions at small times τ. We have worked out explicit solutions for the fields and the energy momentum tensor up to 4th order in an expansion in τ. We generalize the McLerran-Venugopalan model to allow for a systematic treatment of averaged charge densities μ^2 that vary as a function of transverse coordinates. This allows us to calculate radial, elliptic and directed flow of gluon fields. Our results can serve as initial conditions for hydrodynamic simulations of nuclear collisions that include initial flow.

1. Introduction

Nuclear collisions at high energies at the Relativistic Heavy Ion Collider (RHIC) and the Large Hadron Collider (LHC) can create and probe novel forms of QCD matter. One example is a deconfined phase of QCD called Quark Gluon Plasma (QGP). A thermalized QGP is created in nuclear collisions at RHIC and LHC after a time of about 1 fm/c. Relativistic hydrodynamics is the tool of choice to describe the evolution of QGP in this stage [1, 2]. Large uncertainties in the hydrodynamic evolution come from the incomplete understanding of the initial conditions, i.e. the evolution of the nuclear collision prior to 1 fm/c. This is where another new phase of QCD, Color Glass Condensates (CGC) plays an important part. The natural limit of a nucleus at very high energy is characterized by a saturated gluon density in which the gluon field can be described by a quasi-classical field [3, 4]. The earliest phase of the collision of two nuclei can thus be described as the collision of two sheets of color glass [5]. The classical field resulting from the collision then decays and equilibrates to a plasma of quarks and gluons.

To be more precise, in the color glass formalism partons are distinguished by momentum fraction. Sources have large Bjorken-x while the small-x gluons are treated through the classical field $F^{\mu\nu}$. They are related by the Yang-Mills equation $[D_\mu, F^{\mu\nu}] = J^\nu$. Before the collision, the two incoming nuclei on the positive and negative light cone are dominated by transverse chromo-electric and chromo-magnetic fields which are perpendicular to each other [3, 4]. After the collision, the fields in the forward light cone can be parameterized as $A^\pm = \pm x^\pm A$ and a transverse field A^i, $i = 1, 2$. A and A^i depend on the proper time $\tau = \sqrt{t^2 - z^2}$ and transverse coordinates but not on the space-time rapidity η. For small τ, a series expansion $A^\mu = \sum_n \tau^n A^\mu_{(n)}$ in powers of τ can be employed. A few lowest orders in time should give an adequate and analytic description of the gluon fields immediately after the collision [6].

The collision initially ($\tau \to 0$) creates longitudinal chromo-electric and magnetic fields

$$E_0 = ig[A_1^i, A_2^i], \qquad B_0 = -ig\epsilon^{ij}[A_1^i, A_2^j]$$

between the nuclei, where A_1^i and A_2^i are the fields of nucleus 1 or 2 before the collision in light cone gauge. The next order in τ gives us linearly growing transverse electric and magnetic fields

$$E_{(1)}^i = -\frac{1}{2}(\sinh\eta[D^i, E_0] + \cosh\eta\,\epsilon^{ij}[D^j, B_0]), \quad B_{(1)}^i = \frac{1}{2}(\cosh\eta\,\epsilon^{ij}[D^j, E_0] - \sinh\eta[D^i, B_0]).$$

We have calculated analytic expressions for the fields up to fourth order in τ explicitly and will discuss those results elsewhere [7]. The behavior we find is qualitatively consistent with numerical studies of classical field QCD [8].

The energy momentum tensor of the gluon field after the collision can be easily calculated. Up to order $\mathcal{O}(\tau^3)$ — we skip higher orders here for brevity — it can be written in the form

$$T_{\mathrm{f}}^{mn} = \begin{pmatrix} A+C-D & B_1 & B_2 & E/\tau \\ B_1 & A+C & F & B_1'/\tau \\ B_2 & F & A+C & B_2'/\tau \\ E/\tau & B_1'/\tau & B_2'/\tau & (-A-C+D)/\tau^2 \end{pmatrix} \tag{1}$$

in the τ, x, y, η coordinate system. Here $A = (E_0^2 + B_0^2)/2$ is the initial energy density at $\tau = 0$, and B_i and B_i' are the coefficients of the Poynting vector $T^{0i} = B_i \cosh\eta + B_i' \sinh\eta$ and linear in τ ($i = 1, 2$ are transverse indices). C, D, E, $F \propto \tau^2$ are coefficients quadratic in τ. All these coefficients have been calculated analytically [7]. One can check that this energy momentum tensor is boost invariant owing to the original boost-invariant setup of the colliding nuclei.

2. Flow in the McLerran Venugopalan Model

In the McLerran-Venugopalan (MV) model [3, 4], the simplest implementation of color glass, an observable O calculated from sources in the two nuclei has to be averaged over all possible configurations of color source distributions ρ. Each collision samples arbitrary ρ due to the short time scale of the collision compared to internal time scales of the nuclear wave function. The distribution of charge is usually assumed to be Gaussian around zero with an average value μ^2 of the variation. We use the definition

$$\langle \rho_a(\vec{x}_\perp)\rho_b(\vec{y}_\perp)\rangle = \frac{g^2\delta_{ab}}{N_c^2 - 1}\mu^2(\vec{x}_\perp)\delta^2(\vec{x}_\perp - \vec{y}_\perp). \tag{2}$$

We have suppressed the dependence on the longitudinal coordinate for brevity. Note that we have generalized the original expression from the MV model such that μ^2 can be a function of transverse coordinate $\vec{x}_\perp$, instead of being an average charge density that is assumed to be homogeneous. Of course the latter is unrealistic and the generalization needs to be made if long-range transverse dynamics is to be described. The charge density μ^2 of a nucleus is related to the large-x parton distribution in the transverse plane. It can be simulated by sampling the nucleons inside the nucleus, for example using a Woods-Saxon distribution. We have shown explicitly that if μ^2 is varying slowly such that $|\mu^2(\vec{x}_\perp)| \gg m^{-1}|\nabla^i\mu^2(\vec{x}_\perp)| \gg m^{-2}|\nabla^i\nabla^j\mu^2(\vec{x}_\perp)|$, color glass dynamics and long-distance dynamics which are not described by CGC can be safely separated. Here m is an infrared regulator. We find that the generalized two gluon correlation function is [7]

$$\langle A_a^i(\vec{x}_\perp)A_b^j(\vec{x}_\perp)\rangle = \delta_{ab}\frac{g^2\mu^2(\vec{x}_\perp)}{8\pi(N_c^2-1)}\left[\delta^{ij}\ln\frac{Q^2}{(1.47m)^2} + \frac{\nabla^k\nabla^l\mu^2(\vec{x}_\perp)}{m^2\mu^2(\vec{x}_\perp)}\left(\frac{1}{6}\delta^{kl}\delta^{ij} - \frac{7}{12}\delta^{ik}\delta^{jl}\right)\right], \tag{3}$$

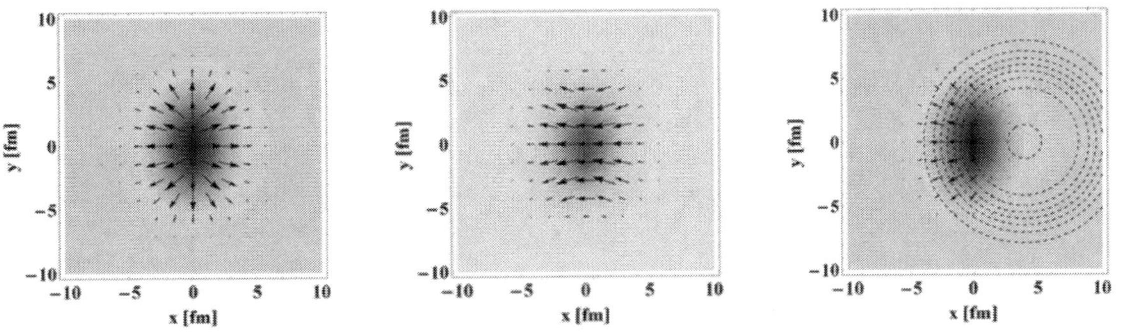

Figure 1. Left: The hydro-like flow component $\vec{B}$ (arrows) and the underlying energy density for Au+Au collisions using Woods-Saxon profiles. Middle: The rapidity-odd flow component $\vec{B}'$. Right: The Poynting vector T^{0i} at $\eta = 1$ together with the energy density. The contour lines indicate the density profile of one of the nuclei.

where the first term is the same as in the original MV model [9]. Q here is the saturation scale.

Let us now focus on the transverse Poynting vector T^{0i} in the energy momentum tensor which describes the energy flow of the gluon field. The term even in η after averaging reads

$$B_i = -\tau \frac{g^6 N_c(N_c^2-1)}{32\pi} \nabla^i (\mu_1^2 \mu_2^2) \ln^2 \frac{Q^2}{(1.47m)^2} \,. \tag{4}$$

This expression gives radial and elliptic flow of energy $\propto -\nabla^i \epsilon$. This is similar to what a hydrodynamic flow field would look like for the same energy density ϵ if the hydrodynamic pressure is a monotonous function of ϵ. The term odd in η is

$$B'_i = -\tau \frac{g^6 N_c(N_c^2-1)}{32\pi} \ln^2 \frac{Q^2}{(1.47m)^2} [\mu_2^2 \nabla^i \mu_1^2 - \mu_2^2 \nabla^i \mu_1^2] \,. \tag{5}$$

Interestingly this term can describe a directed flow of energy. Note that this term does not violate boost-invariance of the energy momentum tensor. On the other hand boost-invariant hydrodynamics would never be able to create such a flow field. Rather one can trace this term back to the QCD analogon of Gauss' Law. This makes this rapidity-odd flow field a curious phenomenon. The flow vector fields $\vec{B}$ and $\vec{B}'$ from two gold nuclei with Woods-Saxon profiles μ^2 colliding with impact parameter $b = 8$ fm are shown in Fig. 1. The net Poynting vector T^{0i} exhibits a clear dipole asymmetry in the transverse plane when $\eta \neq 0$. Flow in color glass at midrapidity has recently also been studied numerically in [10].

3. Hydrodynamic Initial Conditions

Since the classical gluon field only describes the very earliest stage of a nuclear collision we need to translate the energy momentum tensor into initial conditions for hydrodynamics. The dynamics of equilibration is outside the purview of either hydrodynamics nor classical field theory and would require a more complete description of quantum non-abelian dynamics. One can however derive a matching hydrodynamic energy momentum tensor simply using conserved currents both on the classical field side and the hydrodynamic side. For ideal hydrodynamics this was worked out in [11, 12] by just imposing energy and momentum conservation. The flow of energy generally translates into a flow of particles on the hydrodynamic side. The direction of the hydrodynamic transverse flow velocity is parallel to the Poynting vector and its size is a monotonous function of the magnitude of the Poynting vector. A similar correspondence can be worked out in the case of matching to viscous hydrodynamics [7].

4. Directed Flow Phenomenology

Without a further hydrodynamic evolution we can nevertheless check wether our somewhat surprising direct flow result is qualitatively consistent with experiments [13]. The first thing to check is the direction of the flow. It turns out that our directed flow vector $\vec{B}'$ points away from the spectators for a given side $\eta > 0$ or $\eta < 0$ of a collision system with finite impact parameter. This is consistent with experiments. The rapidity dependence of particle directed flow v_1 is also roughly consistent with a $\sinh y$ shape as a function of particle rapidity y [14, 15].

Note that $\vec{B}'$ vanishes for $b = 0$ if the two nuclei are equal. Interestingly, if an asymmetric system like Cu+Au collides with $b = 0$ the flow vector $\vec{B}'$ becomes radial, but points outward in the direction of the smaller nucleus and inward in the direction of the larger nucleus. An asymmetric system with $b \neq 0$ would exhibit a striking blend of both effects. Looking for such a strong dependence of flow on rapidity in central collisions of asymmetric systems could be a signature for the eminence of strong gluon fields early in nuclear collisions.

5. Conclusion

To summarize, if the McLerran Venugopalan model is generalized to allow for variations of the average charge density across the nucleus flow phenomena in early stages of nuclear collisions can be studied. Here we use analytic results for early gluon fields and their energy momentum tensor that can be obtained in an expansion of the Yang-Mills equations around the time of collision. We find radial and elliptic flow patterns similar to hydrodynamic behavior. In addition we also obtain a rapidity-odd term in the flow of gluon energy which resembles directed flow. We have argued that this flow will translate into directed flow of particles in the further course of the collision. The results are in qualitative agreement with results from the STAR experiment.

Acknowledgments

This work is supported by NSF CAREER grant PHY-0847538, JET Collaboration and DOE grant DE-FG02-10ER41682. We thank Joseph Kapusta and Yang Li who contributed to this project.

References

[1] J. Adams *et al.* [STAR Collaboration], Nucl. Phys. **A757**, 102-183 (2005).
[2] K. Adcox *et al.* [PHENIX Collaboration].
[3] L. D. McLerran and R. Venugopalan, Phys. Rev. D **49**, 2233 (1994) [hep-ph/9309289].
[4] L. D. McLerran and R. Venugopalan, Phys. Rev. D **49**, 3352 (1994) [hep-ph/9311205].
[5] A. Kovner, L. D. McLerran and H. Weigert, Phys. Rev. D **52**, 3809 (1995); Phys. Rev. D **52**, 6231 (1995).
[6] R. J. Fries, J. I. Kapusta and Y. Li, arXiv:nucl-th/0604054.
[7] G. Chen, R. J. Fries, J. I. Kapusta and Y. Li, *in preparation*
[8] T. Lappi, Phys. Lett. B **643**, 11 (2006).
[9] J. Jalilian-Marian, A. Kovner, L. D. McLerran and H. Weigert, Phys. Rev. D **55**, 5414 (1997).
[10] B. Schenke, P. Tribedy and R. Venugopalan, Phys. Rev. Lett. **108** (2012) 252301 [arXiv:1202.6646 [nucl-th]].
[11] R. J. Fries, J. I. Kapusta and Y. Li, Nucl. Phys. A **774** (2006) 861 [hep-ph/0511101].
[12] R. J. Fries, J. Phys. G **34** (2007) S851 [nucl-th/0702026 [NUCL-TH]].
[13] G. Chen and R. J. Fries, arXiv:1303.2360 [nucl-th].
[14] B. I. Abelev *et al.* [STAR Collaboration], Phys. Rev. Lett. **101**, 252301 (2008) [arXiv:0807.1518 [nucl-ex]].
[15] J. Adams *et al.* [STAR Collaboration], Phys. Rev. C **73**, 034903 (2006) [nucl-ex/0510053].

Transition From Ideal To Viscous Mach Cones In A Partonic Transport Model

I. Bouras[1], A. El[1], O. Fochler[1], H. Niemi[2], Z. Xu[3] and C. Greiner[1]

[1] Institut für Theoretische Physik, Johann Wolfgang Goethe-Universität, Max-von-Laue-Str. 1, D-60438 Frankfurt am Main, Germany
[2] Department of Physics, P.O. Box 35 (YFL), FI-40014 University of Jyväskylä, Finland
[3] Department of Physics, Tsinghua University, Beijing 100084, China

E-mail: bouras@th.physik.uni-frankfurt.de

Abstract.
Using a partonic transport model we investigate the evolution of conical structures in ultrarelativistic matter. Using two different source terms and varying the transport properties of the matter we study the formation of Mach Cones. Furthermore, in an additional study we extract the two-particle correlations from the numerical calculations and compare them to an analytical approximation. The influence of the viscosity to the shape of Mach Cones and the corresponding two-particle correlations is studied by adjusting the cross section of the medium.

1. Introduction

In relativistic heavy-ion collisions at the relativistic heavy-ion collider (RHIC) and the Large Hadron Collider (LHC) a new state of matter, the Quark-Gluon Plasma (QGP), is supposed to be created. In these collisions highly energetic partons propagate through the hot and dense medium and rapidly lose their energy and momentum as the energy is deposited in the medium. Measurements of two- and three-particle correlations in heavy-ion collisions show a under certain circumstances a suppression of the away-side jet, whereas for lower p_T a double peak structure is observed in the two-particle correlation function. A promising origin of these structures was assumed to be the interaction of fast and high-energetic partons with the soft matter which generates collective motion of the medium in form of Mach cones [2].

For this purpose we investigate the propagation and formation of Mach cones in the partonic transport model BAMPS (Boltzmann Approach of MultiParton Scatterings) [4] in the limit of vanishing mass and very small shear viscosity over entropy density ratio η/s of the matter. Two different scenarios for the jet are used. By adjusting η/s the influence of the viscosity on the profile of the Mach cone and the corresponding two-particle correlation is studied for the first time. The results presented here are based on a recent publication [5].

2. Shock Waves and Mach cones

Shock waves are phenomena which have their origin in the collective motion of matter. In a simplified one-dimensional setup shock waves have already been studied within the framework of BAMPS for the perfect fluid limit [6, 7]. Furthermore BAMPS calculations have demonstrated

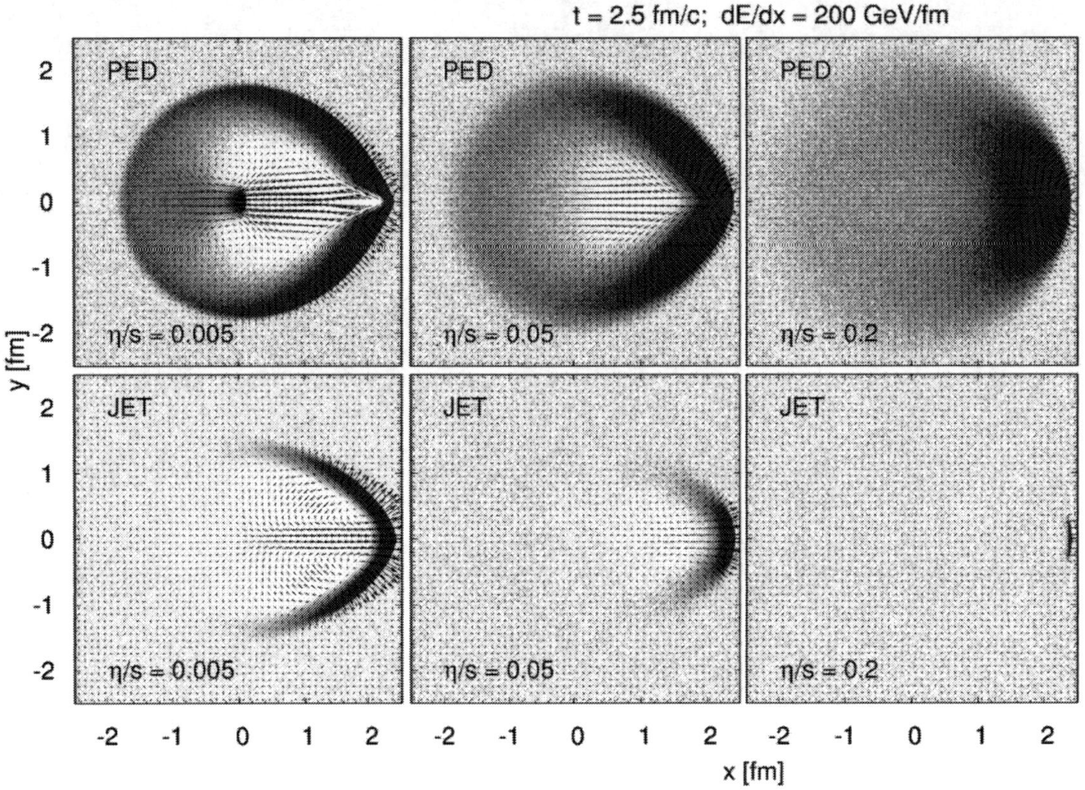

Figure 1. (Color online) Transition from ideal to viscous Mach cones. Shape of a Mach cone shown for different jet scenarios and different viscosity over entropy density ratios, $\eta/s = 0.005$, 0.05 and 0.5. The energy deposition is $dE/dx = 200$ GeV/fm. The upper panel shows the pure energy deposition scenario (PED); the lower panel shows the propagation of a highly energetic jet (JET) depositing energy and momentum in x-direction. Depicted are the LRF energy density within a specific range; as an overlay we show the velocity profile with a scaled arrow length. The results are a snapshot of the evolution at $t = 2.5$ fm/c.

that the shock profile is smeared out when viscosity is large. It was also found that a clear observation of the shock within the short time available in HIC requires a small viscosity.

In the following we study the evolution of "Mach cone"-like structures with different scenarios of the jet-medium interaction by using the parton cascade BAMPS. We focus on investigation of Mach cone evolution in absence of any other effects - i.e. we neglect such effects as initial fluctuations or expansion, which are however relevant in HIC. We use a static box with $T_{\text{med}} = 400$ MeV and binary collisions with an isotropic cross section. Furthermore, we keep the mean free path λ_{mfp} of the medium particles constant in all spatial cells by adjusting the cross section according to $\sigma = 1/(n\lambda_{\text{mfp}})$, where n is the particle density. The related shear viscosity for isotropic binary collisions is given by $\eta = 0.4\, e\, \lambda_{\text{mfp}}$ [8].

The Mach Cones studied here are induced by two different sources. The first of them we refer to as the pure energy deposition scenario (PED) [9]. This is simulated by a moving source depositing momentum end energy isotropically according to the thermal distribution $f(x,p) = \exp(-E/T)$. The second source we refer to as JET. This is simulated by a highly massless particle (jet) which has only momentum in x-direction, i.e. $p_{\text{x}} = E_{\text{jet}}$. After each

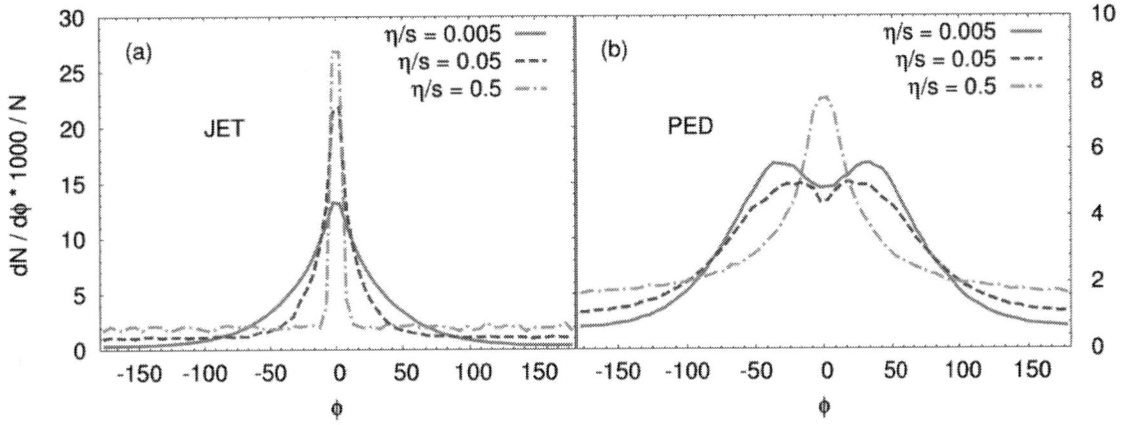

Figure 2. (Color online) Two-particle correlations $dN/(Nd\phi)$ for different viscosities extracted from calculations shown in Fig. 1. The results are shown in the for the JET (a) and PED (b) scenario for $dE/dx = 200$ GeV/fm.

timestep the energy of the jet particle is reset to its initial value. For both scenarios the sources are initialized at $t = 0$ fm/c at the position $x = -0.1$ fm and propagate in x-direction with $v_{\text{source}} = 1$, i.e. with the speed of light.

In Fig. 1 we show the Mach Cone structure for both PED scenario (upper panel) and JET scenario (lower panel) with $\eta/s = 0.005$, 0.05 and 0.5 from left to right, respectively. We show a snapshot at $t = 2.5$fm/c. The energy deposition rate is fixed to $dE/dx = 200$ GeV/fm. Although this rather high energy deposition is not expected to be produced by any known energy-loss process in current heavy-ion collisions, this rate is necessary in order to observe possible double-peak structures in corresponding two-particle correlations.

In both scenarios, PED and JET, for $\eta/s = 0.005$ (left panel), we observe a conical structure, but with obvious differences. The PED case with the isotropic energy deposition induces a spherical shock into back region; this structure is missing in the JET scenario because of the high forward peaked momentum deposition. Another difference is that in the JET scenario a clearly visible head shock appears. This in turn is missing in the PED scenario. Furthermore a (anti)-diffusion wake is induced by the JET (PED) scenario.

Adjusting the shear viscosity over entropy density ratio $\eta/s = 0.05-0.5$ we observe a smearing out of the Mach cone structure. For a sufficient high $\eta/s = 0.5$ the conical structures in both scenarios disappear. This is true for shock fronts as well as for the (anti-) diffusion wake. The difference between the PED and the JET case is that as η/s increases, in the PED scenario the resulting "Mach cone" solution covers approximately the same spatial region regardless of a value of η/s, while in the JET case the structure is concentrated more and more near the projectile as the viscosity increases.

In Fig. 2 we show the two-particle correlations extracted from BAMPS calculations of the Mach Cones shown in Fig. 1. For the JET scenario (a) and sufficiently small $\eta/s = 0.005$ we observe only a peak in direction of the jet. The typical double peak structure, which has been proposed as a possible signature of the Mach cone in HIC, can only be observed for the PED scenario (b) and small η/s. However, the PED scenario has no correspondence in heavy-ion physics. For the JET scenario, which is a simplified model of jets depositing energy and momentum, a double peak structure never appears. This is due to the strong formation of a head shock and diffusion wake.

3. Summary and Conclusions

In this work we investigated the evolution of Mach cones induced by two different source terms, PED and JET, using a partonic transport model. The studies were performed in a static system, neglecting effects coming from initial fluctuations and expansion. The development of Mach cones is observed in case the viscosity of matter is small enough. In addition, the effects of viscosity of the matter were shown by adjusting the shear viscosity over entropy density ratio η/s from 0.005 to 0.5. A clear and unavoidable smearing out of the profile depending on a finite ratio of shear viscosity to entropy density is clearly visible. Investigating the corresponding two-particle correlations we see that Mach cones can not be connected to double peak structures by any realistic picture of jets in HIC. A further investigation using more realistic jets in an expanding system is under way.

Acknowledgements

The authors are grateful to the Center for the Scientific Computing (CSC) at Frankfurt for the computing resources. A.E. was supported by BMBF. This work was supported by the Helmholtz International Center for FAIR within the framework of the LOEWE program launched by the State of Hesse.

References

[1] Fuqiang Wang. Measurement of jet modification at RHIC. *J. Phys.*, G30:S1299–S1304, 2004.

[2] Horst Stoecker. Collective Flow signals the Quark Gluon Plasma. *Nucl. Phys.*, A750:121–147, 2005.

[3] Ioannis Bouras et al. Relativistic Shock Waves and Mach Cones in Viscous Gluon Matter. *J. Phys. Conf. Ser.*, 230:012045, 2010.

[4] Zhe Xu and Carsten Greiner. Thermalization of gluons in ultrarelativistic heavy ion collisions by including three-body interactions in a parton cascade. *Phys. Rev.*, C71:064901, 2005.

[5] I. Bouras, A. El, O. Fochler, H. Niemi, Z. Xu, et al. Transition From Ideal To Viscous Mach Cones In A Kinetic Transport Approach. *Phys.Lett.*, B710:641–646, 2012.

[6] I. Bouras et al. Relativistic shock waves in viscous gluon matter. *Phys. Rev. Lett.*, 103:032301, 2009.

[7] I. Bouras et al. Investigation of shock waves in the relativistic Riemann problem: A comparison of viscous fluid dynamics to kinetic theory. *Phys. Rev.*, C82:024910, 2010.

[8] S. R. de Groot et al. *Relativistic Kinetic Theory: Principles and Applications.* North Holland, Amsterdam, 1980.

[9] Barbara Betz, Jorge Noronha, Giorgio Torrieri, Miklos Gyulassy, Igor Mishustin, et al. Universality of the Diffusion Wake from Stopped and Punch-Through Jets in Heavy-Ion Collisions. *Phys.Rev.*, C79:034902, 2009.

Two-particle correlations in 2D p_t space in Au+Au collisions at $\sqrt{s_{NN}} = 200$ GeV at STAR

Elizabeth W. Oldag (for the STAR Collaboration)

Department of Physics, The University of Texas at Austin, Austin, TX 78712, USA

E-mail: ewingf@physics.utexas.edu

Abstract. Measurements of 2D transverse momentum correlations on (p_{t1}, p_{t2}) from minimum-bias Au+Au collisions at $\sqrt{s_{NN}} = 200$ GeV at STAR are presented and discussed. These correlations, formed from all charged particles with $p_t \geq 0.15$ GeV/c, $|\eta| \leq 1$, and 2π in azimuth, show a broad peak extending from p_t=0.5-4.0 GeV/c. The broad peak is observed in both like- and unlike-sign charge combinations and for same- and away-side relative azimuth angles. Interestingly, the peak in the data for away-side or back-to-back pairs persists even in more-central collisions, remaining at approximately the same p_t for all centralities. These data are compared to theoretical models and the p_t dependence of the same-side angular correlation structure.

1. Introduction

Two-particle correlations provide greater access to the dynamics of heavy-ion collisions than that enabled with single-particle measurements as discussed in [1]. Observations of charged particle pair number correlations in relative angle space ($\eta_\Delta = \eta_1 - \eta_2, \phi_\Delta = \phi_1 - \phi_2$) have shown unexpected trends (e.g. ridge formation and anomalous centrality trends) in heavy-ion collisions relative to N-N binary-collision scaling [1]. Two particle correlations on (p_{t1}, p_{t2}), discussed in this paper, provide complementary information.

2. Two-particle correlation measure

The correlation quantity used in this analysis measures the number of correlated pairs per final-state particle and is defined as $\frac{\Delta\rho}{\sqrt{\rho_{ref}}} = \sqrt{\rho_{ref}}\frac{\rho_{sib}-\rho_{mix}}{\rho_{ref}}$ where ρ_{sib} and ρ_{mix} are the densities of sibling and mixed-event pairs, respectively. The pair density, ρ_{ref}, represents a reference distribution which excludes the physics signal(s) of interest.

In angular space mixed-event pairs provide an accurate representation of ρ_{ref} because the physics signal(s) (e.g. jets, flows) are randomly distributed in the primary (η, ϕ) space. Mixed-event pairs cannot produce statistically significant structures on $(\eta_\Delta, \phi_\Delta)$ due to those processes. However, because dynamical processes (e.g. jets) likely produce particles at similar transverse momentum in different collisions, mixed-event particle pairs on (p_{t1}, p_{t2}) will have a similar structure as that from sibling pairs (from the same event). Given that most interesting angular correlation structures follow (or even exceed) binary collision scaling [1], and may be jet-related, we select a ρ_{ref} that is absent such scaling, or one where multiplicity is proportional to the number of participants (N_{part}). The soft particle spectrum of the two-component model [2]

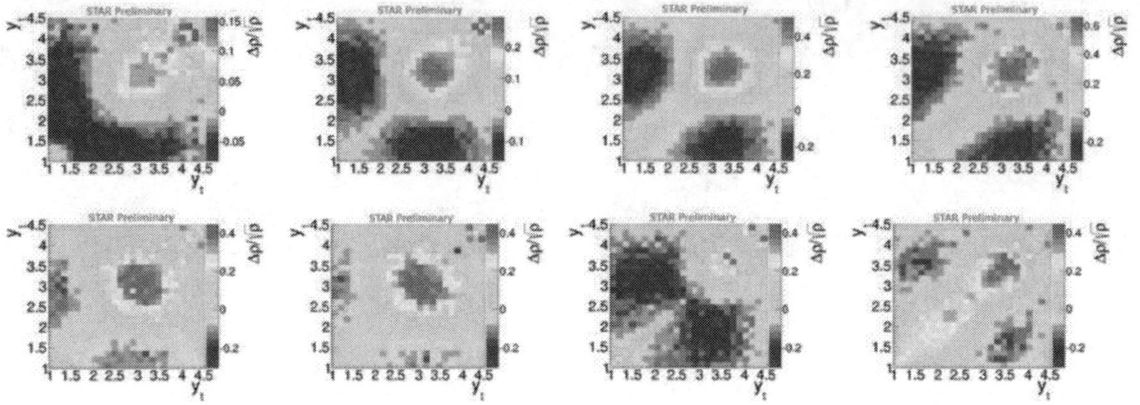

Figure 1. (Color online) Two-dimensional (y_{t1}, y_{t2}) correlations for Au-Au collisions at $\sqrt{s_{NN}}$ = 200 GeV. The upper panels display charge-independent results in centralities 74-84%, 46-55%, 18-28%, and 0-5% (left to right). The lower left [right] two panels are formed from pairs with $|\phi_\Delta| > \pi/2$ (away-side) [$|\phi_\Delta| < \pi/2$ (same-side)] in 0-5% central events. The lower first and third panels display same-sign pairs; the lower second and fourth panels are for unlike-sign pairs.

provides a good estimate; it is represented with a single Levy distribution

$$\frac{d^2N_{ch,soft}}{dy_t d\eta} = 2\pi p_t m_{t,\pi} \frac{N_{part}}{2} \left[\frac{A}{[1 + (m_{t,\pi} - m_\pi)/(nT)]^n} \right] \tag{1}$$

where A=5.81 c/GeV^2, T = 0.169 GeV, n = 13.8, $m_{t,\pi} = \sqrt{p_t^2 + m_\pi^2}$, and m_π is the pion mass.

3. Results

For best visual access the two-particle correlation data will be presented as a function of the transverse rapidity, defined as $y_t = \log[(m_{t,\pi} + p_t)/m_\pi]$. The momentum correlations of all particles with $p_t > 0.15$ GeV/c in the upper panels of Fig. 1 show a bump around (y_{t1}, y_{t2})=(3,3) (p_t=1.4 GeV/c). The amplitude of the bump increases with centrality but remains in approximately the same location. In angular correlations we observe a sharp transition in the features of the correlations around the 46-55% centrality [1] but we observe no sharp transition in momentum space.

The lower panels in Fig. 1 display momentum correlations for the most central 0-5% events for particle pairs distinguished by charge type and azimuthal opening angle. A signal from back-to-back jet fragmentation is expected in the away-side projections (two left panels) and a peak is indeed observed around (y_{t1}, y_{t2})=(3,3). Surprisingly this peak persists in higher centralities, increases in amplitude, as well as remaining in approximately the same location. Interestingly for unlike-sign, same-side pairs a single broad peak is observed in the peripheral bins which then appears to separate into two peaks near the same centrality where a sharp "transition" in the centrality trend occurs in angular correlations [1]. The above centrality trends are represented by the fit model parameters, see Fig. 2. The like-sign, same-side plots show HBT correlations along the diagonal ($y_{t1} = y_{t2}$) as well as a weaker bump at higher momentum.

The main characteristics of the peak at (y_{t1}, y_{t2})=(3,3) in the momentum correlation data are quantified as a function of centrality using a simple fit model within a cut window. The rectangular fitting window, oriented along the $y_{t\Sigma} = y_{t1} + y_{t2}$ and $y_{t\Delta} = y_{t1} - y_{t2}$ axes, was generally defined as $|y_{t\Delta}| \leq 2.0$, $4 < |y_{t\Sigma}| < 8$. This window excluded structure along the edges and at lower y_t. The complete fit model function was defined as $A_0 + A_1\text{exp}[-0.5 * (((y_{t\Sigma} - 2 *$

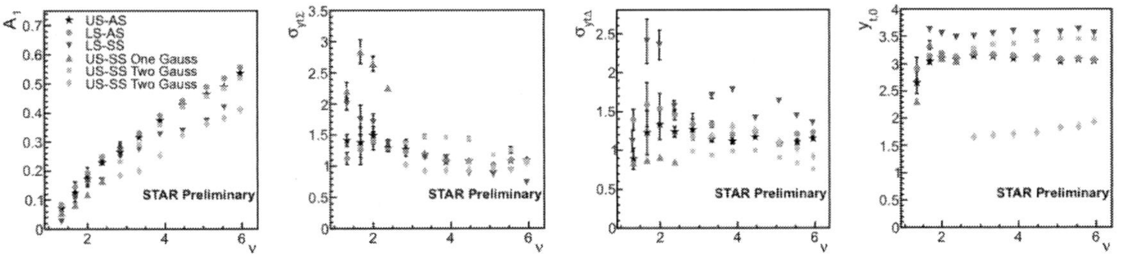

Figure 2. (Color online) Fit parameters for (y_{t1}, y_{t2}) correlation data versus centrality measure $\nu = 2\langle N_{\text{binary}}\rangle / \langle N_{\text{participant}}\rangle$.

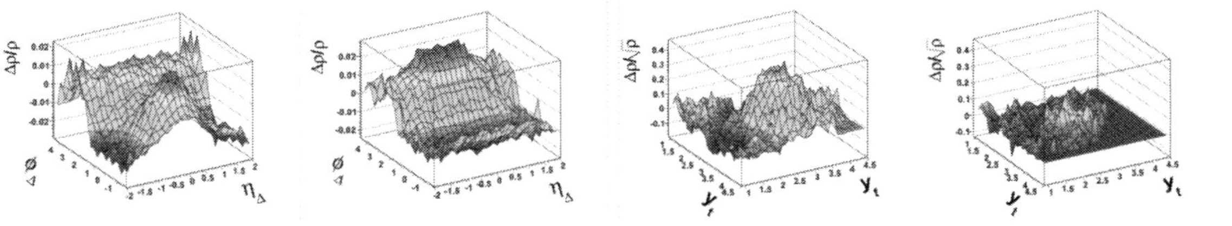

Figure 3. (Color online) Two-particle correlations from HIJING for 75-85% central events. The first and third panels show results with jets on; the second and fourth are with jets off.

$y_{t,0})/\sigma_{yt\Sigma})^2 + (y_{t\Delta}/\sigma_{yt\Delta})^2)]$. For the unlike-sign same-side correlations a second 2D Gaussian at lower momentum was required. The fit results are shown in Fig. 2. The amplitudes of the peaks in all four cases increase monotonically with centrality. The peaks also remain in approximately the same location from most-peripheral to most-central collisions.

These momentum correlations, specifically the peak around (y_{t1}, y_{t2})=(3,3), reveal an important new component of the data with which to test theoretical models. Predictions of two models, HIJING [3] and AMPT [4] are shown in Figs. 3 and 4. HIJING [3], based on the LUND string model and semi-hard jet fragmentation (PYTHIA), describes angular correlation data from peripheral heavy-ion collisions fairly well [1]. HIJING with jets off and on (no jet quenching) suggests that most of the pairs in the peak at (y_{t1}, y_{t2})=(3,3) are from jet fragmentation.

AMPT [4] is a hybrid transport model that includes a period of parton-parton re-scattering. It successfully reproduces the quadrupole (v_2) component in heavy-ion collisions. For this study the partonic re-scattering cross section was increased until the quadrupole amplitude matched the data and then the corresponding correlations in momentum space were observed (Fig. 4). The correlation study was done using partons after the cascade because the hadron correlations produced by AMPT's coalescence algorithm display non-intuitive dependence on the partonic cross section. When the partonic cross section increased to 6 mb the fitted quadrupole amplitude matched the experimental value as well as other general features of the angular correlation data. However, the peak around $(y_{t1}, y_{t2}) \approx (2.5, 2.5)$, evident in the 0 mb case, dissipates and the features deviate from that observed in data.

Measurement of the (y_{t1}, y_{t2}) dependence of 2D angular correlations completes the experimental determination of the six-dimensional two-particle correlations and provides access to the y_t dependence of angular correlation features such as the same-side 2D peak (ridge). In this analysis the (y_{t1}, y_{t2}) range ([1,4.5],[1,4.5]) was divided into 28 unique, square cut bins with sides of length 0.5. The angular correlation of the pairs in each bin was formed and fit with the 11 parameter fit function in [1]. Figure 5 shows the (y_{t1}, y_{t2}) distribution of the volume (number

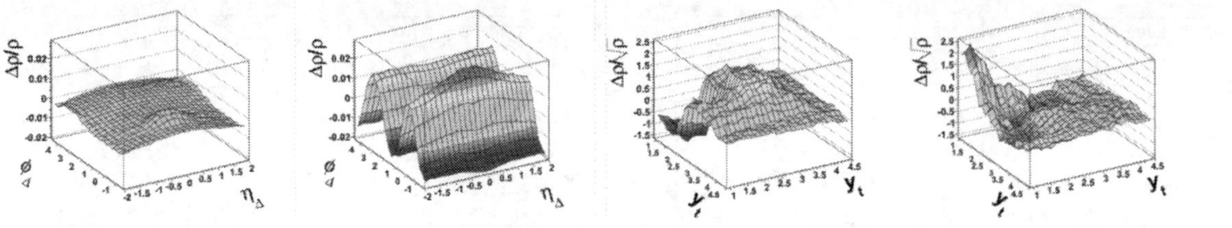

Figure 4. (Color online) Two-parton correlations from AMPT for collisions in the 45-55% centrality bin. The first and third (second and fourth) panels are for events with 0 mb (6 mb) partonic cross section in angular and momentum spaces, respectively.

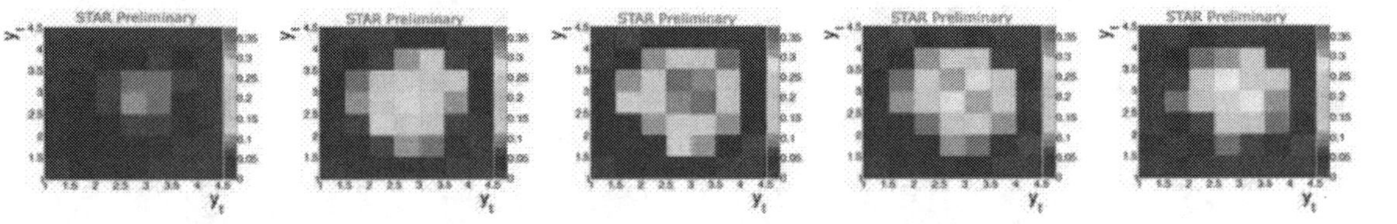

Figure 5. (Color online) The volume of the 2D Gaussian on (y_{t1}, y_{t2}). The left three panels show the volume of the 2D Gaussian from $0 < |\eta_\Delta| < 2/3$ in three centrality bins, 38-46%, 18-28%, and 0-5% respectively. The fourth and fifth panels show the volume in the 0-5% centrality from $2/3 < |\eta_\Delta| < 4/3$ and $4/3 < |\eta_\Delta| < 2$ respectively.

of correlated pairs per final-state particle) of the same-side 2D Gaussian in various $|\eta_\Delta|$ ranges. Most of these correlated pairs are distributed near (y_{t1}, y_{t2})=(3,3), similar to the momentum correlations in Fig. 1.

The y_t distribution of correlated pairs contributing to the same-side ridge as a function of η_Δ for the 0-5% centrality is shown in the three right-most panels in Fig. 5. The peak near (3,3) remains in the same location as $|\eta_\Delta|$ increases. This shows that correlated pairs in the η elongated portion of the same-side angular peak structure come from roughly the same momentum distribution as pairs in the center near $\eta_\Delta = 0$.

4. Conclusion

The momentum dependent results presented here complete the measurement of six-dimensional, two-particle correlations and provide definitive tests for theoretical models. The results show a broad peak extending from 0.5-4.0 GeV/c in all charge and azimuth pair combinations which, as a function of centrality, remains at a fixed position while monotonically increasing in amplitude. HIJING predicts a similar broad peak when jets are turned on. Partons in AMPT do not follow the observed trends when sufficient interaction strength is included to reproduce v_2; the peak is strongly dissipated. In a complementary study the momentum distribution of angular correlation features was investigated. The momentum distribution of correlated pairs that contribute to the same-side 2D Gaussian, commonly referred to as the "ridge" and which may be due to minijets [1], peak near (y_{t1}, y_{t2})=(3,3) (p_t=1.4 GeV/c) and do not soften with increased centrality or at larger $|\eta_\Delta|$.

References

[1] G. Agakishiev *et al.* (STAR Collaboration) 2012 *Phys. Rev. C* **86**, 064902
[2] T.A. Trainor 2008 *Int. J. Mod. Phys. E* **17**, 1499
[3] X.-N. Wang and M. Gyulassy 1991 *Phys. Rev. D* **44** 3501
[4] Zi-Wei Lin *et al.* 2005 *Phys. Rev. C* **72** 064901

Measurement of the distribution of event-by-event harmonic flow in Pb-Pb collisions at $\sqrt{s_{NN}}$ =2.76 TeV with the ATLAS detector

Soumya Mohapatra on behalf of the ATLAS Collaboration
Departments of Physics & Astronomy and Chemistry, Stony Brook University, NY 11794, USA
E-mail: soumya@cern.ch

Abstract. In recent years, the measurement of harmonic flow coefficients v_n has provided important insight into the hot and dense matter created in heavy ion collisions at RHIC and LHC. These coefficients are now understood to reflect the hydrodynamic response of the produced medium to the collision geometry. Due to finite number of nucleons in the system, the collision geometry can fluctuate from one event to another, and hence measuring the full distribution of the event-by-event v_n coefficients can provide better insights on the nature of these fluctuations and possible non-linear effects in the hydrodynamic response. This proceeding presents the first measurements of the event-by-event v_n distributions for n=2-4 in Pb-Pb collisions at $\sqrt{s_{NN}}$ = 2.76 TeV with the ATLAS detector.

Relativistic heavy-ion collisions produce an extremely hot and dense medium commonly termed as the Quark Gluon Plasma. The produced medium expands anisotropically due to asymmetric pressure gradients with larger particle yields in the direction of the largest gradients. The azimuthal anisotropy can be expressed as a Fourier series:

$$\frac{dN}{d\phi} \propto 1 + 2\sum_{n=1}^{\infty} v_n(p_{\mathrm{T}}, \eta)\cos n(\phi - \Phi_n) \tag{1}$$

where, v_n and Φ_n are the magnitude and phase of the nth order harmonic flow [1]. Measurements of the p_{T} , η and centrality dependence of the harmonics v_n have been done via the event-plane (EP) method [1] as well as by multi-particle correlation methods [1, 2]. In these methods, the v_n are obtained by averaging over a large number of events. Due to event by event fluctuations in the initial geometry, the v_n values vary event by event. Averaging over events, the EP and multi-paricle correlation methods measure a mean response of the medium. For example, the EP method gives a value in between the mean and the RMS of the v_n distribution, i.e. $\langle v_n \rangle < v_n^{EP} < \sqrt{\langle v_n^2 \rangle}$ [3]. A deeper understanding of the initial geometry and the expansion mechanism of the produced medium can be obtained by measuring the event-by-event (EbE) v_n distributions.

The large multiplicity in Pb-Pb collisions at the LHC as well as the large acceptance of the ATLAS inner detector [5] covering $|\eta| < 2.5$ allow for the first measurements of the EbE v_n distributions in heavy-ion collisions. Figure 1 shows the azimuthal distribution (black points) of charged tracks with p_{T} >0.5 GeV for three individual events in the (0-5)% centrality class. The red points show the anisotropy in the detector acceptance (arbitrary normalization) obtained by averaging over many events. It is clear that the EbE fluctuations in particle distributions are much larger than the detector acceptance effects.

The v_n measurements presented here were made using 8 μb^{-1} of Minimum Bias Pb-Pb data at $\sqrt{s_{NN}}$ of 2.76 TeV. Details of the results presented here are published in [6].

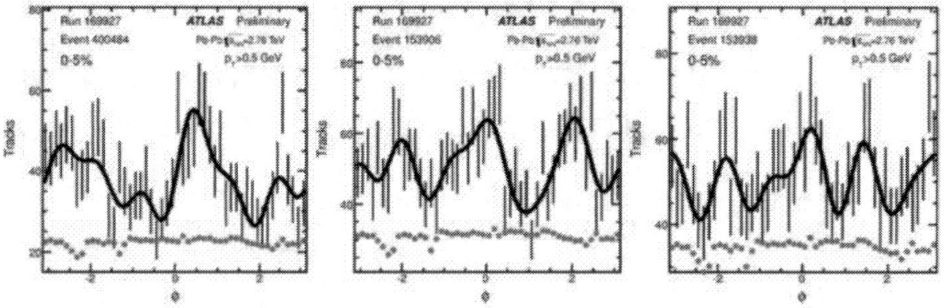

Figure 1. Single track ϕ distributions for three events (from left to right) in the (0-5)% centrality interval. The bars indicate the foreground distributions, the solid curves indicate a Fourier parameterization including first six harmonics and the red points indicate the detector acceptance functions. Charged tracks with $p_{\mathrm{T}} > 0.5\,\mathrm{GeV}$ are used. Figure taken from [6].

The azimuthal distribution of charged tracks in an event can be expanded in a Fourier series to obtain the *observed* flow-vector $(v_{n,x}^{obs}, v_{n,y}^{obs})$ as:

$$\frac{dN}{d\phi} \propto 1 + 2\sum_{n=1}^{\infty} v_n^{obs} \cos n(\phi - \Phi_n^{obs}) = 1 + 2\sum_{n=1}^{\infty} (v_{n,x}^{obs} \cos n(\phi) + v_{n,y}^{obs} \sin n(\phi)) \qquad (2)$$

To account for the detector efficiency effects, each track is weighted by the the inverse of the tracking efficiency. The v_n^{obs} in an event is obtained as $\sqrt{(v_{n,x}^{obs})^2 + (v_{n,y}^{obs})^2}$. Due to the finite number of tracks in the event, the $v_{n,x}^{obs}$ and $v_{n,y}^{obs}$ and hence the v_n^{obs} fluctuate about the true values. The v_n^{obs} distribution needs to be corrected for this smearing using the response function, which gives the probality distribution of v_n^{obs} for a given true v_n. To obtain the response function, each event is divided into two sub-events containing tracks with $\eta > 0$ and $\eta < 0$ respectively. Taking the difference of the flow-vector between the two sub-events, the physical signal cancels out and the resulting distribution $(v_{n,x}^{obs,1} - v_{n,x}^{obs,2}, v_{n,y}^{obs,1} - v_{n,y}^{obs,2})$ is consistent with a 2D Gaussian with identical widths $\delta_{2\mathrm{SE}}$ in the x and y directions. The response function can be obtained by shifting this 2D distribution to $(v_{n,x}, v_{n,y})$ and then projecting it along the radial direction [6]:

$$p(v_n^{obs}|v_n) \propto v_n^{obs} e^{-\frac{(v_n^{obs})^2 + v_n^2}{2\delta^2}} I_0(\frac{(v_n^{obs})^2 + v_n^2}{\delta^2}), \delta = \delta_{2\mathrm{SE}}/2 \qquad (3)$$

where, I_0 is the modified Bessel function of the second kind. This response function is used along with the Bayesian unfolding procedure [4] to calculate the unfolded v_n distributions. Note that this procedure implicitly assumes that the physical flow signal is rapidity independent, which is true when averaged over many events [1], but might not hold event by event.

Figure 2 shows the final probability distributions for $v_2 - v_4$ for several different centrality classes. The shape of the v_2 distribution changes considerably from central to peripheral events while for v_3 and v_4 the change is relatively small. This is expected as the higher order harmonics v_3 and v_4 are produced due to fluctuations in the collision geometry, however v_2 is also driven by the average second order eccentricity, which increases from central to peripheral events. The v_3 and v_4 distributions are well described by radial projections of 2D Gaussian distributions in

$\vec{v_n}$ with $P(|\vec{v_n}|) = (|\vec{v_n}|/\sigma)e^{-|\vec{v_n}|^2/\sigma^2}$, $\sigma = \sqrt{2/\pi}\langle v_n\rangle$ (solid lines) across all centralities. For the v_2 distribution this only works for the (0-1)% central events.

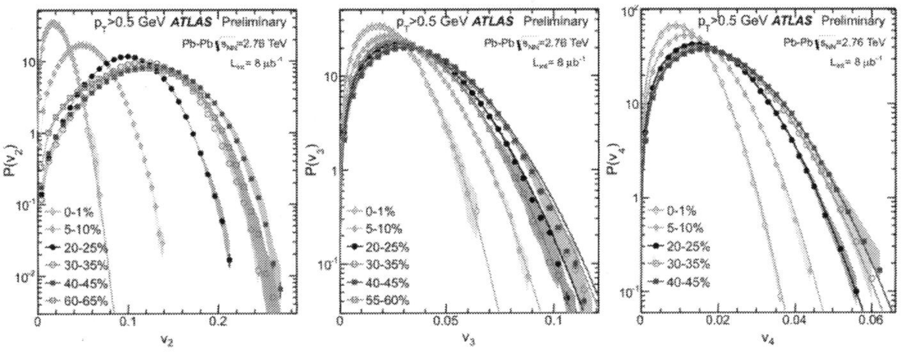

Figure 2. The probability distribution of the EbE v_n in several centrality intervals for n=2, 3 and n=4. The error bars are statistical uncertainties, and the shaded bands are uncertainties on the v_n shape. The solid curves are distributions calculated assuming the v_n are radial projections of 2D Gaussian distributions; they are shown for 0-1% centrality interval for v_2 and all centrality intervals for v_3 and v_4. Figure taken from [6].

Figure 3 compares the EbE distributions for the v_n for three different p_T ranges: $0.5< p_T <1.0$ GeV, $p_T > 1.0$ GeV and $p_T >0.5$ GeV for the (20-25)% centrality class. The v_n distributions for tracks with $p_T >1.0$ GeV are much broader than the ones for $p_T <1.0$ GeV reflecting the fact that the v_n increase with p_T. However, once the three distributions are scaled to the same mean value, their reduced shapes are almost identical as shown in the lower panels. This indicates that the hydrodynamic response to the initial geometry scales linearly with p_T.

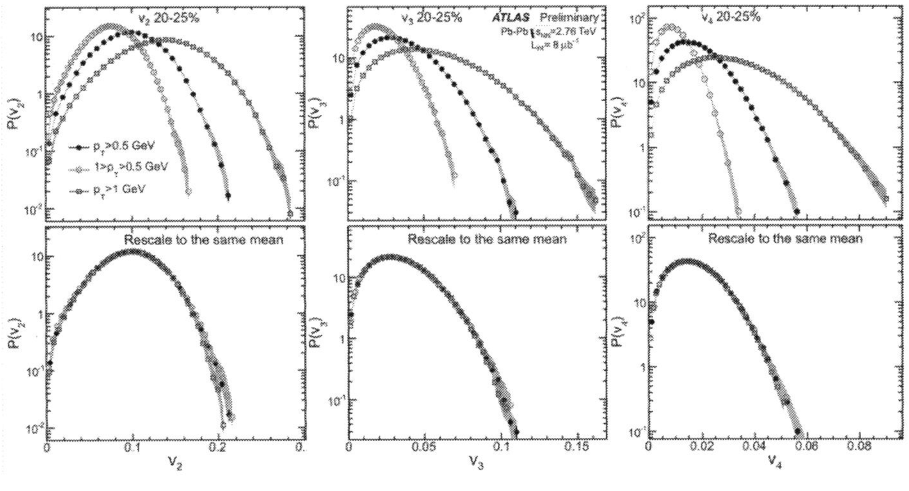

Figure 3. Top panels: The unfolded distributions for v_n in the (20-25)% centrality interval for charged particles in $p_T >0.5$ GeV, $0.5< p_T <1.0$ GeV and $p_T >1$ GeV. Bottom panels: same distributions but rescaled horizontally so the v_n values match that for $p_T >0.5$ GeV. The shaded bands represent the systematic uncertainties on the v_n shape. Figure taken from [6].

Figure 4 shows the mean values ($\langle v_n \rangle$) and RMS widths (σ_{v_n}) of the v_n distributions as well as the ratio of $\sigma_{v_n}/\langle v_n \rangle$ as a function of the number of participating nucleons $\langle N_{\rm part} \rangle$. While the $\langle v_n \rangle$ and σ_{v_n} change with $p_{\rm T}$ the ratio is independent for all three harmonics. The ratio $\sigma_{v_3}/\langle v_3 \rangle$ is independent of $\langle N_{\rm part} \rangle$ and consistent with the value of $\sqrt{4/\pi - 1} = 0.52$ expected from 2D-Gaussian distributions, v_4 shows a similar behavior as well (not shown). For v_2 the Gaussian limit is reached only in most central events and shows a considerable deviation for mid-central and peripheral events, where the average collision geometry has a large second-order eccentricity. For all cases, the v_n measured via the EP method are found to lie in between $\langle v_n \rangle$ and $\sqrt{\langle v_n^2 \rangle}$ within statistical and systematic errors [6].

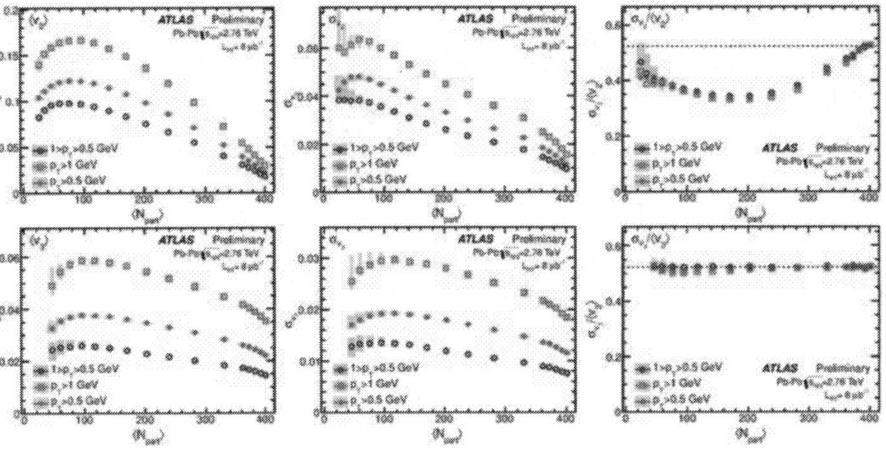

Figure 4. The $\langle N_{\rm part} \rangle$ dependence of $\langle v_n \rangle$ (left column), σ_{v_n} (middle column) and σ_{v_n}/v_n (right column) for n=2 (top row) and n=3 (bottom row). Each panel shows the results for three $p_{\rm T}$ ranges together with the total systematic uncertainties. The dotted lines in the right column indicate the value 0.52 expected for the radial projection of a 2-D Gaussian distribution. Figure taken from [6].

ATLAS has measured the event-by-event distribution of harmonic flow coefficients $v_2 - v_4$ in various centrality bins. The v_2 distribution is consistent with radial projection of 2D Gaussian distributions in most central events, but shows significant deviation for >5% centrality. For v_3 and v_4 the distributions are consistent with 2D Gaussian for all centralities. The reduced shape of the v_n distributions has no $p_{\rm T}$ dependence showing that the hydrodynamic response to the initial geometry is independent of $p_{\rm T}$ up to an overall normalization. These measurements are the first of their kind and provide constraints on the hydrodynamic response as well as initial geometry fluctuations of the produced medium [6].

References

[1] ATLAS Collaboration, Phys. Rev. C 86, 014907 (2012).
[2] ATLAS Collaboration, ATLAS-CONF-2012-118, Measurement of the elliptic flow with multi-particle cumulants in lead-lead collisions at $\sqrt{s_{NN}}$ = 2.76 TeV with the ATLAS detector at the LHC, http://cdsweb.cern.ch/record/1472940
[3] PHOBOS Collaboration, Phys. Rev. C 77, 014906 (2008).
[4] G. D'Agostini, Nucl. Instrum. Meth. A 362, 487 (1995); T. Adye, arXiv:1105.1160 [physics.data-an].
[5] ATLAS Collaboration, JINST 3, S08003 (2008).
[6] ATLAS Collaboration, ATLAS-CONF-2012-114, Measurement of event by event harmonic flow in 2.76 TeV Pb+Pb collisions with ATLAS, http://cdsweb.cern.ch/record/1472935

Transport approach to anisotropic flows from viscous hydro regime to high p_T

Salvatore Plumari[1,2], Armando Puglisi[2], Francesco Scardina[1,2] and Vincenzo Greco[1,2]

[1] Department of Physics and Astronomy, University of Catania, Via S. Sofia 64, I-95125 Catania (Italy)

[2] Laboratorio Nazionale del Sud, INFN-LNS, Via S. Sofia 63, I-95125 Catania (Italy)

E-mail: salvatore.plumari@hotmail.it

Abstract. We discuss the build up of the elliptic flow v_2 in a transport approach at fixed viscosity. We point out that in the intermediate transverse momentum region flows are also sensitive to microscopic details of the cross section, while hydrodynamics suffers from large uncertainties due to the non-equilibrium correction to the distribution function. We show, exploring the temperature dependence of η/s, that a study of v_2 in a wide p_T range allows to understand the difference behind the collective flows at LHC respect to RHIC. In particular the transport approach provides a tool able to naturally describe the rise and fall and saturation of the $v_2(p_T)$.

1. Introduction

In recent years the experimental results accumulated in the ultra relativistic heavy ion collisions before at RHIC program at BNL and more recently at the LHC program at CERN has shown that the elliptic flow v_2 is the largest ever seen in Heavy Ion Collision (HIC) [1, 2]. Viscous hydrodynamics at second order in gradient expansion according to the Israel-Stewart theory [3, 4, 5] is the most commonly used framework to describe the evolution of the fireball created in these collisions. Here all the information is encoded in a few macroscopic parameter like the Equation of State and shear viscosity. The comparison between the experimentally measured v_2 and theoretical calculations have suggested that an almost perfect fluid with a very small shear viscosity to entropy density ratio η/s has been created, with a viscosity of the order of $4\pi\eta/s \approx 1-3$ [3, 5]. Similar conclusions have been obtained from kinetic transport theory [6, 7, 8]. Hydrodynamics however has the fundamental problem of a limited range of validity in η/s and in the transverse momentum $p_T \leq 2$ GeV. At increasing p_T the validity of viscous hydrodynamics breaks because the relative deviation from the equilibrium distribution function $\delta f/f_{eq}$ increases with p_T becoming large already at $p_T \geq 3T \sim 1$ GeV. Due to the sensitivity of the elliptic flow on the ratio η/s it becomes important to know the correct relation between $\eta \leftrightarrow T, \sigma(\theta)$. In literature there are several approximation methods for the calculation of η one of these is the Chapmann-Enskog (CE) [9]. In Ref.[10] it has been shown that CE supplies an analytical expression for η with an accuracy better than 5% while the Relaxation Time Approximation (RTA) underestimates η by a factor 2 and therefore it provides a way to construct a kinetic theory at fixed $\eta/s(T)$. In this proceeding we will show some results for v_2

in the framework of a parton cascade at fixed η/s with a larger accuracy with respect to first tentatives in RTA [7, 4, 8].

2. The parton cascade at fixed η/s

Our approach is a $3+1$ dimensional Monte Carlo cascade [7] for on-shell partons based on the stochastic interpretation of the collision rate [11]. In the following discussion we use the pQCD inspired cross section typically used in parton cascade approaches [6, 7, 11, 12, 13]:

$$\frac{d\sigma}{dt} = \frac{9\pi\alpha_s^2}{2} \frac{1}{(t-m_D^2)^2}\left(1+\frac{m_D^2}{s}\right) \tag{1}$$

where s, t are the Mandelstam variables and m_D is the Debye thermal mass. In the CE approximation the shear viscosity assumes the following form $\eta = g(z, T/m_D)\frac{T}{\sigma_{tot}}$ (see [10]), where σ_{tot} is the total cross section while $g(z, T/m_D)$ is given by the following expression

$$g(z,\overline{a})^{-1} = \frac{8}{5}\frac{z}{K_3^2(z)}\int_1^\infty dy\,(y^2-1)^3\,h(2^{-1}z^{-2}y^{-2}\overline{a}^2)\left[(z^2y^2+1/3)K_3(2zy)-zyK_2(2zy)\right] \tag{2}$$

Where the function h is related to the transport cross section $\sigma_{tr} = \sigma_{tot}h(a)$ with $h(a) = 4a(1+a)[(2a+1)ln(1+1/a)-2]$ while $a = m_D^2/s$ and $\overline{a} = m_D/T$. In our approach we solve the relativistic Boltzmann equation with the constraint that η/s is fixed during the dynamics of the collisions in a way similar to [14] but with an exact local implementation as described in detail in [7]. We evaluate locally in space and time the strength of the cross section $\sigma_{tot}(\rho, T)$ needed to have η/s at the wanted value. Therefore considering that the entropy density for a massless gas is $s = \rho(4-\mu/T)$ with μ being the fugacity, the total cross section assume the following expression $\sigma_{tot} = \frac{g(m_D,T)}{\rho(4-\mu/T)}\frac{T}{\eta/s}$. Within this approach the total cross section has the form $\sigma_{tot} = K(\rho,T)\sigma_{pQCD} > \sigma_{pQCD}$ where $K(\rho,T)$ takes into account the non perturbative effects responsible to keep fixed the ratio η/s. Note that this approach has been shown to recover the viscous hydrodynamics evolution of the bulk system [4, 7], but implicitly assume that also high p_T particles collide with largely nonperturbative cross section [15, 16]. Within this approach the kinetic freeze out is given by the increase of η/s at lower temperature, see red solid line in Fig.2. This naturally realizes a freeze out (f.o.) because it implies a smooth reduction of σ_{tot}.

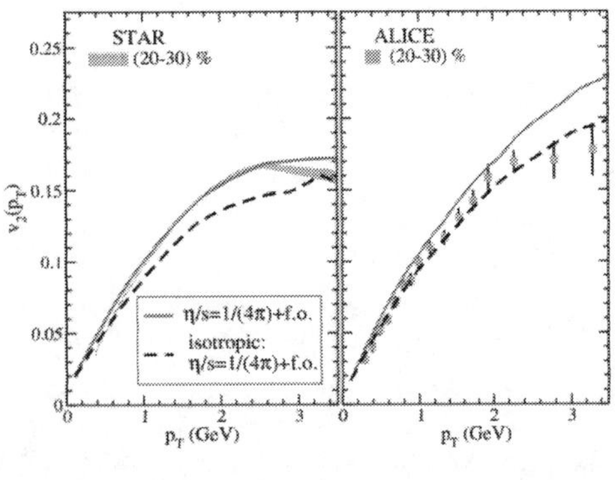

Figure 1. $v_2(p_T)$ at mid rapidity for $20\% - 30\%$ collision centrality for RHIC (left panel) and LHC (right panel), data taken by [2]. The red solid line is the calculation with $4\pi\eta/s = 1$ and the inclusion of freeze out (f.o.) and with an anisotropic cross section with finite m_D while the black dashed line is the case with an isotropic cross section ($m_D \to \infty$).

3. Elliptic flow within a parton cascade at fixed η/s

In our calculation the initial conditions are longitudinal boost invariant. We have assumed parton-hadron duality and the initial $dN/d\eta$ has been chosen in order to reproduce the final

$dN_{ch}/d\eta(b)$ at mid rapidity observed in the experiments at RHIC and LHC energies. The initial condition in coordinate space are given by the Glauber model. In the momentum space at RHIC (LHC) the partons with $p_T \leq p_0 = 2$ GeV (4 GeV) are distributed according to a thermalized spectrum with a maximum temperature in the center of the fireball of $T_0 = 2T_C$ ($T_0 = 3T_C$) while for $p_T > p_0$ we take the spectrum of non-quenched minijets according to standard NLO-pQCD calculations. The initial time is $t_0 = 0.6$ fm/c at RHIC and $t_0 = 0.3$ fm/c at LHC. As shown in Fig.1 in the low p_T region ($p_T < 1.5$ GeV) the v_2 depends on the value of η/s while for higher value of p_T the v_2 becomes an observable sensitive to the microscopical details of the interaction (namely on the value of m_D). For the case shown in the figure with the increase of the thermal debye mass (toward an isotropic cross section), we observe a suppression of the v_2. In order to study the role of η/s in the build up of v_2 we have considered in our simulations three different temperature dependences for η/s, with a constant $4\pi\eta/s = 1$ during all the evolution of the system (black dashed line), with a constant $4\pi\eta/s = 1$ in the QGP phase but with freeze out (f.o.) (red solid line) and finally the case with f.o. and $4\pi\eta/s \propto T^2$ in the QGP phase (blue dot-dashed line), see Fig.2. As shown in Fig. 3, the $\eta/s(T)$ affects the generation of the v_2 in different way between RHIC and LHC. This different behaviour of v_2 can be explained looking at the different life time of the fireball between RHIC and LHC. The life time of the fireball is smaller at RHIC than that at LHC energies, $4-5$ fm/c at RHIC against the about $8-10$ fm/c at LHC. Therefore the elliptic flow at RHIC has not enough time to develop completely and it becomes more sensitive to the value of η/s in the cross over region (for $T \approx T_C$) while at LHC due to a larger life time the v_2 can develop completely and it can discriminate better the value of η/s in the QGP phase. The comparison with data would suggest that at LHC they are compatible with a linear T-increase of η/s. Similar calculations performed in the framework of viscous hydrodynamics (see Ref.[17]) have shown a good description of the data for almost central collissions with a slightly stronger temperature dependence of η/s than that used in our work. As pointed out in [15, 16] the procedure discussed in the previous section doesn't permit

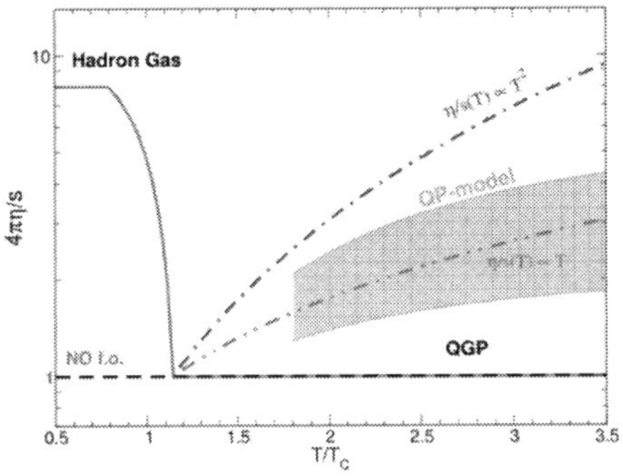

Figure 2. Different temperature dependent parametrizations for η/s. The orange area take into account the quasi-particle model predictions for η/s [18]. The solid, dashed and dot dashed lines are the three different temperature profiles of $\eta/s(T)$ used in our simulations, see Fig.3.

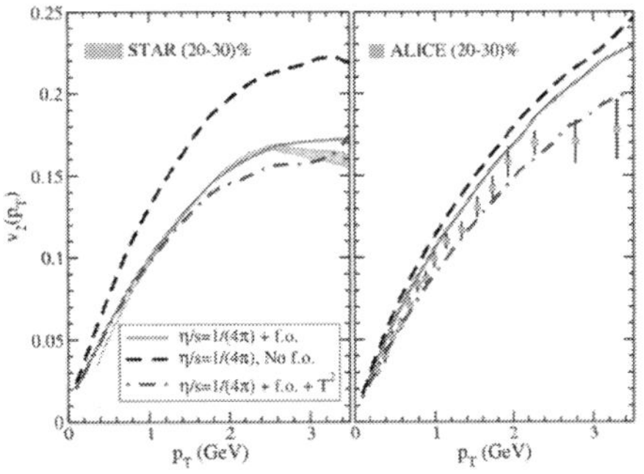

Figure 3. $v_2(p_T)$ for $20\% - 30\%$ collision centrality for RHIC (left panel) and LHC (right panel), data taken by [2]. The black dashed lines is the case with $4\pi\eta/s = 1$ and without f.o., the red line is the case with $4\pi\eta/s = 1$ with f.o. while the blue dot-dashed line refers to the case with f.o. and $4\pi\eta/s \propto T^2$ in the QGP phase.

to recover the pQCD limit for hard collision therefore a natural way to extend it is to consider a K factor that depends on the invariant energy of the collision $K(s)$. This is a way to switch off non-perturbative effects at higher s. We have chosen for $K(s)$ the following exponential

Journal of Physics: Conference Series **446** (2013) 012025 doi:10.1088/1742-6596/446/1/012025

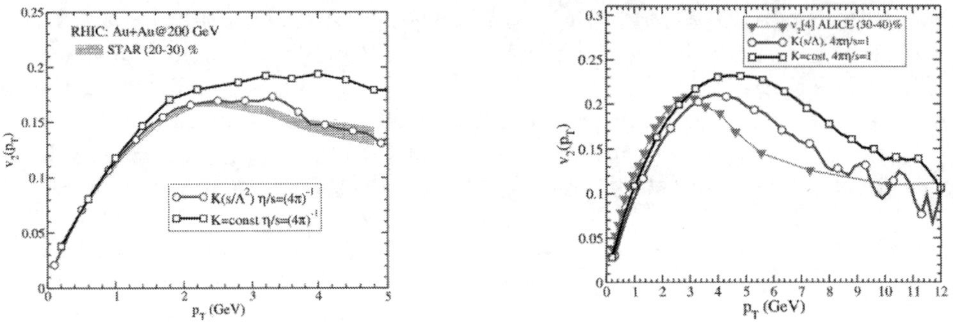

Figure 4. $v_2(p_T)$ for 20% − 30% collision centrality for RHIC (left) and LHC (right) with $4\pi\eta/s = 1$ and f.o., data taken by [2]. The black lines with squares are the cases with $K = const$ while the blue lines with circles are the same calculations but with $K(s/\Lambda)$ with $\Lambda = 3$ GeV.

form $K(s/\Lambda^2) = 1 + \gamma\, e^{-s/\Lambda^2}$, where Λ is a scale parameter that fix the energy scale for the transition to pQCD behaviour (where $K(s) \to 1$). While γ plays the same role of K in the old calculations and it is determined again in order to keep fixed locally the η/s. In Fig.4 we have shown the results for the v_2 obtained using the RTA for the viscosity [15]. The implementation with the CE parametrization is in progress. A comparison between RHIC and LHC shows that the effect of the $K(s/\Lambda^2)$ function is to reduce the v_2 at high p_T for both energies. In general with the inclusion of $K(s/\Lambda^2)$ we can reproduce the shape of the elliptic flow as shown in the experiments in a very wide range of p_T. We can interpret the raise of v_2 as an effect due to the non perturbative behaviour at low p_T while the fall and consequently saturation as given by the disappearence of the non perturbative effect and the restoration of the pQCD behavior.

4. Conclusions

We have investigated within a parton cascade approach the effect of a temperature dependent η/s on the generation of v_2 at RHIC and LHC. Our calculations show that the v_2 at RHIC is sensitive to the f.o. (hadronic phase) while at LHC all the v_2 comes from the QGP phase and the hadronic phase is irrelevant on it. Furthermore we observe that the v_2 in the intermediate p_T region becomes an observable sensitive to the microscopical details of the system.

References

[1] Adams J *et al.* (STAR Collaboration) 2005 *Nucl.Phys.* **A757** 102–183 (*Preprint* nucl-ex/0501009)
[2] Aamodt K *et al.* (ALICE Collaboration) 2010 *Phys.Rev.Lett.* **105** 252302 (*Preprint* 1011.3914)
[3] Romatschke P and Romatschke U 2007 *Phys.Rev.Lett.* **99** 172301 (*Preprint* 0706.1522)
[4] Huovinen P and Molnar D 2009 *Phys.Rev.* **C79** 014906 (*Preprint* 0808.0953)
[5] Song H and Heinz U W 2008 *Phys.Rev.* **C78** 024902 (*Preprint* 0805.1756)
[6] Xu Z and Greiner C 2009 *Phys.Rev.* **C79** 014904 (*Preprint* 0811.2940)
[7] Ferini G, Colonna M, Di Toro M and Greco V 2009 *Phys.Lett.* **B670** 325–329 (*Preprint* 0805.4814)
[8] Plumari S, Baran V, Di Toro M, Ferini G and Greco V 2010 *Phys.Lett.* **B689** 18–22 (*Preprint* 1001.2736)
[9] Wiranata A and Prakash M 2012 *Phys.Rev.* **C85** 054908 (*Preprint* 1203.0281)
[10] Plumari S, Puglisi A, Scardina F and Greco V 2012 *Phys.Rev.* **C86** 054902 (*Preprint* 1208.0481)
[11] Xu Z and Greiner C 2005 *Phys.Rev.* **C71** 064901 (*Preprint* hep-ph/0406278)
[12] Plumari S, Baran V, Di Toro M and Greco V 2011 *J.Phys.Conf.Ser.* **270** 012061 (*Preprint* 1009.2664)
[13] Molnar D and Gyulassy M 2002 *Nucl.Phys.* **A697** 495–520 (*Preprint* nucl-th/0104073)
[14] Molnar D 2008 (*Preprint* 0806.0026)
[15] Plumari S and Greco V 2012 *AIP Conf.Proc.* **1422** 56–61 (*Preprint* 1110.2383)
[16] Plumari S, Puglisi A, Colonna M, Scardina F and Greco V 2012 (*Preprint* 1209.0601)
[17] Niemi H, Denicol G, Huovinen P, Molnar E and Rischke D 2012 *Phys.Rev.* **C86** 014909 (*Preprint* 1203.2452)
[18] Plumari S, Alberico W M, Greco V and Ratti C 2011 *Phys.Rev.* **D84** 094004 (*Preprint* 1103.5611)

Event plane correlation from single-shot hydrodynamics

Derek Teaney and Li Yan*
Department of Physics and Astronomy, Stony Brook University, Stony Brook, New York, 11794, USA
E-mail: *li.yan@stonybrook.edu

Abstract. We apply a nonlinear flow response formalism to the recently measured event plane correlations. We find that the observed event plane correlations can be understood as a result of the averaged effect of linear and quadratic flow response.

1. Introduction

One of the recent progresses made in the studies of high energy heavy-ion collisions was the realization of initial state fluctuations[1, 2]. Due to these fluctuations, bumpy instead of smooth initial geometry dominates the early stage of bulk evolution, leading to new signatures which were later confirmed in experiment. For instance, triangular flow v_3[3] were measured on an event-by-event basis and found to be responsible for the 'ridge' and 'shoulder' structure in the two-particle correlations in heavy-ion collisions[1, 3]. Apart from the magnitude of harmonic flow v_n, the associated angle is also of interest and was recently investigated via the event plane (EP) correlations by ATLAS collaboration[4]. Some two-plane and three-plane correlation patterns were measured and consistent predictions were made via event-by-event hydrodynamics[5]. In this paper, with linear and nonlinear flow response included in a single-shot hydrodynamic calculation[6], we will analyze the origins of these observed EP correlation. The hydrodynamics in our work is solved for $\sqrt{s_{NN}} = 2.76$GeV Pb-Pb collisions at LHC, with parameters set in the same way as in [6]. The initial state with fluctuations is simulated by PHOBOS MC-Glauber model.[7]

2. Methodology

Although event-by-event hydrodynamics has obtained thus far the best predictions of v_n[8], with respect to fluctuating initial state, single-shot hydrodynamics has the advantage of interpreting flow generation. In particular, nonlinear mode mixing in hydrodynamics gives rise to a significant fraction of contribution to v_n, which compensated for the failure of single-shot hydrodynamics (without nonlinear flow generation) for higher order harmonics, $n \geq 4$[6]. Characterizing initial fluctuations with cumulants instead of moments, and including nonlinear mode mixing in hydrodynamics, a correspondence between initial state and final spectrum can be established, such as initial participant plane (PP) angle Φ_n and EP angle Ψ_n, which serves as the basic idea of this paper. In the cumulant definition of ϵ_n and Φ_n[9], nonlinearities are systematically

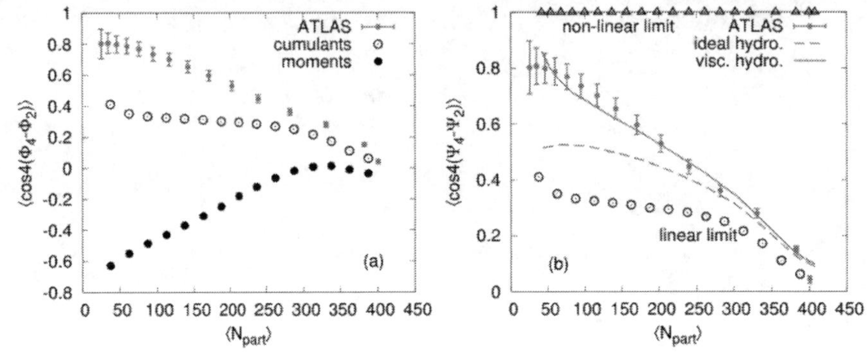

Figure 1. (a) Initial participant plane Φ_2 and Φ_4 correlation, as a function of centrality. The data points are extracted from ATLAS [4] for comparison. (b) Event plane Ψ_2 and Ψ_4 correlation, with predictions made by ideal and viscous hydrodynamics.

subtracted, e.g., for $n = 4$,

$$\epsilon_4 e^{i4\Phi_4} = -\left[\langle r^4 e^{i4\phi_r}\rangle - 3\langle r^2 e^{i2\phi_r}\rangle^2\right] / \langle r^4\rangle, \tag{1}$$

has an extra subtraction term which is proportional to ε_2^2. Subtracting nonlinearities in initial definitions leads to negligible effects for ϵ_n, but pronounced difference for PP angle correlations. As seen in Fig.1(a), $\langle\langle\cos 4(\Phi_4 - \Phi_2)\rangle\rangle$ is positive with cumulant definition, but negative with moment definition. Nevertheless, there still exists a gap between initial state plane correlations and experiment observations, which has to be filled with medium evolution.

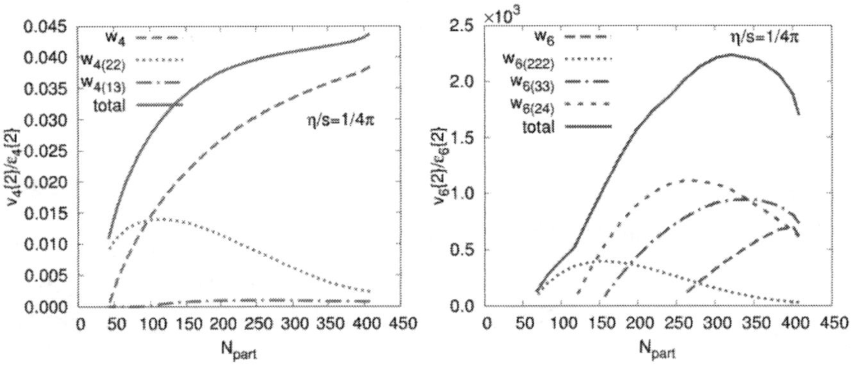

Figure 2. Linear and nonlinear flow response contribution to $v_4\{2\}$ and $v_6\{2\}$.

A systematic way to include linear and nonlinear mode mixing for harmonic flow generation, in a single-shot hydrodynamic calculation, is formulated as follows. In a particle spectrum,

$$\frac{dN}{d\phi_p} = \frac{N}{2\pi}\left[1 + \sum_n v_n e^{in(\phi_p - \Psi_n)} + c.c.\right] = \frac{N}{2\pi}\left[1 + \sum_n z_n e^{in\phi_p} + c.c.\right], \tag{2}$$

for each harmonic order $z_n = v_n e^{in\Psi_n}$, we consider contributions from both linear and nonlinear flow response, then we expect, for example,

$$z_4 = v_4 e^{-i4\Psi_4} = w_4 e^{-i4\Phi_4} + w_{4(22)} e^{-i4\Phi_2}\,. \tag{3}$$

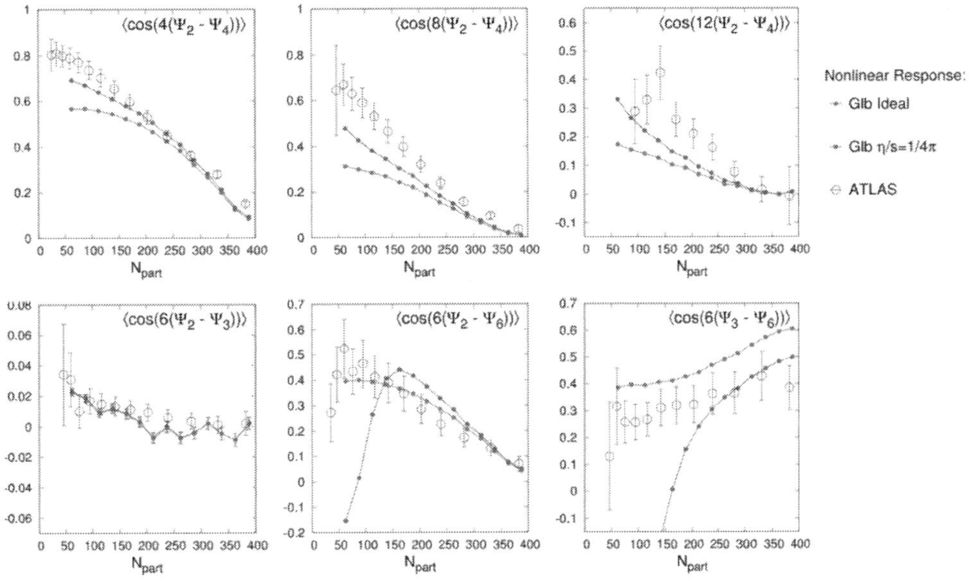

Figure 3. Two-plane correlations of event plane correlations measured at ATLAS[4].

w_4 indicates the linear response to initial fourth order eccentricity ϵ_4, while $w_{4(22)}$ is the quadratic order flow response to ϵ_2^2. Note there should be infinite nonlinear contributions to flow generation, but we only need to consider the dominant ones. For v_4, as shown clearly in Fig.2, the next possible nonlinear contribution $w_{4(13)}$ is negligible. However, for even higher order harmonic flow like v_6, it becomes more involved since there are many more possible nonlinear contributions that are comparably important.

3. Predictions

As an application of (2), given all necessary linear and nonlinear flow generation terms, EP correlations can be formulated. For example,

$$\cos 4(\Psi_4 - \Psi_2) = \frac{\mathrm{Re}[z_2^2 z_4^*]}{|z_2|^2|z_4|} = \frac{\cos 4(\Phi_4 - \Phi_2)w_4 + w_{4(22)}}{|w_4 e^{-i4\Phi_4} + w_{4(22)}e^{-i4\Phi_2}|}. \tag{4}$$

(4) has a typical structure, that in both the numerator and denominator, there is one term determined by pure linear flow response w_4 while the other one contains contribution from nonlinear flow response $w_{4(22)}$. Then in the limit of ignoring the nonlinear flow response, the EP correlation will reduce to its corresponding PP correlation $\langle\langle\cos 4(\Phi_4 - \Phi_2)\rangle\rangle$. On the other hand if we ignore the linear flow response, (4) leads to a complete correlation, *i.e.*, $\langle\langle\cos 4(\Psi_4 - \Psi_2)\rangle\rangle = 1$, as shown in Fig.1(b). Referring these two limits as the linear and nonlinear limits respectively, the observed EP correlations can be understood as an average of these two, with the actual value controlled by the relative magnitudes of the two response coefficients, w_4 and $w_{4(22)}$. Viscous hydrodynamics have indicated that nonlinear flow generation dominates for the case with larger η/s and larger centrality[6], we see indeed $\langle\langle\cos 4(\Psi_4 - \Psi_2)\rangle\rangle$ approaches the nonlinear limit at peripheral bins from viscous hydrodynamics.

Generally more nonlinear terms need to be considered, thus more complex terms appear in the expression of EP correlation, we still have the very similar analyses. The results are shown in Fig.3 and Fig.4, in comparison with ATLAS two-plane and three-plane EP correlations. Except

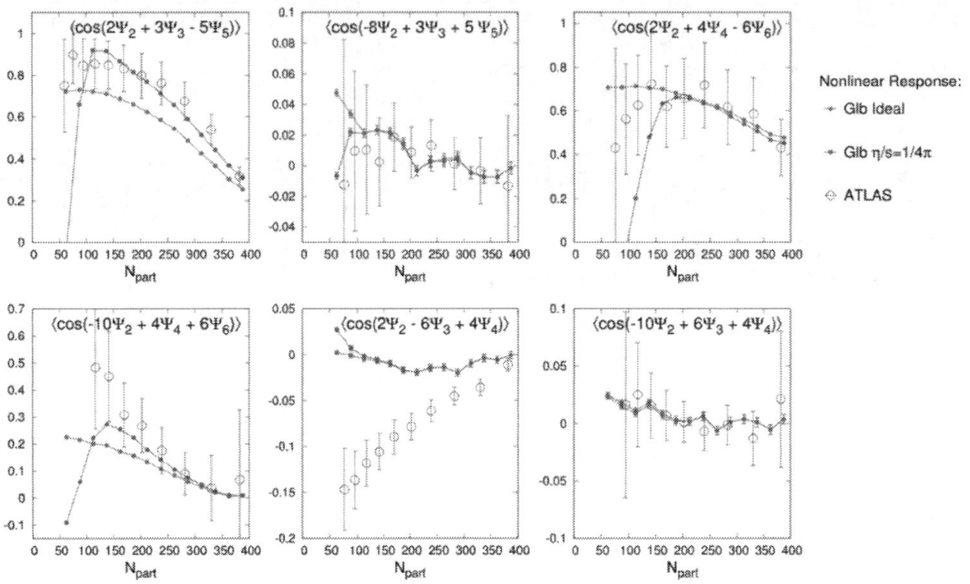

Figure 4. Three-plane correlations of event plane correlations measured at ATLAS[4]. The nonlinear flow generations we considered in these calculations are: $w_{3(12)}$ for v_3, $w_{4(22)}$ for v_4, $w_{5(23)}$ for v_5, and $w_{6(33)}$, $w_{6(24}$, $w_{6(222)}$ for v_6. Linear flow generation w_6 is ignored because of its comparably small contributions.

$\langle\langle\cos(2\Psi_2-6\Psi_3+4\Psi_4)\rangle\rangle$, the measured correlations follow the expected trend from hydrodynamic simulation as a result of averaged effect of linear and nonlinear limits.

4. Summary and discussions

We analyzed EP correlations in the framework of single-shot hydrodynamics, using linear and nonlinear flow response formalism. We found that the observed EP correlations can be effectively interpreted as the average of linear and nonlinear limit. Our results are comparable with the experiment except $\langle\langle\cos(2\Psi_2-6\Psi_3+4\Psi_4)\rangle\rangle$, however the prediction can be improved when more nonlinear contributions are taken into considerations.

Acknowledgements

This work is supported in part by the Sloan Foundation and by the Department of Energy through the Outstanding Junior Investigator program DE-FG-02-08ER4154.

References

[1] Alver B and Roland G 2010 *Phys.Rev.* **C81** 054905 (*Preprint* 1003.0194)
[2] Jia J (ATLAS Collaboration) 2012 (*Preprint* 1209.4232)
[3] Alver B H, Gombeaud C, Luzum M and Ollitrault J Y 2010 *Phys.Rev.* **C82** 034913 (*Preprint* 1007.5469)
[4] Jia J (ATLAS Collaboration) 2012 (*Preprint* 1208.1427)
[5] Qiu Z and Heinz U W 2012 *Phys.Lett.* **B717** 261–265 (*Preprint* 1208.1200)
[6] Teaney D and Yan L 2012 *Phys.Rev.* **C86** 044908 (*Preprint* 1206.1905)
[7] Alver B, Baker M, Loizides C and Steinberg P 2008 (*Preprint* 0805.4411)
[8] Gale C, Jeon S, Schenke B, Tribedy P and Venugopalan R 2013 *Phys.Rev.Lett.* **110** 012302 (*Preprint* 1209.6330)
[9] Teaney D and Yan L 2011 *Phys.Rev.* **C83** 064904 (*Preprint* 1010.1876)

Journal of Physics: Conference Series **446** (2013) 012027 doi:10.1088/1742-6596/446/1/012027

Charged-particle pseudorapidity density and anisotropic flow over a wide pseudorapidity range using ALICE at the LHC

K. Gulbrandsen[1] **for the ALICE Collaboration**

[1] Niels Bohr Institute, University of Copenhagen, Blegdamsvej 17, 2100 Copenhagen Ø, Denmark

E-mail: gulbrand@nbi.dk

Abstract. The pseudorapidity density and anisotropic flow of charged-particles provide fundamental information about global variables and correlations in heavy-ion collisions. The pseudorapidity density is related to the energy available for particle production while the anisotropic flow is related to collective effects from interactions between these particles. Extending these measurements to very forward pseudorapidities yields information about the longitudinal expansion of the system. The first measurements performed at the LHC over more than 8 units of pseudorapidity are presented. The longitudinal scaling of the measurements is analyzed and comparison to models is performed.

Introduction

The goal of colliding heavy-ions is to understand the strong interaction in the hot dense state known as the Quark–Gluon Plasma (QGP). One of the most fundamental characteristics of these collisions is the energy available for particle production, which can be assessed by measuring the charged-particle pseudorapidity density, $dN_{ch}/d\eta$. Further insights can be gained by investigating quantities which probe the collective behavior of the produced particles. Anisotropic flow [1] is one signature of this collectivity arising from multiple interactions between particles which transfers the initial spatial anisotropy of the collision into a momentum anisotropy of the produced particles. It is quantified as the measurement of the flow harmonics, v_n, of the particle distribution [2]. Extending the measurements to forward pseudorapidity allows for an assessment of the physics in the fragmentation region and clarifies the role of fluctuations in the initial energy in the anisotropic flow measurement.

Detector and data analysis

The measurements are carried out using the ALICE detector [3] at CERN. Three main sub-detectors, each with full (2π) azimuthal coverage, are utilized in these measurements. At mid-rapidity, the Silicon Pixel Detector (SPD) [3] is used. This detector has 2 cylindrical layers of silicon at radii of 3.9 and 7.6 cm from the beampipe with pseudorapidity coverages of $|\eta| < 2$ and $|\eta| < 1.4$, respectively. With almost 10 million total channels, the SPD provides excellent spatial resolution in both η and φ for registering particles produced in the collision. The detector's close proximity to the beampipe reduces the number of detected secondary particles.

At forward rapidities the Forward Multiplicity Detector (FMD) [4] and the VZERO [5] are used. The FMD is divided into three sections at different positions along the beam axis with each section made of one or two rings of silicon strip sensors. Each ring has 10240 silicon strips with high η resolution and 20 or 40 azimuthal sectors for resolving φ. The nominal acceptance of the FMD is $-3.4 < \eta < -1.7$ and $1.7 < \eta < 5.0$. The VZERO has a nominal acceptance of $-3.7 < \eta < -1.7$ and $2.8 < \eta < 5.1$. This detector is composed of scintillator counters whose signal is proportional to the number of particles incident upon the detector. The detector has 4 radial segments for η resolution and 8 azimuthal segments for φ resolution.

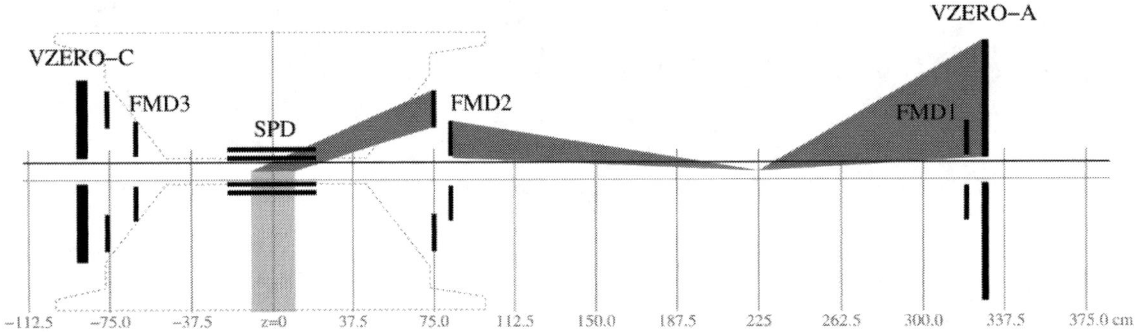

Figure 1. Drawing of the cross-section of the sub-detectors used for this analysis and locations of the nominal ($z = 0$) and 'satellite' interaction points. The dashed line indicates where dense support material exists.

Figure 1 shows a schematic cross-section of these detectors and the trajectories which particles would follow from different interaction vertices. From the nominal vertex at $z = 0$, few secondary particles are detected in the SPD, but for FMD2, FMD3, and VZERO-C many secondary particles produced in support material are detected. This effect requires a very accurate material description and a correction dependent on simulations. An alternate method was, therefore, developed to measure $dN_{ch}/d\eta$ where only FMD2, FMD1, and VZERO-A were used to extend the mid-rapidity measurement of $dN_{ch}/d\eta$ from the SPD. 'Satellite' vertices spaced in intervals of 37.5 cm from $z = 0$ created via beam debunching during injection and acceleration [6] were used to avoid large corrections due to secondary particle production and also to extend the η coverage for the $dN_{ch}/d\eta$ measurement to $-5.0 < \eta < 5.5$. The collision rate with displaced vertices is about 1000 times smaller than at the nominal vertex position which does not allow for a precision measurement of v_n with this method.

Pseudorapidity density

Figure 2(a) shows the result of the $dN_{ch}/d\eta$ analysis using the SPD with nominal vertex collisions for the region of $|\eta| < 2$ and the FMD and VZERO detectors with 'satellite' vertex collisions. The results are compared to three models. The AMPT model [7] tuned to the measured mid-rapidity $dN_{ch}/d\eta$ [8] fails to reproduce the shape of the $dN_{ch}/d\eta$ distribution. UrQMD [9] predicts the $dN_{ch}/d\eta$ shape in the region of $|\eta| < 1$ in the two most central bins, but does not properly describe the distribution at forward η. The CGC based model [10, 11] reproduces the $dN_{ch}/d\eta$ shape within the limited pseudorapidity range of the prediction, $|\eta| < 2$.

Figure 2(b) shows the distribution of $dN_{ch}/d\eta$ per participant pair as a function of $\eta - y_{beam}$. The most central ALICE data are compared to lower energy data from RHIC [12, 13]. Near beam rapidity the lower energy distributions overlap, suggesting the validity of longitudinal scaling in the fragmentation region [14, 15]. The ALICE acceptance does not extend far enough to show

Hot Quarks 2012 IOP Publishing
Journal of Physics: Conference Series **446** (2013) 012027 doi:10.1088/1742-6596/446/1/012027

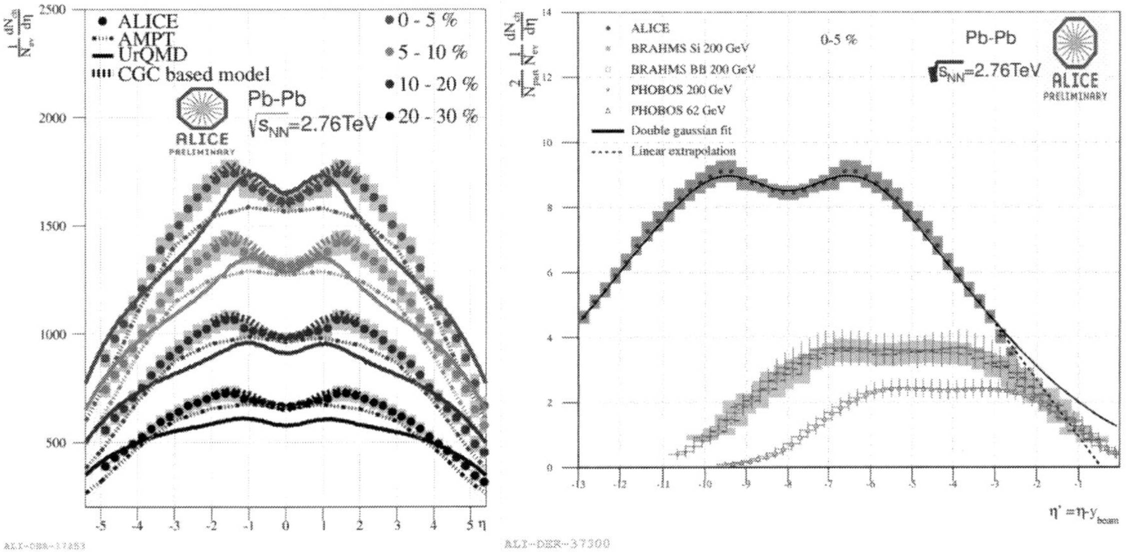

(a) (b)

Figure 2. (a) $dN_{ch}/d\eta$ versus η measured by ALICE in Pb–Pb collisions at $\sqrt{s_{NN}} = 2.76$ TeV in 4 different centrality bins. The data are symmetrized in the region of $|\eta| < 5$ to reduce systematic errors. Model comparisons are shown [7, 8, 9, 10, 11]. (b) $dN_{ch}/d\eta$ per participant pair versus $\eta - y_{beam}$. The most central ALICE data are compared to lower energy measurements [12, 13]. The full drawn line is an extrapolation of the distribution from a fit to the difference of two Gaussian distributions centered at $\eta = 0$, but with different widths. The dashed line is a straight line fit to the forward pseudorapidity points of the measured distribution.

an overlap with the lower energy data, but extrapolations are consistent with the preservation of longitudinal scaling in the fragmentation region.

Anisotropic flow

The data were analyzed using the cumulant method [16] to extract the values of the flow harmonics, v_n. Different multiparticle correlators were used to understand the effect of flow fluctuations. The analysis was carried out using only collisions occurring around the nominal vertex region ($|z| < 10$ cm). The inner layer of the SPD was used for the measurement in the range $|\eta| < 2$ and the FMD (all three sections) extended this measurement to the range $-3.4 < \eta < 5.0$. Figure 3(a) shows the result of the analysis. The large value of v_2 is due to the harmonic's sensitivity to the large anisotropy in the initial geometrical shape present in the 30–40% most central collisions. The smaller value of v_3 is attributed to the harmonic's dominant sensitivity to fluctuations in the initial geometrical shape. The essentially constant difference between $v_2\{2\}$ and $v_2\{4\}$ implies that the effect of flow fluctuations is similar at mid-rapidity and forward rapidity. The data are compared to the tuned AMPT model [8]. With this tuning, the model reproduces the shape of measured flow harmonics as a function of η.

Figure 3(b) shows v_2 versus $\eta - y_{beam}$ compared to lower energy data [17]. Close to beam rapidity, $|\eta - y_{beam}| \sim 3$, v_2 measured at RHIC and LHC seems to follow a universal dependence, hinting that longitudinal scaling in the fragmentation region seen at RHIC energies [17] also holds at the LHC.

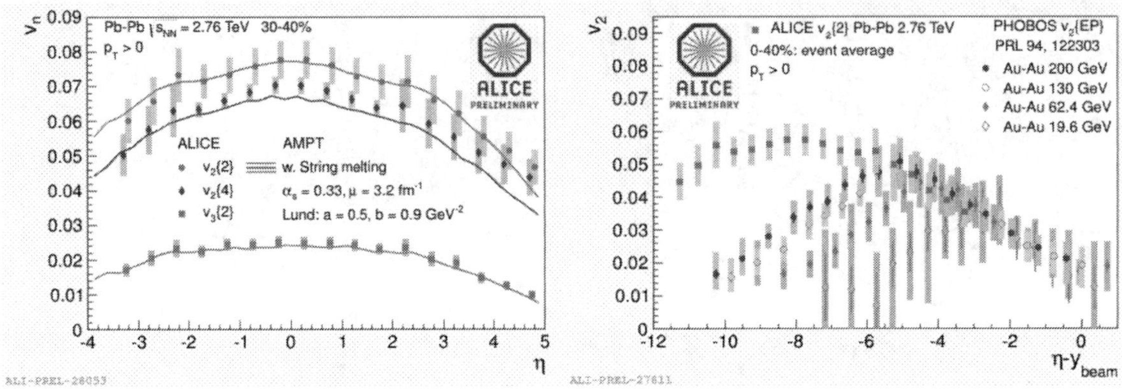

(a) (b)

Figure 3. (a) v_2 and v_3 versus η for the 30–40% centrality class measured with 2 and 4 particle cumulants [16]. The data are compared to the AMPT model [7] with the settings shown [8]. (b) The measurement of v_2 versus $\eta - y_{\rm beam}$. The data are compared to lower energy results [17].

Summary

The charged-particle pseudorapidity density, elliptic flow (v_2), and triangular flow (v_3) have been measured over more than 8 units of pseudorapidity. The available models do not predict the full $dN_{ch}/d\eta$ shape. A large value of v_2 is seen for mid-central collisions (30-40%), consistent with the anisotropy of the initial geometrical shape of the collision, while v_3 is smaller, consistent with its sensitivity to fluctuations in the shape. The contribution of flow fluctuations to the flow signal appears to be similar over a wide range of pseudorapidity. The measured $dN_{ch}/d\eta$ and elliptic flow appear to exhibit longitudinal scaling in the fragmentation region up to $\sqrt{s_{NN}} = 2.76$ TeV.

References

[1] Ollitrault J Y 1992 *Phys. Rev. D* **46**(1) 229–245
[2] Voloshin S and Zhang Y 1996 *Z.Phys.* **C70** 665–672 (*Preprint* `hep-ph/9407282`)
[3] Aamodt K *et al.* (ALICE Collaboration) 2008 *Journal of Instrumentation* **3** S08002
[4] Christensen C H *et al.* (ALICE Collaboration) 2007 *Int.J.Mod.Phys.* **E16** 2432–2437 (*Preprint* `0712.1117`)
[5] Cortese P *et al.* (ALICE Collaboration) 2004 CERN-LHCC-2004-025
[6] Jeff A *et al.* 2012 CERN-ATS-2012-088
[7] Lin Z W, Ko C M, Li B A, Zhang B and Pal S 2005 *Phys. Rev. C* **72**(6) 064901
[8] Xu J and Ko C M 2011 *Phys. Rev. C* **83**(3) 034904
[9] Mitrovski M, Schuster T, Graf G, Petersen H and Bleicher M 2009 *Phys.Rev.* **C79** 044901 (*Preprint* `0812.2041`)
[10] Albacete J L and Dumitru A 2010 (*Preprint* `1011.5161`)
[11] Albacete J L, Dumitru A and Nara Y 2011 *J.Phys.Conf.Ser.* **316** 012011 (*Preprint* `1106.0978`)
[12] Alver B *et al.* (PHOBOS Collaboration) 2011 *Phys. Rev. C* **83**(2) 024913
[13] Bearden I G *et al.* (BRAHMS Collaboration) 2002 *Phys. Rev. Lett.* **88**(20) 202301
[14] Benecke J, Chou T T, Yang C N and Yen E 1969 *Phys. Rev.* **188**(5) 2159–2169
[15] Back B B *et al.* (PHOBOS Collaboration) 2003 *Phys. Rev. Lett.* **91**(5) 052303
[16] Bilandzic A, Snellings R and Voloshin S 2011 *Phys.Rev.* **C83** 044913 (*Preprint* `1010.0233`)
[17] Back B *et al.* (PHOBOS Collaboration) 2005 *Phys.Rev.Lett.* **94** 122303 (*Preprint* `nucl-ex/0406021`)

Journal of Physics: Conference Series **446** (2013) 012028 doi:10.1088/1742-6596/446/1/012028

Measurement of Direct-Photon Elliptic Flow in Pb-Pb Collisions at $\sqrt{s_{\mathrm{NN}}} = 2.76\,\mathrm{TeV}$

Daniel Lohner (for the ALICE Collaboration)

Physikalisches Institut, Ruprecht-Karls-Universität Heidelberg, Heidelberg, Germany

E-mail: Daniel.Lohner@cern.ch

Abstract. We present the first measurement of the direct-photon elliptic flow $v_2^{\gamma,\mathrm{dir}}$ in Pb-Pb collisions at $\sqrt{s_{\mathrm{NN}}} = 2.76\,\mathrm{TeV}$ with data taken by the ALICE experiment at the LHC. The measurement provides evidence for a non-zero $v_2^{\gamma,\mathrm{dir}}$ for $1 < p_{\mathrm{T}} < 3\,\mathrm{GeV}/c$ with a magnitude similar to the observed charged pion $v_2^{\pi^\pm}$. In order to explain the large inverse slope parameter T_{eff} of the low p_{T} direct-photon spectrum observed at LHC and RHIC, recent hydrodynamical descriptions of the direct-photon production include a substantial portion of thermal photons from the hot plasma phase. As a consequence of the early production time, $v_2^{\gamma,\mathrm{dir}}$ is expected to be small compared to hadrons. A large $v_2^{\gamma,\mathrm{dir}}$ might lend support for a significant direct-photon emission from late stages of the system evolution where hadron flow has developed.

1. Introduction

A unique tool for the study of the collision evolution in nucleus-nucleus collisions is the measurement of photons. Besides photons from hadron decays also direct photons are emitted at every stage of the system evolution. Since photons interact only weakly with the strongly coupled medium they carry undistorted information of the system at their production time [1]. Theoretical descriptions of direct-photon production in nucleus-nucleus collisions include prompt photons, fragmentation photons, thermal photons and photons from parton-medium interactions [2, 3]. Prompt photons originate from primary, hard interactions of partons including quark-antiquark-annihilation ($q+\bar{q} \rightarrow g+\gamma$) and quark-gluon compton scattering ($q+g \rightarrow q+\gamma$). The production of prompt photons can be described by next-to-leading-order (NLO) perturbative QCD. Fragmentation photons are produced in the fragmentation of hard scattered quarks or gluons (e.g. $q+q \rightarrow q+q+\gamma$). Thermal photons are emitted by the hot thermalized medium through scattering of particles (e.g. $q+\bar{q} \rightarrow g+\gamma$) during the QGP phase and hadronic interactions (e.g. $\pi^+ + \pi^- \rightarrow \rho_0 + \gamma$) in the hot hadron gas phase.

The direct-photon spectrum can be calculated from the measured inclusive photon spectra by subtraction of all contributions from hadron decays. Measurements at RHIC [4] and LHC [5] yield a significant contribution of direct photons to the inclusive photon spectra. Figure 1 shows the direct-photon spectrum in the 0–40 % most central Pb-Pb collisions at $\sqrt{s_{\mathrm{NN}}} = 2.76\,\mathrm{TeV}$. At transverse momenta above $4\,\mathrm{GeV}/c$ the signal is described by binary scaled NLO (pQCD) calculations for pp [5]. At low p_{T} the spectrum shows a significant excess above the pQCD prediction which can be described by an exponential function with an inverse slope parameter of $T_{\mathrm{eff}}^{\mathrm{LHC}} = (304 \pm 51^{stat+sys})\,\mathrm{MeV}$ [5].

Journal of Physics: Conference Series **446** (2013) 012028 doi:10.1088/1742-6596/446/1/012028

The PHENIX collaboration reports an inverse slope parameter of $T_{\rm eff}^{\rm RHIC} = (220 \pm 19^{stat} \pm 19^{sys})$ MeV in 0–20 % Au-Au collisions at $\sqrt{s_{\rm NN}} = 0.2$ TeV [4]. Assuming a thermal origin of low p_T direct photons the inverse slope parameter can be interpreted as an effective temperature of the source integrated over the whole system evolution. A high effective temperature compared to the critical temperature for deconfinement (150–170 MeV [6, 7]) implies an early production time of direct photons during the system evolution.

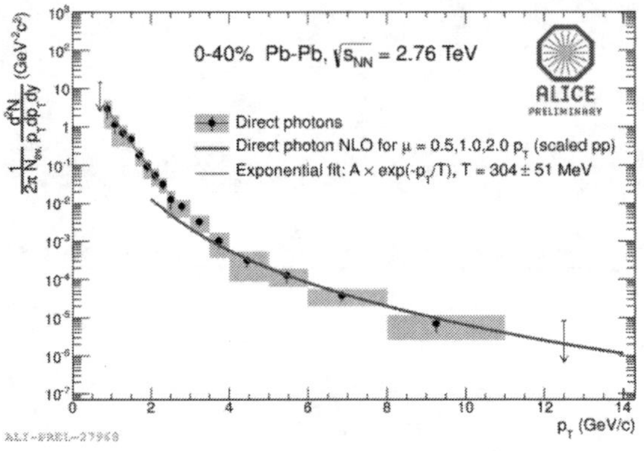

Figure 1: Direct-photon invariant yield in 0–40 % Pb-Pb collisions with NLO pQCD predictions and exponential fit [5].

A fingerprint of the thermalization of the system created in nucleus-nucleus collisions is the observation of azimuthally anisotropic particle emission. The azimuthal anisotropy is studied via a Fourier decomposition with harmonic coefficients v_n. In non-central collisions the initial asymmetry of the overlap zone leads to a dominant second harmonic v_2 known as elliptic flow. At small transverse momenta ($p_{\rm T} \lesssim 3$ GeV/c) the hadron v_2 can be understood in terms of the collective expansion of the medium. At higher momenta the anisotropy of hadrons is thought to be caused by the in-medium energy loss of partons.

While measurements of hadrons quantify the system evolution at the kinetic freeze-out, direct photons carry the system information at their production time. Consequently, a measurement of $v_2^{\gamma,\rm dir}$ allows to put additional constraints on their production time. Hydrodynamical descriptions of thermal photon production include a substantial portion of photons emitted in the hot plasma phase where flow has not yet been developed. Other sources of direct photons have either small or zero elliptic flow. Thus, a generic model expectation is that the $v_2^{\gamma,\rm dir}$ is smaller than for charged particles [8, 9].

2. Analysis Method

The analysis is based on about 2×10^7 minimum bias Pb-Pb collisions at $\sqrt{s_{\rm NN}} = 2.76$ TeV recorded in 2010 with the ALICE experiment at the LHC. Photons were detected at mid rapidity ($|\eta| \leq 0.8$) via their conversion in the detector material using the tracking and particle identification capabilities of the Inner Tracking System (ITS), the Time Projection Chamber (TPC) and the Time-Of-Flight (TOF) detector. Photon candidates were reconstructed using a Kalman filter based secondary vertex finding algorithm. A set of selection criteria on kinematic variables and particle identification allows to separate photons efficiently from other sources of secondary vertices (combinatorial background, Λ, $K_{\rm s}^0$, etc) [10]. Centrality and event plane angle were determined using the VZERO detector ($-3.7 \leq \eta \leq -1.7$ and $2.8 \leq \eta \leq 5.1$).

The inclusive photon $v_2^{\gamma,\rm inc}$ was studied in small centrality classes in order to minimize systematic multiplicity effects (0–5 %, 5–10 %, 10–20 %, 20–30 %, 30–40 %). The direct-photon spectrum was measured in the 0–40 % centrality class [5]. In order to minimize systematic uncertainties the ratio $\frac{N^{\gamma,\rm inc}}{N^{\gamma,\rm bg}}$ was calculated via the double ratio R:

$$R = \frac{\left(\frac{dN^{\gamma,\rm inc}/dp_{\rm T}}{dN^{\pi^0}/dp_{\rm T}}\right)}{\left(\frac{dN^{\gamma,\rm bg}/dp_{\rm T}}{dN^{\pi^0}/dp_{\rm T}}\right)_{MC}} = \frac{N^{\gamma,\rm inc}}{N^{\gamma,\rm bg}} \quad (1)$$

The inclusive photon spectrum was obtained from Pb-Pb data. The decay photon spectrum was calculated in a cocktail simulation based on the measured neutral pion spectrum. The contribution of η's and other mesons was estimated by transverse mass (m_{T}) scaling. In order to estimate the decay photon elliptic flow $v_2^{\gamma,\mathrm{bg}}$, the initial hadron azimuthal distributions were parametrized with the measured charged pion $v_2^{\pi^\pm}$ [11] scaled in transverse kinetic energy ($KE_{\mathrm{T}} = m_{\mathrm{T}} - m_0$). Finally, the direct-photon $v_2^{\gamma,\mathrm{dir}}$ was calculated from the double ratio R, inclusive photon $v_2^{\gamma,\mathrm{inc}}$ and decay photon $v_2^{\gamma,\mathrm{bg}}$:

$$v_2^{\gamma,\mathrm{dir}} = \frac{R v_2^{\gamma,\mathrm{inc}} - v_2^{\gamma,\mathrm{bg}}}{R - 1} \tag{2}$$

3. Results

Figure 2 shows the inclusive photon $v_2^{\gamma,\mathrm{inc}}$ for several centrality classes in Pb-Pb collisions. The magnitude of $v_2^{\gamma,\mathrm{inc}}$ is comparable to hadronic v_2 [11] and decreases with increasing centrality. Figure 3 shows the v_2 components for decay photons from various hadrons. The magnitude of decay photons is also comparable to hadrons but the values are shifted in transverse momentum p_{T} depending on the mass of the mother particles. The decay kinematics can even result into negative values for v_2 at small transverse momenta. About 80% of the decay photons come from neutral pion decays ($\pi^0 \rightarrow \gamma\gamma$) and about 18 % from the η meson ($\eta \rightarrow \gamma\gamma$).

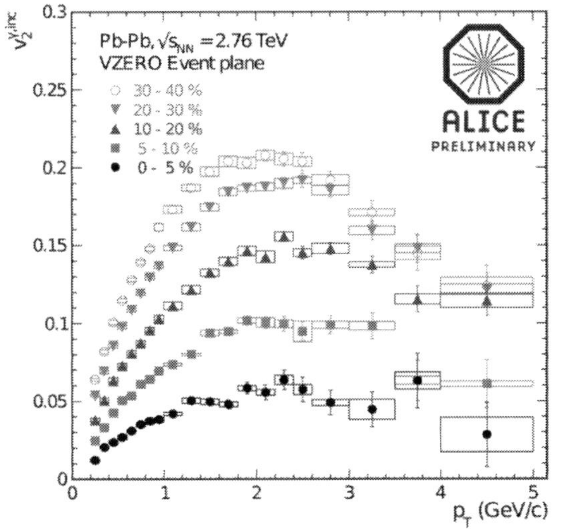

Figure 2: Inclusive photon $v_2^{\gamma,\mathrm{inc}}$ for several centrality classes in Pb-Pb collisions.

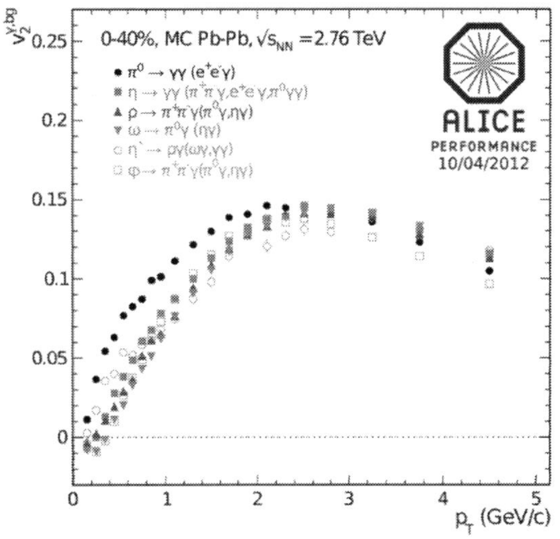

Figure 3: Simulation of decay photon $v_2^{\gamma,\mathrm{bg}}$ from different hadron decays.

The accumulated $v_2^{\gamma,\mathrm{bg}}$ is shown in Fig. 4 in comparison to the inclusive photon $v_2^{\gamma,\mathrm{inc}}$ in 0–40 %. Below a transverse momentum p_{T} of 3 GeV/c decay and inclusive photons are consistent within uncertainties which allows for two interpretations: either direct and decay photons have a similar elliptic flow ($v_2^{\gamma,\mathrm{dir}} \approx v_2^{\gamma,\mathrm{bg}}$) or there are no direct photons ($R = 1$). Above 3 GeV/c the inclusive photon $v_2^{\gamma,\mathrm{inc}}$ is significantly smaller than $v_2^{\gamma,\mathrm{bg}}$ which can only be explained by a direct-photon contribution with smaller $v_2^{\gamma,\mathrm{dir}}$ compared to $v_2^{\gamma,\mathrm{bg}}$. Systematic uncertainties of the decay photon $v_2^{\gamma,\mathrm{bg}}$ include very conservative assumptions for a breaking of m_{T} and KE_{T} scaling, a possible deviation between neutral and charged pion v_2 and the unknown centrality dependence of direct-photon production. Consequently, the estimate of the $v_2^{\gamma,\mathrm{bg}}$ is the dominant

contribution of systematic uncertainty of the $v_2^{\gamma,\mathrm{dir}}$ measurement.
Figure 5 shows the first measurement of $v_2^{\gamma,\mathrm{dir}}$ in 0–40 % Pb-Pb collisions at $\sqrt{s_{\mathrm{NN}}} = 2.76\,\mathrm{TeV}$.

4. Summary and Conclusions

The results provide evidence for a non-zero $v_2^{\gamma,\mathrm{dir}}$ for $1 < p_{\mathrm{T}} < 3\,\mathrm{GeV}/c$ with a magnitude similar to the observed charged pion $v_2^{\pi^\pm}$ [11]. Similar results were reported by the PHENIX collaboration [12]. Recent hydrodynamical calculations [8, 9] include a substantial portion of thermal photons from the hot plasma phase and also a sizable fraction from other sources in order to describe the observed direct-photon spectra. However, the emission from early stages of the system evolution yields a small $v_2^{\gamma,\mathrm{dir}}$ compared to hadrons. Thus, the observed large $v_2^{\gamma,\mathrm{dir}}$ might lend support for a significant emission from late stages of the system evolution where the hadron flow has built up.

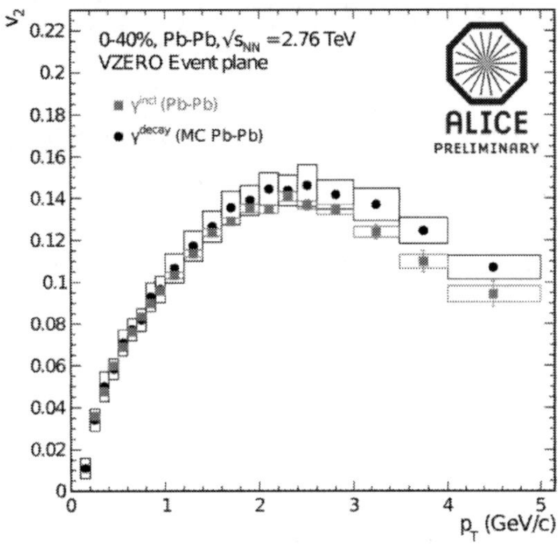

Figure 4: Inclusive photon $v_2^{\gamma,\mathrm{inc}}$ and decay photon $v_2^{\gamma,\mathrm{bg}}$ in 0–40 % Pb-Pb collisions.

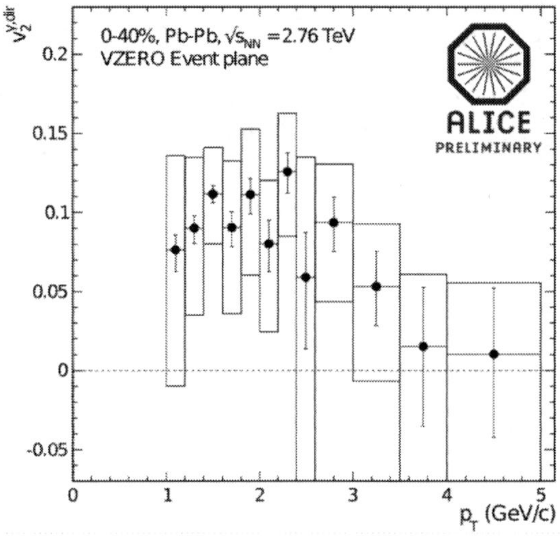

Figure 5: Direct-photon $v_2^{\gamma,\mathrm{dir}}$ in 0–40 % Pb-Pb collisions.

References

[1] Chatterjee R, Frodermann E S, Heinz U and Srivastava D K 2006 *Phys. Rev. Lett.* **96**(20) 202302
[2] Liu F M and Werner K 2009 *J.Phys.* **G36** 035101 (*Preprint* 0712.3619)
[3] Gale C 2010 *Landolt Börnstein* **23** 445 (*Preprint* 0904.2184)
[4] Adare A *et al.* (PHENIX Collaboration) 2010 *Phys.Rev.Lett.* **104** 132301 (*Preprint* 0804.4168)
[5] Wilde M (for the ALICE Collaboration) 2012 (*Preprint* 1210.5958)
[6] Aoki Y, Borsanyi S, Durr S, Fodor Z, Katz S D *et al.* 2009 *JHEP* **0906** 088 (*Preprint* 0903.4155)
[7] Cheng M *et al.* 2006 *Phys.Rev.* **D74** 054507 (*Preprint* hep-lat/0608013)
[8] Holopainen H, Rasanen S and Eskola K J 2011 *Phys.Rev.* **C84** 064903 (*Preprint* 1104.5371)
[9] van Hees H, Gale C and Rapp R 2011 *Phys.Rev.* **C84** 054906 (*Preprint* 1108.2131)
[10] Abelev B *et al.* (ALICE Collaboration) 2012 *Phys.Lett.* **B717** 162–172 (*Preprint* 1205.5724)
[11] Aamodt K *et al.* (ALICE Collaboration) 2010 *Phys. Rev. Lett.* **105** 252302
[12] Adare A *et al.* (PHENIX Collaboration) 2012 *Phys.Rev.Lett.* **109** 122302 (*Preprint* 1105.4126)

Journal of Physics: Conference Series **446** (2013) 012029 doi:10.1088/1742-6596/446/1/012029

Fluctuating Hydrodynamics Confronts the Rapidity Dependence of Transverse Momentum Fluctuations

Rajendra Pokharel[a], Sean Gavin[a] and George Moschelli[b]
a) Wayne State University, 666 W Hancock, Detroit MI 48084, USA, b) Frankfurt Institute for Advanced Studies, Johann Wolfgang Goethe University 60438 Frankfurt am Main, Germany.
E-mail: rajpol@wayne.edu

Abstract.
Interest in the development of the theory of fluctuating hydrodynamics is growing. Early efforts suggested that viscous diffusion broadens the rapidity dependence of transverse momentum correlations. That work stimulated an experimental analysis by STAR. We study the hydrodynamic evolution using second order causal viscous hydrodynamics including Langevin noise. We obtain a deterministic evolution equation for the transverse momentum density correlation function. We use the latest theoretical equations of state and transport coefficients to compute the STAR observables. The results are in excellent accord with the measured broadening. In addition, we predict features of the distribution that can distinguish 2nd and 1st order diffusion.

1. Introduction

Relativistic Heavy Ion Collider (RHIC) experiments have provided evidence of formation of deconfined state of quark and gluons, or quark-gluon plasma (QGP). It was shown that QGP behaves like a most perfect liquid known. The main reason behind this conclusion was the successful use of relativistic ideal hydrodynamic model to describe the flow data. The quantity used to indicate the perfectness of QGP is the ratio of viscosity to entropy density η/s. Of course, for perfect fluid it is zero. Current consensus is that for QGP it is in the range between the conjectured lower bound (the "KSS" bound) of $1/4\pi$ and ~ 0.3 (in units where $\hbar = k_B = c = 1$). This range is much smaller than the value ($= 0.7$) of superfluid liquid helium. In Ref.[1] Gavin and Abdel-Aziz propose an independent method for estimating η/s using rapidity distribution of transverse momentum correlations. They calculate the difference in the widths of these distributions between peripheral and central collisions, using first order diffusion of p_t fluctuations. They estimate η/s in the range 0.08 - 0.3, which is the same as that obtained from the flow data. Recently STAR has measured [2] the widths of correlations and p_t covariance profile in rapidity. Here we show that causally constrained diffusion equation with temperature dependent η/s is required to explain the data. We start with a brief discussion of diffusion of transverse momentum fluctuation in Section 2, before applying the concepts to two-particle p_t correlations in Section 3. We then present the results and discuss them in Section4.

2. Diffusion of Transverse Flow Fluctuations

Equations of dissipative relativistic hydrodynamics are derived from the basic principles of conservations of charges and of energy-momentum. Depending on the order of gradients included in the dissipative part of the energy momentum tensor, relativistic hydrodynamics is said to be of first or second order. The first order is the relativistic version of the famous Navier-Stokes theory and the second order is the well known Israel-Stewart theory. If we take small fluctuations on quantities (like the flow u^μ) over equilibrium values, we can see that transverse modes decouple from the longitudinal modes. The longitudinal modes are propagating sound modes while the transverse modes are solely the diffusing shear modes. As we are more interested in shear viscosity we focus on the transverse modes. Note that shear viscosity is the most important mode of dissipation in QGP. Starting with relativistic hydrodynamic equations and constitutive relation for shear tensor (see [3], for example, for such equations and relations) and then linearizing, one arrives at [4] diffusion equations for a transverse momentum current:

$$\frac{\partial \delta T_{0y}}{\partial t} = \nu \nabla_z^2 \delta T_{0y}, \qquad \tau_\pi \frac{\partial^2 \delta T_{0y}}{\partial t^2} + \frac{\partial \delta T_{0y}}{\partial t} = \nu \nabla_z^2 \delta T_{0y} \tag{1}$$

These two equations follow from the first order and second order theories, respectively. Here $T_{0y} \approx (\epsilon_0 + p_0)\delta u^y$. δu^y is a transverse fluctuation in flow, ϵ_0 and p_0 are energy density and pressure at equilibrium. Note that the diffusion coefficient $\nu = \eta/(\epsilon_0 + p_0) = \eta/Ts$ contains the ratio η/s and thus encapsulates the information on viscosity. In [5] Hirano and Gyulassy have put together the latest information on η: a combination of results from perturbative QCD, kinetic theory and $\mathcal{N} = 4$ Super Yang-Mills (SYM) at infinite coupling strengths, in the respectively applicable temperature ranges. This information has been used here for shear viscosity. For entropy density we have looked at values from lattice QCD calculations, specifically s95p-v1 from [6, 7].

We make use of Bjorken boost invariant coordinates (τ, x, y, η), where τ and η are proper time and space-time rapidity: $\tau = \sqrt{t^2 - z^2}$ and $\eta = (1/2)\ \ln((t+z)/(t-z))$. [1] The entropy production equation with boost invariance can be written as [3] $\frac{ds}{d\tau} + \frac{s}{\tau} = \Phi/\tau$, where $\Phi = 4\eta/3\tau$ for the first order case, and for second order case $\tau_\pi \frac{d\Phi}{d\tau} + \left(1 + \frac{\tau_\pi}{2\tau} + \frac{1}{2}\eta T \frac{d}{d\tau}(\frac{\tau_\pi}{\eta T})\right)\Phi = \frac{4\eta}{3\tau}$, where $\Phi = \pi_0^0 - \pi_z^z$.

The first equation of (1) is a regular diffusion equation. There is a well-known problem with the first order theory: this diffusion equation violates causality and is not suitable for relativistic particles or fluid cells. It has been shown that second order theory gives relativistically consistent diffusion equation. Equation (1), which follows from second order theory, does not violate causality. The relaxation time τ_π in the second order diffusion equation restores causality. For details on causality and diffusion see [3], [8] and the references therein.

3. Two-Particle Transverse Momentum Correlations

Correlation function of two-particle transverse momentum current is $r = \langle T_1^{0y} T_2^{0y}\rangle - \langle T_1^{0y}\rangle\langle T_2^{0y}\rangle$, where $T_1^{0y} \equiv T^{0y}(\mathbf{x}_1)$ and $T_2^{0y} \equiv T^{0y}(\mathbf{x}_2)$ are momentum currents of particle pairs at points $\mathbf{x}_1$ and $\mathbf{x}_2$, respectively. Considering small perturbations about the average, $T^{0y} = \langle T^{0y}\rangle + \delta T^{0y}$ and treating them as stochastic quantities, we can write $r = \langle \delta T_1^{0y} \bar{T}_2^{0y}\rangle + \langle \bar{T}_1^{0y} \delta T_2^{0y}\rangle + \langle \delta T_1^{0y} \delta T_2^{0y}\rangle$. The bars over the letters represents the same average - the average over the ensembles of events. The last term becomes important in stochastic methods. This term vanishes in the absence of noise. Using the second of (1) and the expression for r and manipulating the terms we can [4] subtract away the noise term to obtain

$$\tau_\pi \frac{\partial^2 \Delta r}{\partial t^2} + \frac{\partial \Delta r}{\partial t} = \nu(\nabla_{z1}^2 + \nabla_{z2}^2)\Delta r. \tag{2}$$

[1] Note that the same letter η is used for shear viscosity and for the longitudinal space-time rapidity here. The meaning should be clear from the context of its use.

We can get similar equation for the first order case [1]. Here $\Delta r = r - r_{eql}$.
If we use Bjorken boost invariance, (2) we can obtain this equation in terms of τ and η:

$$\tau_\pi \frac{\partial^2 \Delta r}{\partial \tau^2} + \frac{\partial \Delta r}{\partial \tau} = \frac{\nu}{\tau^2}(\nabla^2_{\eta 1} + \nabla^2_{\eta 2})\Delta r \tag{3}$$

Clearly, η_1 and η_2 are space-time rapidities for particle 1 and particle 2. This equation and its first order version [1] (i.e., (1) without the first term) are the equations we numerically solve. The diffusion coefficient and the relaxation time are both temperature (and time) dependent.

4. Results and Discussion

The main observables we compute from the numerical solutions (Δr) are the width σ of the correlation function and an important observable $\mathcal{C}$ given by [1] $\mathcal{C} = \langle N\rangle^{-2}\int \Delta r_g(\mathbf{x_1}, \mathbf{x_2})dx_1dx_2 = \langle N\rangle^{-2}\langle\sum_{i\neq j} p_{ti}p_{tj}\rangle - \langle p_t\rangle^2$, where i labels particles from each event. It can be seen that $\mathcal{C}$ connects Δr to experimentally measurable p_t-covariance. The relaxation time is $\tau_\pi = \beta\frac{\eta}{Ts} = \beta\nu$. Suggested value of β are 5-6 [9] and 6.32 [10]. Here we have used $\beta = 6$. Initial (thermalization) time is taken to be $\tau_0 = 1$ fm/c. Initial Δr is taken to be a gaussian in relative rapidity $\Delta\eta$, with initial width $\sigma_0 = 0.54$, making it consistent with experimental data for the most peripheral collisions. We have assumed a fixed temperature freezout, at 150 MeV. Glauber model has been used to connect impact parameters and multiplicity of the collisions. The freezeout proper time is assumed as $\tau_F - \tau_0 \propto (R - R_0)^2$, with R the rms participant radius. Freezeout time for most central collisions is taken to be 9 fm/c. The computed σ and $\mathcal{C}$ are compared with the values measured recently by STAR [2].

Figure 1 shows widths versus centralities computed using both first and second order theory with two different equations of state - EOS I and EOS II. EOS I is the one which contains the entropy density calculated from lattice QCD. EOS II uses the standard entropy density, summarized in [5]. We see that second order calculations agree very well with the experimental data. We also note that the choice of EOS makes a very small difference. However, there is significant difference between the first and second order diffusion, except for a few most central cases. The deviation of the first order results from data relates to how the relaxation time compares with the evolution time [4]. This clearly shows that one needs causally constrained hydrodynamic evolutions to better explain the experimental data. Figure.1 also compares results with NeXSPHeRIO calculations [11]. NeXSPHeRIO uses non-viscous (ideal) hydrodynamics in evolution of initial correlations. It reproduces most qualitative features of the two-particle correlations. It however does not reproduce the broadening of the rapidity width with increasing centrality, as one cane see in Fig. 1. We attribute that to the absence of viscosity in its hydrodynamic. Other explanations involving collective initial state behavior are also possible.

Figure 2 shows observable $\mathcal{C}$ vs $\Delta\eta$ for various centralities from our numerical calculations and the values measured by STAR [2]. The STAR measured values have error bars in the offsets (ridge), which we have not shown here. The plots show good agreement of the numerical results with the experiments. The single point at $\Delta\eta = 0$ in the most central case here has been attributed to the track merging and is just a detector artifact. One important observation we make here is the flattening of the peak (and to some extent double humps) in $\mathcal{C}$. The reason for the double hump (or rather the flattened peak) is a second order diffusion effect. We note that the second order diffusion equation has propagating wave part as well. Thus there is competing wave and diffusion effects, depending on the size of τ_π and ν. Wavefronts propagate in opposite directions and diffusion fills in the space in between. This can be seen more effectively in coordinate space (not shown here). In rapidity space, width saturates since the effective diffusion coefficient ν/τ^2 in (3) decreases rapidly with time. The first order diffusion obviously does not show such humps and flattening - there are no propagating waves. STAR data for other centralities [12] show the humps exactly at the same centralities. Shapes show fair agreement with the calculations. Here we have restricted ourselves with the published data only for the three centralities shown in Fig2. Finally, we just want to mention that the order of entropy production equations does not make practical difference in our results. Note that we have relatively large β in $\tau_\pi = \beta\eta/Ts$.

It would be interesting to measure the observable $\mathcal{C}$ from the p-A [13] and Pb-Pb [14] collisions data from LHC. In the pA case, it would be interesting to see whether there is a broadening with respect to pp. Experiments have identified a ridge in pA [13]. If hydrodynamics is applicable in the pA system, then viscous diffusion would broaden the rapidity width of $\mathcal{C}$.

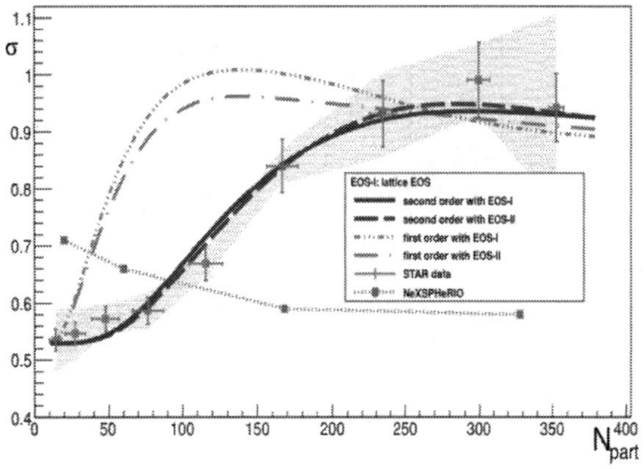

Figure 1. Second order, first order, with both EOS [4] and NeXSPHeRIO [11] calculations of width vs STAR [2] measurements.

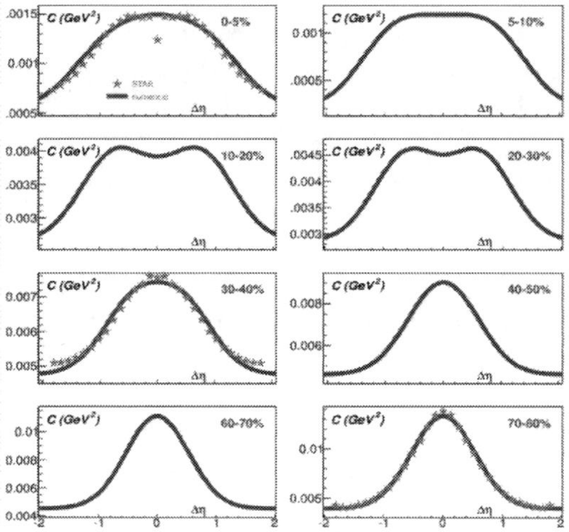

Figure 2. $\mathcal{C}$ vs $\Delta\eta$. Comparison with STAR data at different centralities. [2].

5. Acknowledgement

We would like to thank C. Pruneau and M. Sharma for a great deal of communication and discussions of the experimental results and for generously providing STAR data. RP would like to thank P. Huovinen for his helpful responses to questions on lattice data. This work was supported by U.S. NSF grant PHY-1207687.

References

[1] Gavin S and Abdel-Aziz M 2006 *Phys. Rev. Lett.* **97** 162302 (*Preprint* nucl-th/0606061)
[2] Agakishiev H *et al.* (STAR Collaboration) 2011 *Phys. Lett.* **B704** 467–473 (*Preprint* 1106.4334)
[3] Muronga A 2004 *Phys. Rev.* **C69** 034903 (*Preprint* nucl-th/0309055)
[4] Pokharel R, Gavin S and Moschelli G in preparation
[5] Hirano T and Gyulassy M 2006 *Nucl. Phys.* **A769** 71–94 (*Preprint* nucl-th/0506049)
[6] Huovinen P and Petreczky P 2010 *Nucl. Phys.* **A837** 26–53 (*Preprint* 0912.2541)
[7] https://wiki.bnl.gov/hhic/index.php/Lattice_calculatons_of_Equation_of_State
[8] Aziz M A and Gavin S 2004 *Phys. Rev.* **C70** 034905 (*Preprint* nucl-th/0404058)
[9] York M A and Moore G D 2009 *Phys. Rev.* **D79** 054011 (*Preprint* 0811.0729)
[10] Hong J, Teaney D and Chesler P M 2011 (*Preprint* 1110.5292)
[11] Sharma M, Pruneau C, Gavin S, Takahashi J, de Souza R D *et al.* 2011 *Phys.Rev.* **C84** 054915 (*Preprint* 1107.3587)
[12] Sharma M and Pruneau C private communications
[13] D. Velicanu's presentation, Hot Quarks 2012
[14] Aamodt K *et al.* (ALICE Collaboration) 2010 *Phys. Rev. Lett.* **105** 252302 (*Preprint* 1011.3914)

Journal of Physics: Conference Series **446** (2013) 012030 doi:10.1088/1742-6596/446/1/012030

Observation of near-side angular correlations in pPb collisions at 5.02TeV with CMS

Dragos Velicanu

24-522 Heavy Ion Group, 77 Massachusetts Avenue, Cambridge, MA, USA

E-mail: velicanu@mit.edu

Abstract.
The first measurement of two-particle correlations in pPb collisions at a center-of-mass energy of 5.02 TeV is presented. The correlations are studied in azimuth, ϕ , and pseudorapidity, η , as a function of charged-particle transverse momentum, p_T , and multiplicity. A total of two million collisions are used in this analysis. A long range ($2 < |\Delta\eta| < 4$), near-side ($\Delta\phi \approx 0$) ridge-like structure emerges in the two-particle $\Delta\eta-\Delta\phi$ correlation functions of high-multiplicity events. This is the first time such a correlation has been seen in proton-nucleus collisions, which which resembles the correlations seen in a broad energy range of nucleus-nucleus collisions and in high-multiplicity proton-proton collisions at $\sqrt{s} = 7$ TeV. The correlation strength rises approximately linearly with multiplicity and has a maximum in the range of $p_T = 1 - 1.5$ GeV/c .

1. Introduction

These proceedings present two-particle correlations in pPb collisions at $\sqrt{s_{NN}} = 5.02$ TeV with the Compact Muon Solenoid (CMS) detector at the Large Hadron Collider (LHC). Collective behavior in the medium can give rise to two-particle correlations. These can be studied using two-dimensional $\Delta\eta - \Delta\phi$ correlation functions. Of particular interest are the long-range (large $\Delta\eta$) correlations that are less affected by correlations such as fragmentations from jets and resonance decays. These long-range correlations are related to the medium properties and its evolution and have been studied at many collision energies in proton-proton (pp), proton-nucleus (pA), and nucleus-nucleus collisions (AA). Recently measurements of 7 TeV pp collisions have revealed the existence of long-range, near-side ($\Delta\phi \approx 0$) correlations in high multiplicity collisions [1]. Diverse theoretical models have been proposed to explain the origin of the ridge in pp collisions (see Ref. [2] for a recent review). The first comparison of pPb and pp correlations, presented in these proceedings, provides information for understanding the ridge correlation signal observed in high-multiplicity pp collisions. In section 2 the two-particle correlation function is described, then in section 3 the quantitative ridge measurements are presented.

2. The Two-Particle Correlation Function

The two-particle correlations analysis was performed in bins of track multiplicity in order to match the analysis strategy in high-multiplicity pp collisions [1], and using the procedure established in [3, 4]. Charged particles, which have their trajectories reconstructed in the silicon tracker, are denoted as "tracks", and multiplicity, $N_{\rm trk}^{\rm offline}$, is defined as the number of tracks having $|\eta| < 2.4$ and $p_T > 0.4$ GeV/c and passing a set of quality cuts described in more detail in [5]. For each multiplicity bin, the "trigger" particles are defined to be all charged particles in a given p_T range originating from the primary vertex. Charged-particle pairs are formed by associating all other tracks in the same p_T range as the trigger particle, where a minimum of two particles is required in each p_T interval for each collision. The definition of per-trigger-particle associated yield is

$$\frac{1}{N_{\rm trig}}\frac{{\rm d}^2{\rm N}^{\rm pair}}{{\rm d}\Delta\eta\,{\rm d}\Delta\phi} = B(0,0)\times\frac{S(\Delta\eta,\Delta\phi)}{B(\Delta\eta,\Delta\phi)}, \tag{1}$$

where $\Delta\eta$ and $\Delta\phi$ are the pair differences in η and ϕ. The signal distribution, $S(\Delta\eta,\Delta\phi)$, is the per-trigger-particle yield of same event particle pairs,

$$S(\Delta\eta,\Delta\phi) = \frac{1}{N_{\rm trig}}\frac{{\rm d}^2{\rm N}^{\rm same}}{{\rm d}\Delta\eta\,{\rm d}\Delta\phi}\,. \tag{2}$$

The background distribution is constructed the same way as the signal, except associating particles from different events than the trigger particle, but from events with the same longitudinal position along the beam axis and multiplicity range.

The 2-D two-particle correlation function for low (a) and high (b) multiplicity pPb events is shown in Figure 1 for charged-particle pairs with $1 < p_{\rm T} < 3$ GeV/c. Both the high and low

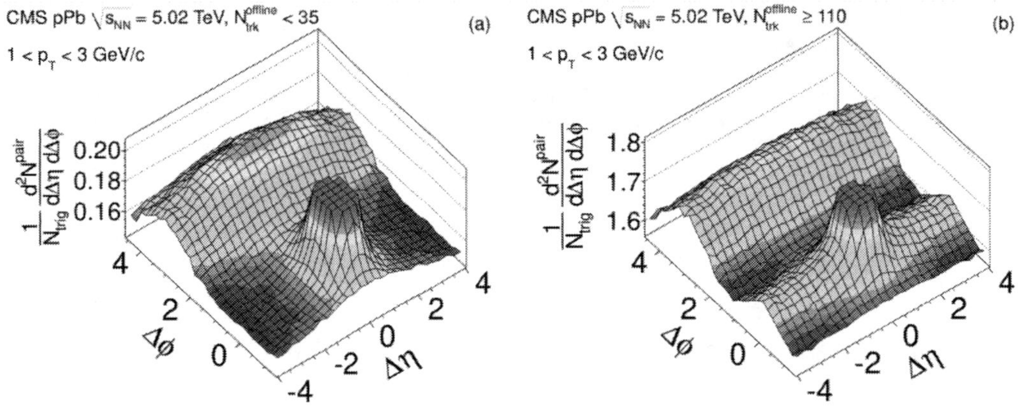

Figure 1. 2-D two-particle correlation functions for charged-particle pairs with $1 < p_{\rm T} < 3$ GeV/c from 5.02 TeV pPb collisions are shown with low-multiplicity ($N_{\rm trk}^{\rm offline} < 35$) (a) and high-multiplicity ($N_{\rm trk}^{\rm offline} \geq 110$) (b). The near side jet peak is truncated.

multiplicity correlation functions exhibit a peak at $(\Delta\eta,\Delta\phi) = (0,0)$ originating from particle pairs produced by a jet, and an elongated structure at $\Delta\phi \approx \pi$ from particle pairs produced by back-to-back jets. However, the high-multiplicity correlation function also has a pronounced ridge at $\Delta\phi \approx 0$ extending at least 4 units in $|\Delta\eta|$.

3. Results

To further investigate these long-range near-side pA correlations and to provide a quantitative comparison to measurements in pp collisions, one-dimensional (1D) $\Delta\phi$ distributions are formed by averaging the two-dimensional signal and background distributions over the range of $2 < |\Delta\eta| < 4$ [1, 3, 4]. The minimum of the correlation function in the azimuthal range of $0.1 < |\Delta\phi| < 2$ is first found and the corresponding value, $C_{\rm ZYAM}$, is subtracted from the 1-D correlation function [6].

Figure 2 shows the 1D distributions for $\sqrt{s_{NN}}$ = 5.02 TeV pPb (solid), $\sqrt{s_{NN}}$ = 7 TeV pp (open), with $p_{\rm T}$ increasing horizontally and multiplicity increasing from top to bottom. Also shown are predictions, for pPb collisions modeled by HIJING [7] (dashed curves), and by hydrodynamics [8] (solid curves) in the multiplicity bins that best match the bins of the theoretical predictions. The color-glass condensate framework also predicted a ridge structure, but in this case it arises from initial state gluon correlations rather than from collective behavior in the produced matter [9].

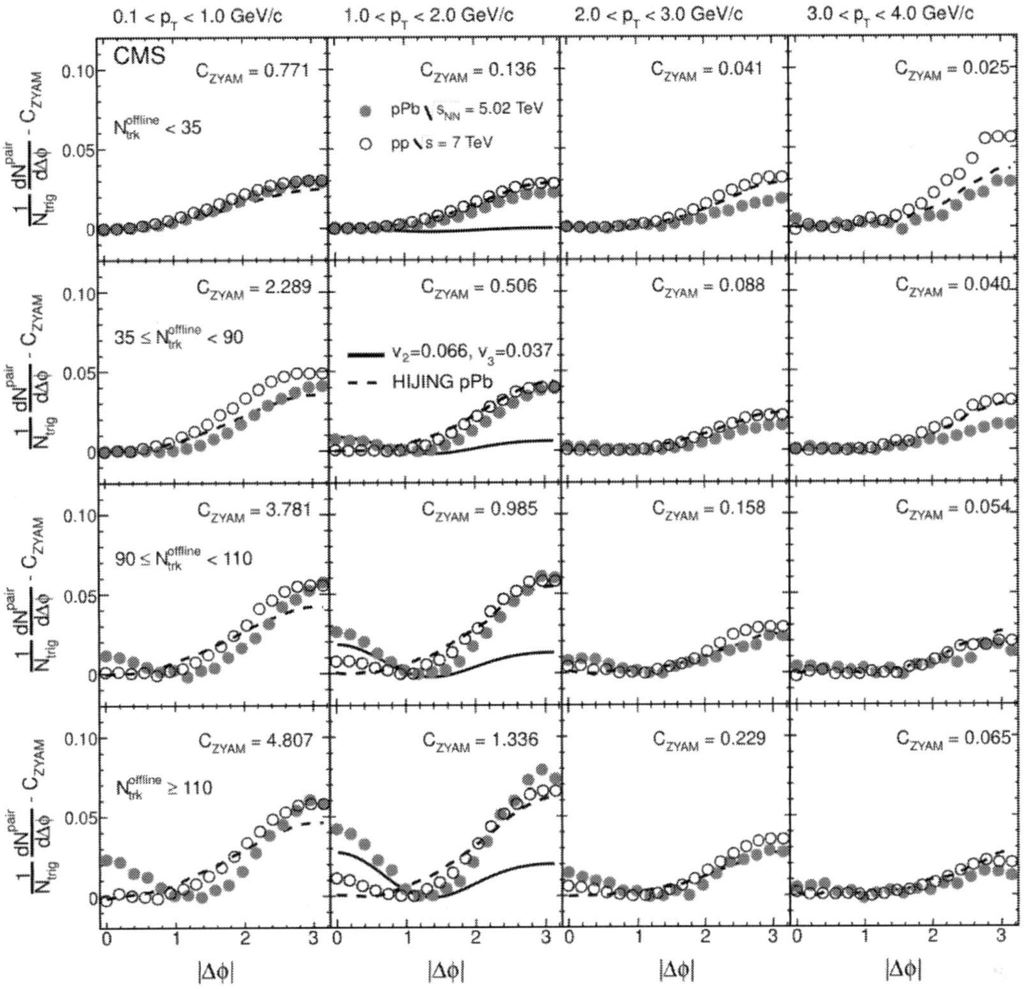

Figure 2. Zero-yield-at-minimum (ZYAM) correlated yield in different p_T and multiplicity bins averaged over $2 < |\Delta\eta| < 4$ as a function of $|\Delta\phi|$. Shown are 5.02 TeV pPb data (solid circles) and 7 TeV pp data (open circles). Statistical uncertainties are smaller than the marker size. The subtracted ZYAM constant is listed in each panel.

In low multiplicity events the correlation functions from both pPb and pp collisions have minima at $\Delta\phi = 0$ and a maximum at $\Delta\phi = \pi$ for all p_T ranges. However in high-multiplicity collisions a second maximum emerges at $\Delta\phi = 0$, corresponding to the two-dimensional ridge, for both the pp and pPb collisions systems, which is most pronounced in the $1 < p_T < 2$ GeV/c range.

The strength of the ridge yield can be further quantified in finer multiplicity and p_T bins by integrating the correlated yield from Fig. 2 over $|\Delta\phi| < 1.2$ and normalizing by the width of the p_T interval. The resulting pPb ridge-yield is shown in Fig. 3 in ten p_T bins for high-multiplicity collisions 3(a) and in twelve multiplicity classes fixing $1 < p_T < 2$ GeV/c in 3(b). The corresponding pp data (open circles) is also shown. The error bars show the statistical uncertainties and the shaded boxes the systematic.

Both pp and pPb data show a maximum ridge yield in the range of $1 < p_T < 2$ GeV/c. At low multiplicity the ridge yield is zero due to the ZYAM procedure, indicating the ridge correlations are absent or smaller than other negative correlations. At high-multiplicity the yield rises almost linearly with multiplicity once it turns on at a multiplicity of around $N_{trk}^{offline} = 50$,

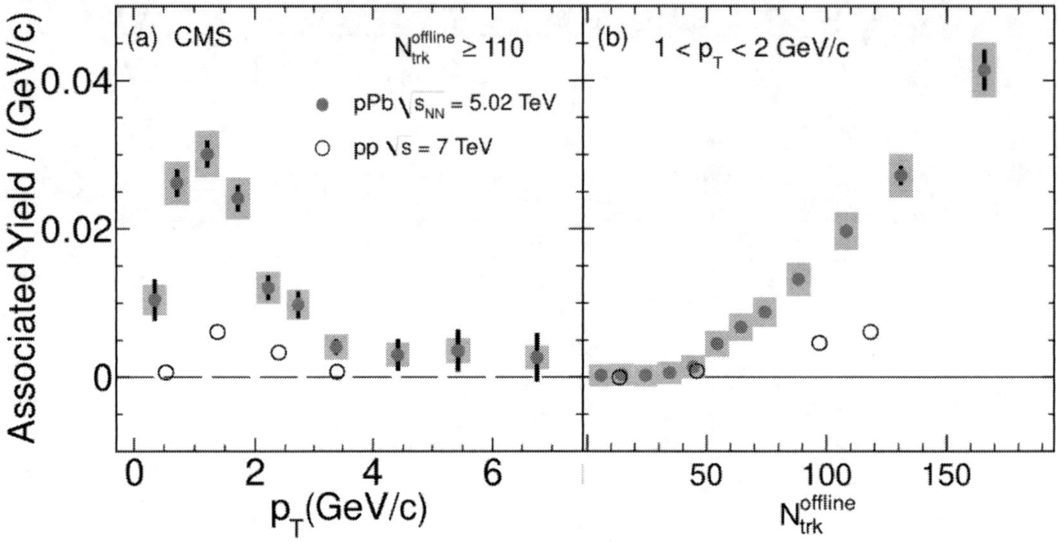

Figure 3. Ridge associated yield in different p_T and multiplicity bins averaged over $2 < |\Delta\eta| < 4$ and integrated over $|\Delta\phi| < 1.2$. Shown are 5.02 TeV pPb data (solid circles) and 7 TeV pp data (open circles). Panel (a) shows the p_T dependence of the associated yield in high-multiplicity events. Panel (b) shows the multiplicity dependence of the associated yield in the $1 < p_T < 2$ GeV/c range. The error bars are statistical uncertainties and the shaded areas are systematic uncertainties.

which is slightly higher than the mean multiplicity in minimum bias events..

4. Summary

A ridge-like structure has been observed in two-particle angular correlations in high-multiplicity pPb collisions with the CMS experiment at the LHC. This structure is qualitatively similar to the ridge observed in high-multiplicity pp collisions at $\sqrt{s} = 7$ TeV and in AA collisions for a wide range range of center-of-mass energies. The effect is strongest in the intermediate transverse momentum range $1 < p_T < 1.5$ GeV/c. The dependence of the effect on multiplicity and p_T in pPb collisions is similar to that seen in pp data, but the magnitude of the ridge yield is significantly larger.

References

[1] V. Khachatryan *et al.* Observation of long-range near-side angular correlations in proton-proton collisions at the LHC. *JHEP*, 09:091, 2010.
[2] W. Li. Observation of a ridge correlation structure in high multiplicity proton-proton collisions: A brief review. *Mod. Phys. Lett. A*, 27:1230018, 2012.
[3] S. Chatrchyan *et al.* Long-range and short-range dihadron angular correlations in central PbPb collisions at a nucleon-nucleon center of mass energy of 2.76 TeV. *JHEP*, 07:076, 2011.
[4] S. Chatrchyan *et al.* Centrality dependence of dihadron correlations and azimuthal anisotropy harmonics in PbPb collisions at 2.76 TeV. *Eur. Phys. J. C*, 72:2012, 2012.
[5] S. Chatrchyan *et al.* Observation of long-range, near-side angular correlations in ppb collisions at the lhc. *Physics Letters B*, 718(3):795 – 814, 2013.
[6] N. N. Ajitanand, J. M. Alexander, P. Chung, W. G. Holzmann, M. Issah, Roy A. Lacey, A. Shevel, A. Taranenko, and P. Danielewicz. Decomposition of harmonic and jet contributions to particle-pair correlations at ultrarelativistic energies. *Phys. Rev. C*, 72:011902, 2005.
[7] M. Gyulassy and X. Wang. HIJING 1.0: A Monte Carlo program for parton and particle production in high-energy hadronic and nuclear collisions. *Comput. Phys. Commun.*, 83:307, 1994.
[8] P. Bożek. Collective flow in p-Pb and d-Pb collisions at TeV energies. *Phys. Rev. C*, 85:014911, 2012.
[9] K. Dusling and R. Venugopalan. Evidence for BFKL and saturation dynamics from di-hadron spectra at the LHC, arXiv:1210.3890, 2012.

Journal of Physics: Conference Series **446** (2013) 012031 doi:10.1088/1742-6596/446/1/012031

Event-by-Event di-hadron correlations in Pb-Pb 2.76 TeV collisions from the ALICE experiment.

Anthony R. Timmins

Department of Physics, University of Houston, 617 Science and Research Building 1, Houston, TX 77204

E-mail: anthony.timmins@cern.ch

Abstract. The large multiplicities at the LHC may permit flow harmonics to be determined on an event-by-event basis in Pb-Pb collisions. We extract these harmonics from event-by-event di-hadron correlations. Within a fine centrality bin, we find the correlation function varies substantially on an event-by-event basis beyond expectations from statistical fluctuations. This implies large fluctuations in the initial conditions occur for a given impact parameter. Such large fluctuations lead to some events being highly triangular or highly elliptical, where the angular correlation function is completely dominated by the respective second and third Fourier harmonics. We will show unfolded v_2 distributions for various centralities, and implications for our understanding of the initial conditions.

1. Introduction

Relativistic heavy-ion collisions aim to create the QGP (Quark-Gluon Plasma), a unique state of matter where quarks and gluons can move freely over large volumes in comparison to the typical size of a hadron. Spatial inhomogeneities in the initial state can create pressure gradients in the QGP, which induce angular correlations known as azimuthal flow. It is convenient to express the single particle density of produced particles as a Fourier series [1]:

$$\frac{dN}{d\phi} \propto 1 + 2\sum_{n=1}^{\infty} v_n cos(n(\phi - \psi_n)) \tag{1}$$

where v_n and ψ_n are the magnitude and direction of order n harmonic flow. The magnitude v_n is typically extracted from azimuthal correlations averaged over an event ensemble, where the second order (elliptic flow) is usually the dominant contribution. It has been proposed that event-by-event fluctuations in the initial state give rise to the odd orders [2]. They also lead to variance in v_n (σ_{vn}) for all orders, that can be extracted from data via two and four particle cumulants [3]. These fluctuations should be also evident in the event-wise pair distributions. All orders of v_n can be extracted without the need to identify ψ_n using the pair density:

$$\frac{dN}{d\Delta\phi} \propto 1 + 2\sum_{n=1}^{\infty} V_{n\Delta} cos(n\Delta\phi) \tag{2}$$

where $\Delta\phi$ is the angular difference between the two particles in a pair. If factorisation occurs, $V_{n\Delta}$ can be expressed as the product of v_n so $V_{n\Delta} = v_n^2$, and this has been demonstrated for

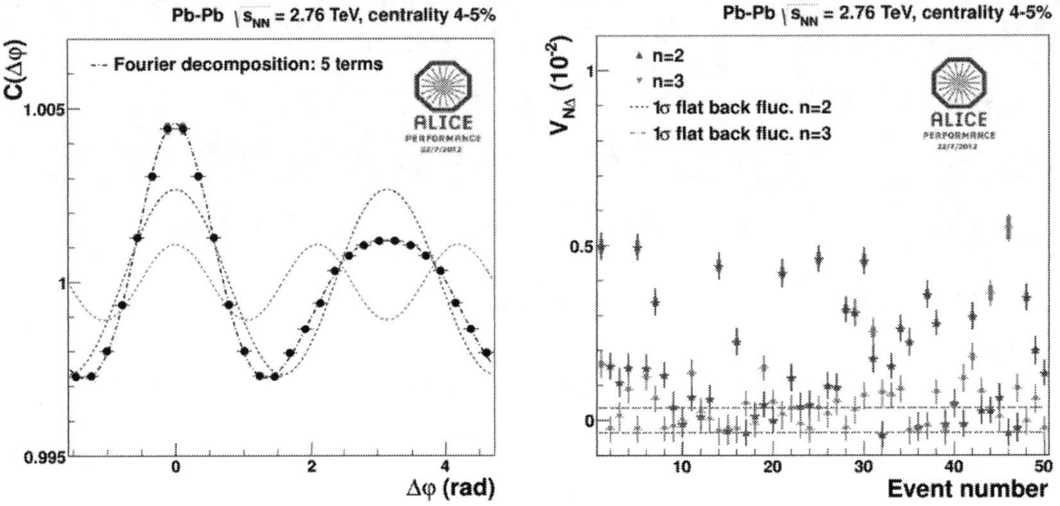

Figure 1. Left Panel: Event averaged $\Delta\phi$ distributions for Pb-Pb $\sqrt{s_{NN}} = 2.76$ TeV collisions. The dot-dash lines show the sum of a Fourier decomposition (with $n = 2$ and $n = 3$ contributions below). Right Panel: Event wise $V_{n\Delta}$ extracted from 50 events in chronological order. Dashed lines show $\pm 1\sigma$ statistical fluctuations for events with true $v_n = 0$.

these energies at low p_T [4]. Measurements of di-hadron correlations from ALICE show v_2 and v_3 manifest themselves as a nearside "ridge" and awayside "double hump" for very central collisions [4, 5, 6]. We present measurements of event-by-event di-hadron correlations in Pb-Pb $\sqrt{s_{NN}} = 2.76$ TeV collisions. The large multiplicities at the LHC and near full p_T acceptance of the ALICE detector favour such measurements. Values of $V_{2\Delta}$ and $V_{3\Delta}$ are extracted from these distributions, and we will show events with large values of $V_{2\Delta}$ and $V_{3\Delta}$. Measurements of event wise $V_{2\Delta}$ allow for the full v_2 distributions to be obtained, which will also be shown.

2. The correlation function

We use charged hadrons with transverse momentum $p_T > 0.15$ GeV/c from the ALICE TPC to form the pair correlation function. This is defined as follows:

$$C(\Delta\phi)_i = \frac{N_{pairs}(\Delta\phi)_i}{\langle N_{pairs}\rangle} \tag{3}$$

where $N_{pairs}(\Delta\phi)_i$ is the number of pairs in bin i, and $\langle N_{pairs}\rangle$ is the average number of pairs summed over all bins. It therefore quotes the relative deviation from the mean number of pairs. The main advantage of the correlation function is that pair inefficiencies due to the detector (e.g. tracking inefficiencies) in the numerator and denominator cancel to leading order. To extract $V_{n\Delta}$ from $C(\Delta\phi)$, a definition used in a previous ALICE publication will be adopted [4]. It is given by:

$$V_{n\Delta} = \sum_i C_i cos(n\Delta\phi) / \sum_i C_i \tag{4}$$

where n is the harmonic of interest and C_i is the value of the correlation function in a particular $\Delta\phi$ bin. The key advantage of this method is that the calculation of $V_{n\Delta}$ is invariant with respect to the choice of normalisation of the correlation function. The left panel of figure 1 shows the event averaged correlation function obtained for Pb-Pb centrality 4-5%. A Fourier

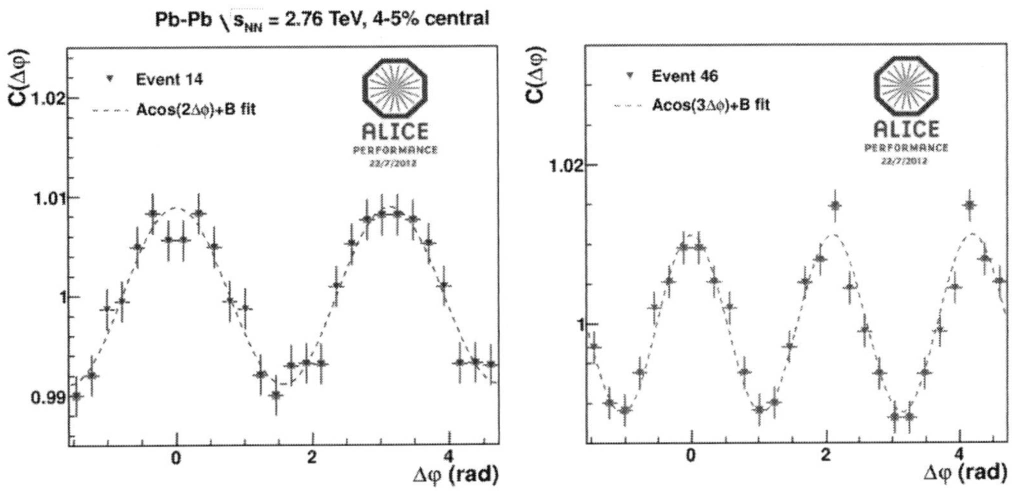

Figure 2. $\Delta\phi$ distributions from events 14 and 46 in the right panel of figure 1.

decomposition is performed, and each coefficient is extracted using equation 4. The Fourier series is shown by the black lines. It was found that the first 5 terms are needed for a good description of the data for this p_T range, and one can obverse the $n = 2$ contribution is larger than the $n = 3$ contribution.

3. event-by-event $V_{N\Delta}$

Rather than obtaining $C(\Delta\phi)$ and $V_{n\Delta}$ from an event ensemble, we investigate these measurements in single events. In the right panel of figure 1, event-wise values of $V_{2\Delta}$ and $V_{3\Delta}$ are shown for 50 events that occur for centrality of 4-5% chronologically. The dashed lines show expected $\pm 1\sigma$ levels of background fluctuations for each order, due to statistical fluctuations. There are clearly many events with $V_{2\Delta}$ higher than expected from just background fluctuations. Such an observation indicates large fluctuations in the initial conditions, even within a small centrality range. The $V_{2\Delta}$ fluctuations are bigger than the $V_{3\Delta}$ fluctuations, and this is consistent with the fact $\langle V_{2\Delta} \rangle$ is bigger than $\langle V_{3\Delta} \rangle$ (shown in the left panel). This is also inconsistent with background fluctuations, where we find all orders of n fluctuate equally (the red and blue dashed lines overlap). In figure 2, the $\Delta\phi$ distributions for event 14 and event 46 from the left panel of figure 1 are shown. It is observed that these events are completed dominated by elliptical ($n = 2$) and triangular ($n = 3$) flow respectively, and are in contrast with the distribution seen in the left panel of figure 1.

4. Unfolded v_2 distributions

In figure 3, we show v_2 distributions as a function of centrality. These are obtaining by unfolding raw $V_{2\Delta}$ distributions and applying the transformation $f(v_2) = f(\sqrt{V_{2\Delta}})$, where $f(V_{2\Delta})$ is the unfolded $V_{2\Delta}$ distribution. The unfolding procedure [7] is needed to remove the effects of the non-flow and trivial statistical smearing. The statistical smearing is determined by sampling a $dN/d\phi$ distribution with a known input $V_{2\Delta}$, then extracting the event-wise reconstructed $V_{2\Delta}$. The sampling is repeated, so one obtains the distribution of reconstructed $V_{2\Delta}$ for the given input. The contribution of non-flow is determined by fitting the pseudo-rapidly dependence of $V_{2\Delta}$ with a Gaussian (non-flow) and constant (flow). The systematic uncertainties (shown by the boxes) are due to the uncertainty in the non-flow contribution (evaluated by changing the

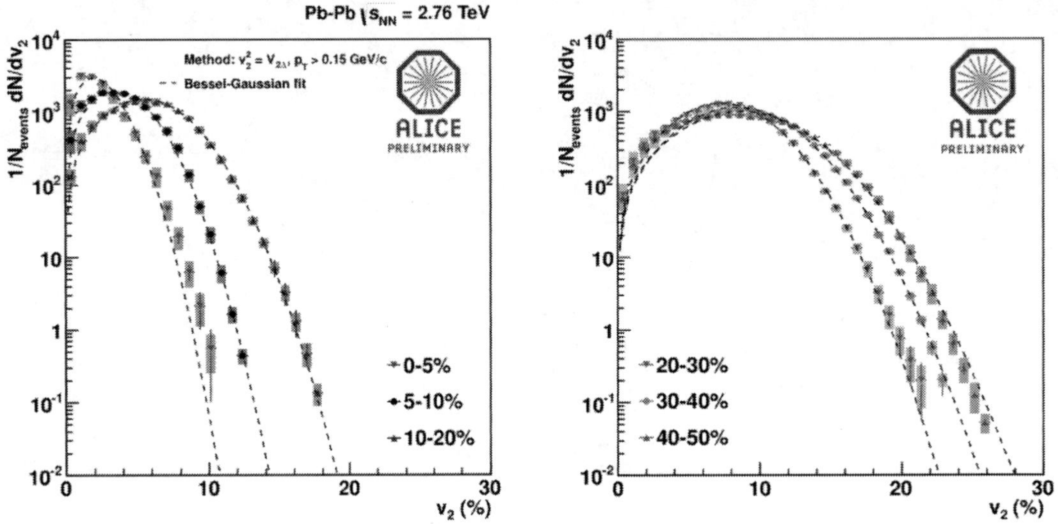

Figure 3. Unfolded v_2 distributions as a function of centrality, subject to a $\Delta\eta > 0.8$ gap.

$\Delta\eta$ cut), statistical smearing hypothesis (evaluated by changing the number of particles used) and centrality selection. The v_2 distributions are expected to reflect the participant eccentricity distributions [8], therefore provide constraints on models of the initial conditions. We fit these distributions with a Bessel-Gaussian distribution, which is the relevant p.d.f if the components of the eccentricity vector (ε_x,ε_y) are subject to Gaussian fluctuations [9]. These fits generally describe the data well, bar central collisions. We found (not shown) that the unfolded $\langle v_2 \rangle$ and $\sqrt{\langle v_2^2 \rangle + 2\sigma^2}$ are consistent with ALICE measurements of $v_2\{4\}$, and $v_2\{2\}$ respectively[5], which is expected if the v_2 distributions are Bessel-Gaussian. The widths of the unfolded distributions were also found to be consistent with the widths obtained via $\sqrt{0.5(v_2\{2\}^2 - v_2\{4\}^2)}$.

5. Summary

We have investigated event-by-event di-hadron correlations, and found some events dominated by $n = 2$ (elliptical events), and others dominated by $n = 3$ (triangular events). The observations are beyond what is expected from trivial statistical fluctuations. We have presented unfolded v_2 distributions and found they are generally described by a Bessel-Gaussian distribution. Theses distributions can be used as constraints for models of initial state eccentricities.

6. References

[1] A. M. Poskanzer and S. A. Voloshin, Phys. Rev. C **58** (1998) 1671
[2] J. Takahashi et al., Phys. Rev. Lett. **103** (2009) 242301; J. B. Alver and G. Roland, Phys. Rev. C **81** (2010) 054905; B. H. Alver, C. Gombeaud, M. Luzum and J. Y. Ollitrault, Phys. Rev. C **82** (2010) 034913
[3] N. Borghini, P. M. Dinh and J. Y. Ollitrault, Phys. Rev. C **64** (2001) 054901; A. Bilandzic, R. Snellings, S. A. Voloshin, Phys. Rev. C **83** (2010) 044913
[4] K. Aamodt et al. (ALICE Collaboration) Phys. Lett. B **708** (2012) 249264
[5] K. Aamodt et al. (ALICE Collaboration) Phys. Rev. Lett. **107** (2011) 032301
[6] A. Timmins (ALICE collaboration) J. Phys. G: Nucl. Part. Phys. **38** (2011) 124093
[7] G. D'Agostini, Nucl. Instrum. Meth. A **362** (1995) 487-498
[8] H. Niemi, G. S. Denicol, H. Holopainen, and P. Huovinen, *arXiv:1212.1008v1*
[9] S. A. Voloshin, A. M. Poskanzer, A. Tang, G. Wang, Phys. Lett. B **659** (2008) 537

Effective model for deconfinement at high temperature

Vladimir Skokov

Department of Physics, Brookhaven National Laboratory, Upton, NY, 11973 USA

E-mail: vskokov@quark.phy.bnl.gov

Abstract. In this talk I consider the deconfining phase transition at nonzero temperature in a SU(N) gauge theory, using a matrix model. I present some results including the position of the deconfining critical endpoint, where the first order transition for deconfinement is washed out by the presence of massive, dynamical quarks, and properites of the phase transition in the limit of large N. I show that the model is soluble at infinite N, and exhibits a Gross-Witten-Wadia transition.

1. The matrix model

In my talk, I addressed the properties of the thermodynamics of pure $SU(N)$ gauge theory in the so-called semi-QGP region for temperatures about or a few times that for the deconfining phase transition T_d.

A commonly accepted order parameter for the deconfinement phase transition is the Wilson line defined by

$$\mathrm{L}(\vec{x}) = \mathrm{P}\exp\left(ig\int_0^\beta A_0(\vec{x},\tau)d\tau\right). \tag{1}$$

An effective theory is constructed by expanding about a constant background field for the vector potential,

$$A_0^{ij} = \frac{2\pi T}{g}\, q_i\, \delta^{ij}, \tag{2}$$

where the field A_0 was diagonalized by a proper gauge transformation and where $i,j = 1\cdots N$ and the eigenvalues q_i are subject to the $SU(N)$ constraint $\sum_i^N q_i = 0$. In the matrix model q_i are the fundamental variables characterizing the transition. It is assumed that after integrating out the other components of gluon field A_i, that we obtain an effective potential for q_i ($\tilde{V} = V/N^2$) [1, 2, 3, 4, 5, 6]:

$$\tilde{V}_{eff}(q) = -d_1(T)\tilde{V}_1(q) + d_2(T)\tilde{V}_2(q), \tag{3}$$

$$\tilde{V}_n(q) = \sum_{i,j=1}^{N} |q_i - q_j|^n \left(1 - |q_i - q_j|\right)^n. \tag{4}$$

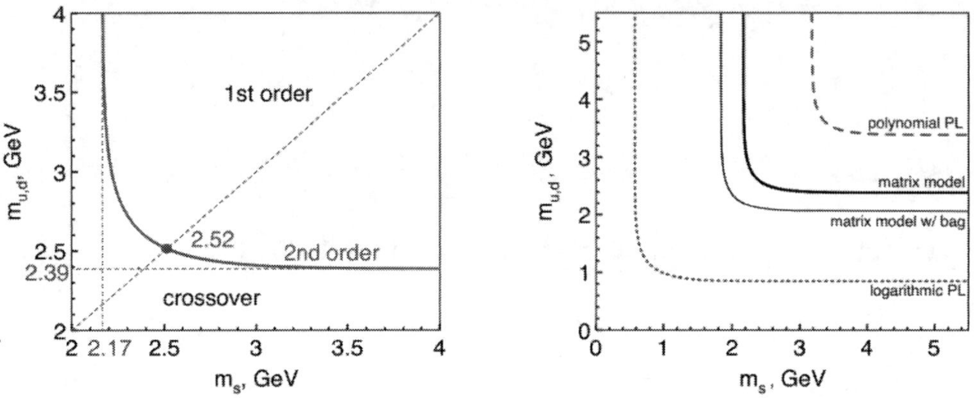

Figure 1. The left panel: The phase diagram for the deconfining phase transition in the matrix model, assuming $T_d \sim 270$ GeV. The right panel: The phase diagram of the deconfining phase transition for the matrix model without/with bag constant (bold/thin solid line) and the Polyakov loop models with the polynomial (dashed line) and logarithmic (dotted line) potentials. The lines correspond to the second order deconfining phase transition.

The potential includes both perturbative ($\tilde{V}_2$) and non-perturbative ($\tilde{V}_1$) contributions. The temperature-dependent functions d_1 and d_2 are given by

$$d_1(T) = \frac{2\pi}{15}\, c_1\, T^2\, T_d^2,\ d_2(T) = \frac{2\pi}{3}\left(T^4 - c_2 T^2 T_d^2\right). \tag{5}$$

The constants are fixed by comparison to the simulations of SU(3) on the lattice [4]. They are not important for further discussion.

2. Critical point of the deconfinement phase transition

When dynamical quarks are added one needs to compute the one-loop perturbative analogous contribution which they make to the q-dependent potential. This has been computed in Ref. [1]. For quarks of mass m, the one loop potential is given by

$$\ln\det(\gamma^\mu\partial_\mu + q\delta^{\mu 4} + im) = 2\ln\det\Big[\Big(\partial_0 + 2\pi T q\Big)^2 - \vec{\partial}^2 + m^2\Big], \tag{6}$$

As fermions, in the Matsubara formalism, the frequencies are odd multiples of πT. The straightforward calculations a single flavor of massive quark lead to the contribution to the potential

$$V^{qk}_{\rm pert}(\mathbf{q}) = \frac{2\,m^2\,T^2}{\pi^2}\sum_{n=1}^{\infty}\frac{(-1)^n}{n^2}\,K_2\left(\frac{n\,m}{T}\right)\sum_{i=1}^{N}\cos(2\pi n q_i)\,. \tag{7}$$

For the pure glue theory, by a $Z(3)$ rotation we can assume that the Polyakov loop is real. With dynamical quarks, this remains true if all of the quark masses are real and the quark chemical potential vanishes. In the left panel of Fig. (1) the results are presented for the matrix model. The critical temperature changes very little from the pure gauge theory,

$$T_{\rm de} = .995\, T_d\,. \tag{8}$$

The results are compared for different models in the right panel of Fig. (1). For the logarithmic Polyakov loop model [7], the masses are light, $m^{de} \sim 1$ GeV. The temperature for the deconfining critical endpoint is significantly less than for the pure glue theory,

$$T_{de} = 0.90\, T_d\ . \tag{9}$$

A polynomial Polyakov loop model [8, 9] gives a very large mass, $m^{de} \sim 3.5$ GeV. The critical temperature is very close to the pure glue theory, $T_{de} \sim 0.996\, T_d$.

The sensitivity of T_{de} and m^{de} to the models' assumptions may allow to single out the model with the most appropriate description of lattice $SU(3)$ simulations. The detailed description can be found in Ref. [5].

3. Large N limit

In the infinite N limit it is convenient to introduce a continuous variable $x = \frac{i}{N}$. Labeling the eigenvalue $q_i \to q(x)$, one gets

$$\rho(q) = \lim_{N\to\infty} \frac{1}{N} \sum_i^N \delta(q - q_i) = \int_0^1 dx \delta[q - q(x)] = \frac{dx}{dq}\ . \tag{10}$$

At finite N, the identities

$$\sum_i^N 1 = N, \ \sum_i^N q_i = 0 \tag{11}$$

become, at infinite N,

$$\int dq \rho(q) = 1, \quad \int dq \rho(q) q = 0. \tag{12}$$

The potential is proportional to N^2,

$$\begin{aligned}\tilde{V}_n(q) = N^2 V_n(q) &= N^2 \int dx\, dy\, |q(x) - q(y)|^n \left(1 - |q(x) - q(y)|\right)^n \\ &= N^2 \int dq\, dq'\, \rho(q)\, \rho(q')\, |q - q'|^n \left(1 - |q - q'|\right)^n .\end{aligned} \tag{13}$$

This representation transforms the potential into a polynomial in q.

The minimum of Eq. (13) was found in Ref. [6]:

$$\rho(q) = 1 + b\cos dq, \quad -q_0 < q < q_0, \tag{14a}$$

$$d = 2\sqrt{\frac{3d_2}{d_1}}. \tag{14b}$$

$$\cot(dq_0) = \frac{d}{3}\left(\frac{1}{2} - q_0\right) - \frac{1}{d\,(1/2 - q_0)}, \tag{14c}$$

$$b^2 = \frac{d^4}{9}\left(\frac{1}{2} - q_0\right)^4 + \frac{d^2}{3}\left(\frac{1}{2} - q_0\right)^2 + 1. \tag{14d}$$

For $T > T_d$, $d > 2\pi$ and $q_0 < \frac{1}{2}$. The eigenvalues do not span the full range between $-\frac{1}{2}$ and $\frac{1}{2}$. The density is discontinuous at the end points $\rho(\pm q_0) > 0$. For $T = T_d^+$, $q_0 = \frac{1}{2}$ and $d = 2\pi$, the density is continuous for all values of q in $[-1/2, 1/2]$. In particular it vanishes at the end points $\rho(\pm q_0) = 0$. For $T < T_d$, the theory is in confined phase, with a uniform distribution of eigenvalues, Eq. (14a) with $q_0 = \frac{1}{2}$ and $b = 0$.

The Polyakov loop jumps at the transition from $L_c = 0$ to $L_d = 0.5$. Assuming that $\delta d \sim T_d - T$, one obtains that as $T \to T_d^+$,

$$L(T) - \frac{1}{2} \sim (T_d - T)^\beta \quad , \quad \beta = 2/5\ . \tag{15}$$

Thus, near the transition $L(T)$ exhibits a power like behavior which is characteristic of a second order phase transition, although $L(T_d^+) \neq 0$. Therefore, the system exhibits a Gross-Witten-Wadia transition. That is, it exhibits aspects of both first order and second order phase transitions; thus it can be termed "critical first order" [10].

Other critical exponents besides β can be calculated, as was demonstrated in Ref. [6], they satisfy the usual Griffths scaling relation,

$$2 - \alpha = \beta(1 + \delta)\ . \tag{16}$$

The critical first order transition described above is clearly special to infinite N. At finite N, one expects a first order phase transition, and a smoothing of the critical behavior. This leads to a natural question: how large must N be to see such putative critical behavior? The present model can be solved numerically for $\infty > N \geq 4$. In Ref.[6] the behavior of the numerical solution for the specific heat, divided by $N^2 - 1$, was shown for different values of N. To observe the divergence in the specific heat, moderate values of N do not suffice. Instead, it is necessary to go to rather large values, $N \geq 40$. However, the recent results of Ref. [11] imply that to detect a manifestation of the Gross-Witten-Wadia phase transitions, it is enough to calculate the interface tension up to $N = 10$.

The present matrix model suggests that *very* near T_d, a novel transition may arise at large N.

4. Acknowledgement

I thank my co-authors, R. Pisarski, S. Lin and K. Kashiwa, with whom I have been working on the subject of this talk. This work was supported by Contract No. DE-AC02-98CH10886 with the U.S. Department of Energy.

[1] Meisinger P N, Miller T R and Ogilvie M C 2002 *Phys.Rev.* **D65** 034009 (*Preprint* hep-ph/0108009)
[2] Meisinger P N and Ogilvie M C 2002 *Phys.Rev.* **D65** 056013 (*Preprint* hep-ph/0108026)
[3] Dumitru A, Guo Y, Hidaka Y, Altes C P K and Pisarski R D 2011 *Phys.Rev.* **D83** 034022 (*Preprint* 1011.3820)
[4] Dumitru A, Guo Y, Hidaka Y, Altes C P K and Pisarski R D 2012 *Phys.Rev.* **D86** 105017 (*Preprint* 1205.0137)
[5] Kashiwa K, Pisarski R D and Skokov V V 2012 *Phys.Rev.* **D85** 114029 (*Preprint* 1205.0545)
[6] Pisarski R D and Skokov V V 2012 *Phys.Rev.* **D86** 081701 (*Preprint* 1206.1329)
[7] Fukushima K 2004 *Phys.Lett.* **B591** 277–284 (*Preprint* hep-ph/0310121)
[8] Pisarski R D 2000 *Phys.Rev.* **D62** 111501 (*Preprint* hep-ph/0006205)
[9] Scavenius O, Dumitru A and Lenaghan J 2002 *Phys.Rev.* **C66** 034903 (*Preprint* hep-ph/0201079)
[10] Dumitru A, Lenaghan J and Pisarski R D 2005 *Phys.Rev.* **D71** 074004 (*Preprint* hep-ph/0410294)
[11] Lin S, Pisarski R D and Skokov V V 2013 (*Preprint* 1301.7432)

Journal of Physics: Conference Series **446** (2013) 012033 doi:10.1088/1742-6596/446/1/012033

Initial State in RHIC and Ground-State Properties of Nuclei

Peter Filip

Institute of Physics, Slovak Academy of Sciences, Dúbravská cesta 9, Bratislava, Slovakia

E-mail: peter.filip@savba.sk

Abstract. The influence of ground-state properties of nuclei on the initial state geometry of strongly interacting matter in relativistic heavy ion collisions is discussed. Self-orientation effect in very-high-multiplicity collisions of deformed nuclei is explained. Deformation and wave function of Au, Cu nuclei are discussed in connection to the initial excentricity simulations. Suggestion for colliding selected isotopes of Sm, Zr and Ru nuclei is presented.

1. Introduction

Elliptic flow in Au+Au collisions measured by RHIC detectors at Brookhaven National Laboratory has become a very important quantity for studying properties of the hot strongly interacting matter created in relativistic interactions of nuclei. The detailed understanding of the elliptic flow phenomenon is thus of primary importance. Although the hydrodynamic picture of the partonic matter expansion in Au+Au collisions allows us to reproduce the elliptic flow data well, there still remain some issues to be understood.

For example, the elliptic flow strength measured in Cu+Cu collisions [1] at RHIC is 2x larger than predictions from our models (which reproduce the observables measured in Au+Au collisions). Is there something substantial in the initial state of dense partonic matter and its expansion in Cu+Cu collisions which we do not take properly into account?

In this contribution we discuss the influence of ground-state properties of nuclei (intrinsic deformation, and shape vibration) on the asymmetric geometry of the initial state, which is believed to be primarily responsible for the final strength of the measured elliptic flow.

2. Initial eccentricity in collisions of deformed nuclei

Motivation for studying deformation effects in relativistic collisions of heavy nuclei came partially from the suggestion [2], that a sample of central Au+Au collisions studied at RHIC/BNL contains a small admixture of events with anomalously large (single-event) elliptic flow values. Assuming deformation of ^{197}Au nucleus, such events could be understood as originating from central collisions of suitably oriented non-spherical Au nuclei. A quantitative study was done [3] with the Optical Glauber Model (OGM) and also with full MC Glauber simulation [4] to investigate effects of deformation in heavy ion collisions. A short summary of these studies is given here and then we suggest the experimental verification of our understanding of the initial state geometry and the elliptic flow origin with carefuly selected collision systems.

2.1. Collisions of Prolate nuclei

In collisions of nuclei with strong prolate ($\beta_2 \approx 0.3$) deformation an effective *self-orientation* effect can take place in the sample of very-high-multiplicity (VHM) ultra-central collisions. The behaviour comes from binary (nucleon-nucleon) interaction contributions to secondary charged particles multiplicity N_{ch}. If one assumes that secondary particle multiplicity in relativistic collisions of nuclei at RHIC can be described by two-component model of particle production [5]

$$dN_{ch}/d\eta = (1-x) \cdot n_{pp}\frac{N_{part}}{2} + x \cdot n_{pp}N_{coll} \qquad (1)$$

then central collisions of longitudinally oriented prolate nuclei have significantly larger $dN_{ch}/d\eta$ due to higher N_{coll} value if compared to other orientations of colliding nuclei (see Fig.1). Thus, in the sample of very-high-multiplicity (VHM) events, collisions of longitudinally oriented prolate nuclei will appear more frequently, and the effective self-orientation occurs. Longitudinal orientation of colliding prolate nuclei increases the achieved transverse energy density [6]. The overlap of longitudinaly oriented nuclei in the transverse plane becomes more spherical at small impact parameters, which gives smaller eccentricity values. A sudden drop of the initial eccentricity in the sample of most central high-multiplicity collisions of prolate nuclei has been predicted [4]. This should be experimentally observable as a decrease (cusp) of elliptic flow strength v_2, if measured as a function of charged particles multiplicity ($dN_{ch}/d\eta$). When the number of participants N_{part} is used for studying v_2 dependence on centrality, all orientations of colliding deformed nuclei get mixed together in the most central bin, and self-orientation effects will not be observed via v_2 behavior.

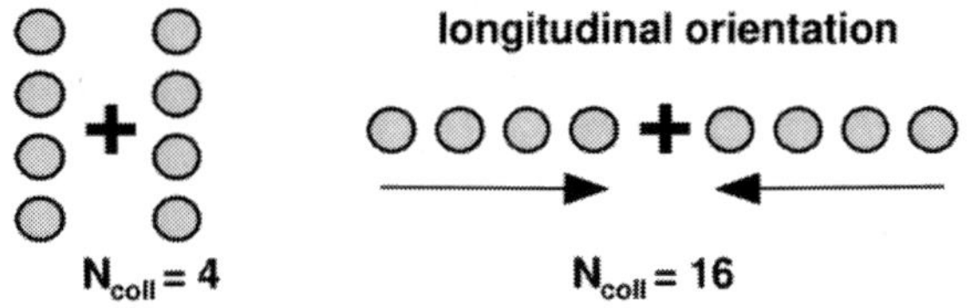

Figure 1. Number of nucleon-nucleon collisions for different orientations of two extremely prolate nuclei.

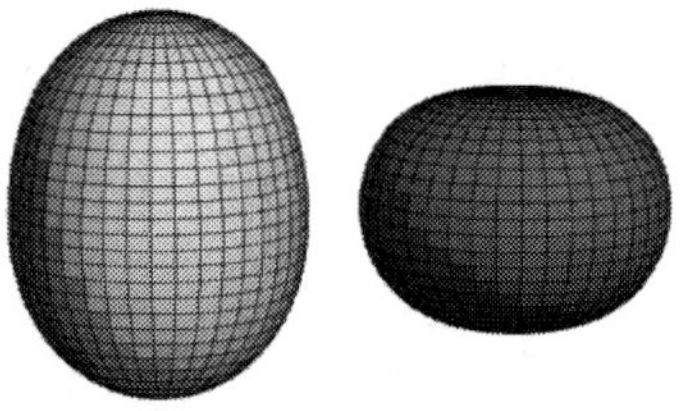

Figure 2. ^{165}Ho,^{116}Cd shapes.

We emphasize here, that the prediction of the effective self-orientation is dependent on the particle production mechanism used in Eq.(1). Additionally a Poissonian (or negative binomial) high-multiplicity fluctuation tail of more frequent collisions with lower average $< N_{ch} >$ values can significantly contaminate the sample of VHM collisions. Therefore, a sudden drop of v_2 value for VHM collisions of prolate nuclei needs to be experimentally verified. Rare-earth elements (e.g. ^{165}Ho) or some nuclei heavier than ^{208}Pb have suitably large $\beta_2 \approx 0.3$ prolate deformation.

2.2. Collisions of Oblate nuclei

For nuclei with strong oblate deformation [7] (β_2≈-0.24 for ^{116}Cd) the effective self-orientation can also happen. The highest binary collisions number N_{coll} and maximal multiplicities N_{ch} are achieved for oblate nuclei when oriented orthogonally to the beam. This configuration gives larger eccentricity than in other orientations, and therefore the increased elliptic flow is expected in VHM collisions of oblate nuclei. In optimal case (large static oblate deformation) a small rise of the elliptic flow strength v_2 can be expected [4] for the VHM tail of e.g. Cd+Cd collisions.

Journal of Physics: Conference Series **446** (2013) 012033 doi:10.1088/1742-6596/446/1/012033

3. Ground-state of Au-197 nucleus

The spectroscopic quadrupole moment of Au nucleus Q=+0.58 [8] confirms the non-sphericity of the charge distribution in ^{197}Au. Theoretical calculations predict ^{197}Au to be oblate [7] with β_2=−0.13 and recent GDR measurements confirm this (see Refs in [4]). Positive sign of Q measured via muonic spectroscopy [8] may correspond to oblate J=3/2 nucleus observed in quantum substate s_z=1/2. Indeed, known formula $Q(s_z) = Q_o[3s_z^2 - J(J+1)]/[(J+1)(2J+3)]$ gives $Q(1/2) > 0$ while $Q_o < 0$. However, an odd number of protons in a stable Au nucleus makes theoretical calculations complicated, and ground-state wave function of ^{197}Au is not well known. Assuming that stable Pt and Hg nuclei behave similarly as ^{197}Au one can speak about the probability distribution of Au nucleus to be found (observed) with positive or negative β_2 value (shape coexistence is expected to occur in the region of Pt-Au-Hg nuclei).

Typical probability distribution of quadrupole deformation β_2 calculated for ground-state wave function of ^{196}Hg nucleus [9] is reproduced in Fig.3. Although the average deformation is oblate ($\langle\beta_2\rangle$<0) a significant probability of finding such nucleus in the prolate shape configuration exists. This type of behavior probably needs to be taken into account in initial eccentricity simulations of central Au+Au collisions at RHIC. Axial vibrations might also be significant.

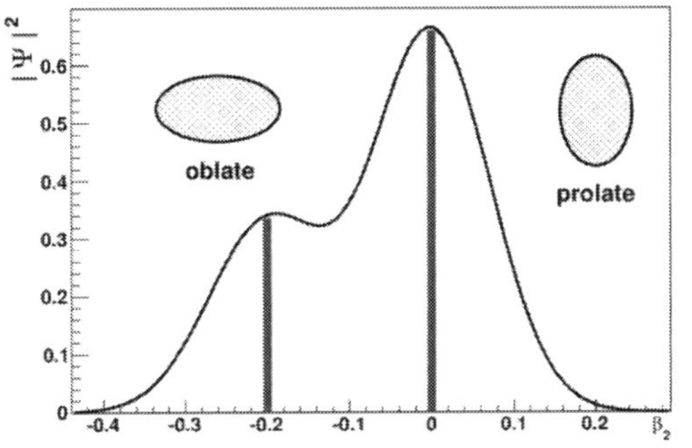

Figure 3. Probability of vibrating oblate nucleus to have deformation with given β_2.

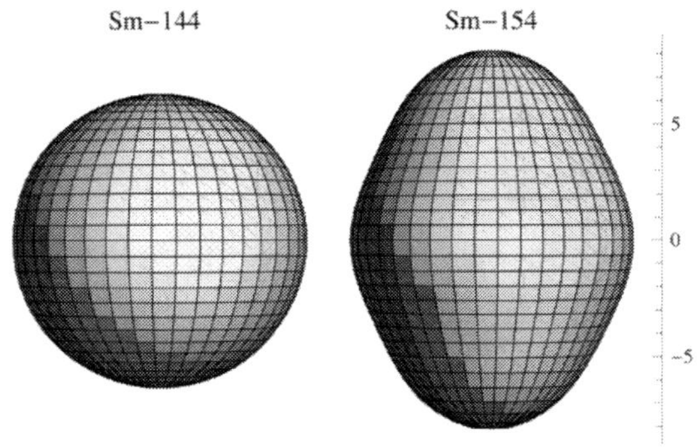

Figure 4. ^{144}Sm and ^{154}Sm nuclei.

4. Ground-state vibration

Even for spherical nuclei with zero static quadrupole deformation, one should consider a possibility of significant shape vibrations yielding average *dynamic* quadrupole moment to be important for the initial-state geometry in VHM collisions. In molecular physics, quantum vibrations are well understood and the amplitude of ground-state shape vibrations of C_{60} molecule was calculated [10]. Also, for muonic heavy hydrogen molecule the vibrational wave function needs to be known for estimating the fusion probability [11] in such system. Since the zero-point-energy (and amplitude) of molecular vibrations is more significant for lighter molecules, some lighter nuclei (e.g. ^{63}Cu) may vibrate more significantly than heavier ones.

5. Suggested collision systems

Verification of our understanding of the relation between initial eccentricity and observed elliptic flow strength (and fluctuations) can be done using collisions of carefully selected nuclei with known properties. Very interesting might be a comparison of ^{144}Sm+^{144}Sm and ^{154}Sm+^{154}Sm collisions. Samarium element allows one to study collisions of spherical and strongly deformed nuclei in a relativistic collider using the same setup of the ion source. Shape of spherical ^{144}Sm and prolate ^{154}Sm isotopes (both stable) is shown in Fig.4. Fluctuations of the initial eccentricity and elliptic flow v_2 should significantly increase at given centrality $(dN_{ch}/d\eta)$ if strongly deformed ^{154}Sm nuclei are collided instead of ^{144}Sm. The expected increase of the elliptic flow fluctuation width σ_{v_2} for collisions of strongly deformed nuclei can be estimated as $\sigma_{v_2} \approx \lambda \cdot \sqrt{\tilde{\sigma}_\varepsilon^2 + \sigma_\beta^2}$ where $\tilde{\sigma}_\varepsilon$ is the eccentricity fluctuation for spherical nuclei collisions originating from all other sources exept a deformation, and σ_β denotes the eccentricity fluctuation width due to deformation β_2 of nuclei. Factor $\lambda \approx 0.2$ comes from hydrodynamic simulations of the partonic matter expansion.

Also, cusp-like decrease of v_2 strength in VHM central collisions of ^{154}Sm nuclei and slight increase of v_2 in VHM collisions of ^{116}Cd nuclei should be observable, if our understanding of the elliptic flow origin, charged multiplicty generation and of the initial eccentricity are correct.

Another interesting possibility is to compare collisions of ^{96}Zr+^{96}Zr and ^{96}Ru+^{96}Ru nuclei (proton number differs by 10%). Dependence of various observables on the magnetic field strength (created by the charge of spectators) can be studied using these nuclei. One has to keep in mind, however, that ^{96}Ru nucleus is almost spherical ($\beta_2 = 0.05$) while ^{96}Zr is prolate ($\beta_2 = 0.22$) [7]. It is therefore important to understand and subtract deformation effects from the relevant observables studied in ^{96}Zr+^{96}Zr system (in comparison with ^{96}Ru+^{96}Ru collisions) before chiral magnetic effect [12] or other isospin-dependent phenomena are claimed to exist. Collisions of Zr+Zr and Ru+Ru nuclei have been studied by FOPI collaboration [13] at GSI.

6. Conclusions

For a detailed understanding of the initilal eccentricity and the elliptic flow in central collisions of Au and U nuclei, one needs to take into account their static deformation and ground-state vibrational wave function. The explanation of the elliptic flow strength in Cu+Cu collisions [1] at RHIC is still awaited. We suggest ground-state properties (e.g. strong shape vibrations) in ^{63}Cu nucleus might partially explain the observed discrepancy with theoretical models.

Experimental study of the relativistic collisions of carefully selected nuclei (^{96}Zr, ^{96}Ru, ^{144}Sm and ^{154}Sm) at RHIC and LHC colliders as well as on SPS at CERN and future facilities NICA, FAIR could allow us to verify and improve our understanding the elliptic flow phenomenon.

Deformation effects we have discussed here are expected to occur mainly in very central (VHM) collisions. We suggest that detailed understanding of such collisions is important. Theoretical calculations predict that extreme magnetic fields (B>10^{14}T) created in non-central collisions of nuclei may influence the critical temperature T_c of QCD phase transitions [14]. It can thus happen, that in the most central collisions (B≈0) the hot partonic matter created may have different properties (Equation of State) than observed in non-central collisions.

Acknowledgements: This work was supported by the Slovak Research and Development Agency under the contract No. APVV-0177-11 and by the grant VEGA No. 1/0171/11.

References

[1] Agakishiev and STAR collaboration 2012 *Phys. Rev.* C **85** 014901
[2] LHE group at JINR Dubna 2007, personal communication
[3] Filip P 2008 *Phys. of Atom. Nuclei* **71** 1609
[4] Filip P, Lednicky R, Masui H, Xu N 2009 *Phys. Rev.* C **80** 054903
[5] Kharzeev D, Nardi M 2001 *Phys. Lett.* B **507** 121
[6] Shuryak E V 2000 *Phys. Rev.* C **61** 034905
[7] Möller J, Nix J R, Florkowski W, Swiatecki W J 1995 *At. Data Nucl. Data Tables* **59** 185
[8] Itano W M 2006 *Phys. Rev.* A **73** 022510
[9] Köppel Th J, Faessler A 1983 *Nucl. Phys.* A **403** 263
[10] Böhm M C, Ramírez R 1995 *Jour. Phys. Chem.* **99** 12401
[11] Jackson J D 1957 *Phys. Rev.* **106** 330
[12] Fukushima K, Kharzeev D E, Warringa H J 2008 *Phys. Rev.* D **78** 074033
[13] Reisdorf W and FOPI collaboration 2007 *Nucl. Phys.* A **781** 495
[14] Mizher A J, Chernodub M N, Fraga E S 2010 *Phys. Rev.* D **82** 105016

Heavy-quark production in heavy-ion collisions

Andrea Dainese
INFN – Sezione di Padova, via Marzolo 8, 35131 Padova, Italy
E-mail: andrea.dainese@pd.infn.it

Abstract. We present an overview of the phenomenology and experimental results on open heavy-flavour and quarkonium production in heavy-ion collisions at RHIC and at the LHC.

1. Introduction: what's special about heavy quarks

Heavy-flavour hadrons, containing a charm (c) or beauty (b) quark and a light quark, and quarkonia, bound states of a $c\overline{c}$ or $b\overline{b}$ pair, are effective probes of the properties of the hot and dense strongly-interacting medium formed in high-energy heavy-ion collisions.

Heavy quarks are produced in the early stage of the collision in primary partonic scatterings with large virtuality Q and, thus, on temporal and spatial scales, $\Delta\tau \sim \Delta r \sim 1/Q$, which are sufficiently small for the production to be unaffected by the properties of the medium, in the case of nucleus–nucleus collisions. In fact, the minimum virtuality $Q_{\min} = 2\,m_{\rm Q}$ in the production of a $\rm Q\overline{Q}$ pair implies a space-time scale of $\sim 1/(2\,m_{\rm Q}) \simeq 1/3\ \rm GeV^{-1} \simeq 0.1$ fm (for charm), to be compared to the expected lifetime of a few fm for the hot and dense medium at RHIC and LHC. In contrast to light quarks and gluons that can be produced or annihilated during the entire partonic phase of the medium, heavy quarks preserve their flavour and mass identity while traversing the medium and can be tagged via the measurement of heavy-flavour hadrons in the final state of the collision.

The two historical pillars of the high scientific relevance of heavy-flavour and quarkonium production studies in heavy-ion collisions are:

- The dead cone effect in radiative energy loss [1, 2]. This effect is predicted to suppress medium-induced gluon radiation off heavy quarks at small angles, thus reducing the quenching of heavy-flavour with respect to light-flavour hadrons.
- The colour screening effect on loosely-bound quarkonium states [3]. This mechanism would determine a suppression in the yield of charmonium and bottomonium states with a sequential pattern according to the binding energy of each state.

In the next section we will describe these concepts in more detail, along with briefly summarizing some other aspects of the rich phenomenology of heavy quarks. The rest of this writeup gives a (subjective and incomplete) overview of the most relevant observations made so far by the experimental collaborations at RHIC and LHC.

2. Heavy-quark production from proton–proton to nucleus–nucleus collisions

Given the large virtualities that characterize the production of heavy quarks, the production cross section of heavy-flavour hadrons in proton–proton collisions can be calculated in the framework of collinear factorization and perturbative QCD (pQCD). The ingredients of the calculations are: the parton distribution functions (PDFs) of the proton, evolved with pQCD starting from initial conditions based on deep inelastic scattering ep data; the partonic hard cross section, computed as a perturbative series of the strong coupling constant α_s; the non-perturbative fragmentation functions, which are parametrized from e^+e^- collision data. The state-of-the-art calculations, at fixed-order next-to-leading-log (FONLL) [4], describe well the production cross sections of open heavy-flavour hadrons (D and B mesons) measured at RHIC ($\sqrt{s}$ = 0.2 and 0.5 TeV), Tevatron (1.96 TeV) and LHC (2.76 and 7 TeV). In the case of quarkonia, it is not possible to separate between the partonic hard cross section and the hadronization function, because of the colour neutralization that is required to bind the $Q\overline{Q}$ pair. This part of the process cannot be parametrized from e^+e^- data and it introduces an additional uncertainty in the calculations. The Colour Singlet + Colour Octet calculations at next-to-leading order, among others, provide a good description of the data [5].

For hard processes, in the absence of nuclear and medium effects, a nucleus–nucleus (AA) or proton–nucleus (pA) collision would behave as a superposition of independent nucleon–nucleon (NN) collisions. The heavy-flavour differential yields would then scale from pp to AA (or pA) proportionally to the number $N_{\rm coll}$ of inelastic NN collisions (binary scaling):

$$d^2N^{h_Q}_{\rm AA(pA)}/dp_{\rm T}dy = N_{\rm coll} \cdot d^2N^{h_Q}_{\rm pp}/dp_{\rm T}dy\,. \tag{1}$$

Several effects break down binary scaling. They are usually divided into two classes:

- Initial-state effects, such as nuclear shadowing, the modification of the parton distribution functions in the nucleus due to gluon recombination at small x. Nuclear shadowing is expected to be the dominant effect at RHIC and LHC energies. Other effects include $k_{\rm T}$-broadening and nuclear absorption of quarkonium states. Initial-state effects (or cold nuclear matter effects) can be studied by comparing pp and pA collisions. Due to space limitations, experimental results on this study are not discussed in this writeup.
- Final-state effects, due to the interaction of the produced partons with the medium formed in the collision. Partonic energy loss by gluon radiation, elastic collisions with the medium constituents and colour screening in the medium are the most relevant effects for the heavy-flavour sector. These effects are expected to depend on the properties of the medium (gluon density, temperature and volume) and should, therefore, provide information on such properties.

The modification of the binary scaling is quantified using the nuclear modification factor

$$R_{\rm AA}(p_{\rm T}, y) = \frac{1}{\langle N_{\rm coll}\rangle} \cdot \frac{d^2N^{h_Q}_{\rm AA}/dp_{\rm T}dy}{d^2N^{h_Q}_{\rm pp}/dp_{\rm T}dy}\,. \tag{2}$$

Due to the QCD nature of parton energy loss, quarks are predicted to lose less energy than gluons (that have a higher colour charge) and, in addition, the dead-cone effect is expected to reduce the energy loss of massive quarks with respect to light partons [1,2]. Therefore, one should observe a pattern of gradually decreasing $R_{\rm AA}$ suppression when going from the mostly gluon-originated light-flavour hadrons (e.g. pions) to D and to B mesons [6]: $R^{\pi}_{\rm AA} < R^{\rm D}_{\rm AA} < R^{\rm B}_{\rm AA}$. A comparative study of the attenuation of various probes allows one to test the consistency of the interpretation of quenching effects as due to energy loss in a deconfined medium and to investigate the properties (density) of such a medium. Heavy quarks with moderate momenta ($p_{\rm T} < 10$ GeV/c) could be slowed down by in-medium interactions to the point of being sensitive to the collective expansion of the system and consequently might also reach thermal equilibrium with the medium constituents [7]. In addition, in this momentum range, hadronization may predominantly occur inside the medium via the mechanism of coalescence with other quarks present in the system, rather than via fragmentation in the vacuum outside the medium [7,8]. The study of this effect may allow the degree of thermal equilibration of the partonic system formed in the collision to be assessed.

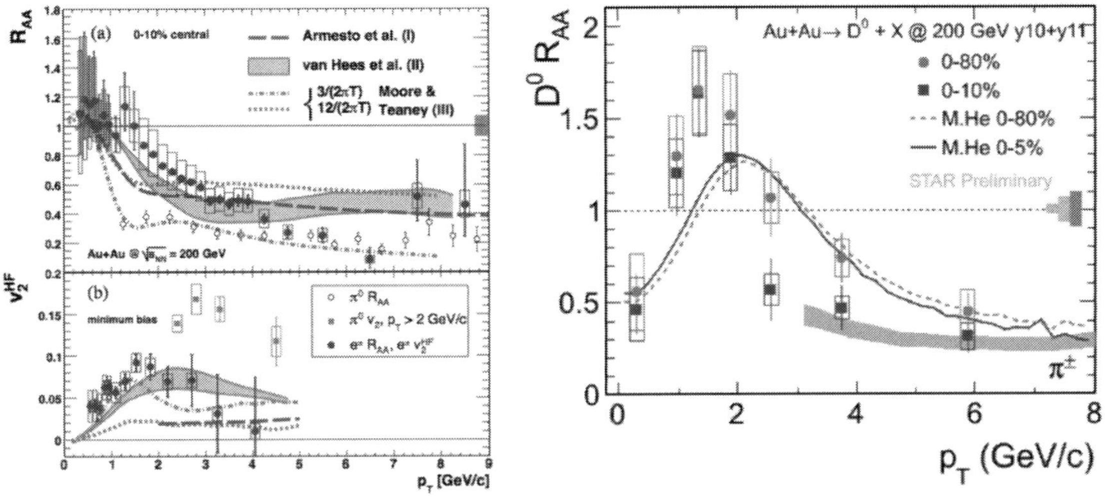

Figure 1. Open heavy-flavour production in Au–Au collisions at top RHIC energy, $\sqrt{s_{NN}}$ = 200 GeV. Left: R_{AA} (top) and v_2 (bottom) of electrons from heavy-flavour decay, measured by PHENIX and compared with the same quantities for pions and with model calculations [13]. Right: R_{AA} of D^0 mesons in central and minimum-bias collisions, measured by STAR [15] and compared with model calculations.

A significant J/ψ suppression beyond the estimated cold nuclear matter effects was first observed at the SPS [9]. At the higher centre-of-mass energies of RHIC and LHC, the most interesting items in the quarkonia sector are: (a) understanding the interplay between colour-screening-induced suppression and statistical regeneration for charmonium production in a medium containing a large number of $c\bar{c}$ pairs [10]; (b) measuring medium effects on the bottomonium states. On this point, the predicted suppression pattern as a function of the medium temperature is particularly interesting: the Υ would melt at $T \approx 2\,T_c$ (where T_c is the critical temperature of the QGP phase transition), a temperature that would be reached only at the LHC, while the Υ' would melt at the same temperature as the J/ψ, $T \approx 1.2\,T_c$ (see e.g. [11, 12]). It is thus important to measure also the Υ', because, at variance with the J/ψ, it is expected to have a small regeneration probability and it would be very useful to disentangle J/ψ suppression and regeneration. Measuring the in-medium dissociation probability of the different quarkonium states at RHIC and LHC would provide an estimate of the initial medium temperature. Hence, the idea of "quarkonium thermometer".

3. Open heavy-flavour production in nucleus–nucleus collisions at RHIC and LHC

In central Au–Au collisions at centre-of-mass energy $\sqrt{s_{NN}}$ = 200 GeV the PHENIX and STAR Collaborations measured a suppression of a factor up to 3–5 ($R_{AA} \approx 0.2$–0.3) at $p_T \simeq 5$ GeV/c for electrons from heavy-flavour hadron decays [13, 14]. This suppression is compatible with that of pions of similar transverse momentum. Figure 1 (left) shows in the upper panel the R_{AA} measured by PHENIX. The STAR Collaboration has recently presented also preliminary results for reconstructed $D^0 \to K^-\pi^+$ decays [15]. The R_{AA} is shown in the right-hand panel of Fig. 1 for the central and minimum-bias collisions. For $p_T > 3$ GeV/c the suppression of D mesons is consisted with that of pions. At $p_T \approx 1$–2 GeV/c, R_{AA} exhibits an enhancement above unity. A model calculation implementing charm quark transport [16] in the medium captures the main features of the data.

At the LHC, the much larger production cross sections (of about a factor 10 for charm and a factor 30 beauty) and the presence of dedicated silicon vertex detectors in the ALICE, ATLAS

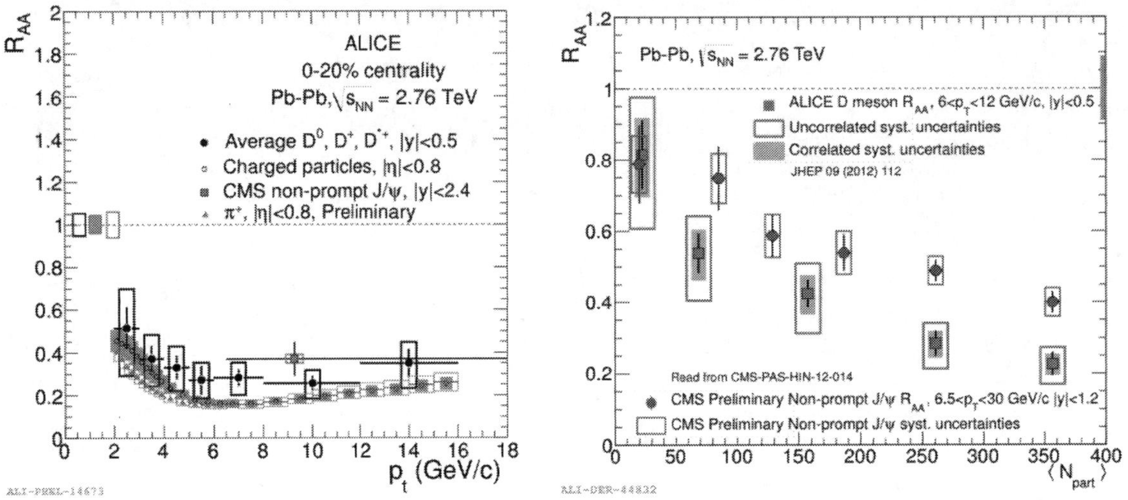

Figure 2. Charm and beauty production in Pb–Pb collisions at the LHC. Left: ALICE measurements of average R_{AA} of D mesons in the 0–20% centrality class [17], compared to that of charged particles [18], π^+, and non-prompt J/ψ from B decays, measured by CMS [19] in the same centrality class. Right: Average R_{AA} of D mesons as a function of centrality (ALICE [17]), compared to that of non-prompt J/ψ from B decays (CMS [20]).

and CMS experiments, allowed for heavy-flavour production measurements already in the first Pb–Pb run in 2010. These measurements are now being extended and improved with data from the 2011 run. The nuclear modification factor of prompt D mesons (average of D^0, D^+ and D^{*+}) in Pb–Pb collisions, measured by the ALICE Collaboration, is shown in Fig. 2: as a function of p_T in central collisions (left) and as a function of centrality for $6 < p_T < 12$ GeV/c (right) [17]. A strong suppression is observed in central collisions, reaching a factor about 4 for $p_T > 5$ GeV/c. R_{AA} is compatible with that of charged pions and charged particles [18], although there seems to be a tendency for $R_{AA}^{D} > R_{AA}^{\pi}$ at low p_T. In the figure, the nuclear modification of non-prompt J/ψ from B decays, measured by the CMS Collaboration [19, 20], is also shown. The comparison with D mesons indicates a different suppression for charm and beauty hadrons in central collisions, consistent with the expectation $R_{AA}^{D} < R_{AA}^{B}$. However, the different p_T and rapidity ranges still prevent from a clear conclusion on this point.

A large suppression is also observed for heavy-flavour decay muons, at forward rapidity by ALICE [21] and at central rapidity by ATLAS [22], and electrons at central rapidity by ALICE [23,24]. Both lepton species exhibit a suppression of a factor of about 2–3 up to 15 GeV/c in central collisions, as shown in Fig. 3.

The results on heavy-flavour suppression at the LHC are described reasonably well by model calculations that implement various combinations of in-medium interaction mechanisms for heavy quarks, including radiative and collisional processes, as well as in-medium formation and dissociation of heavy-flavour hadrons [6, 16, 25–31].

The possible in-medium thermalization of heavy quarks can be studied by measuring the azimuthal anisotropy of heavy-flavour hadrons and of their decay leptons. The anisotropy is quantified by the second coefficient of the Fourier expansion of the azimuthal distribution, the so-called elliptic flow coefficient v_2. Heavy-flavour anisotropy is observed by PHENIX in Au–Au collisions at RHIC with v_2 values of up to ≈ 0.13 for electrons from heavy-flavour decays [13] (see Fig. 1, bottom-left, and Fig. 4, left). A comparable anisotropy is observed by ALICE at

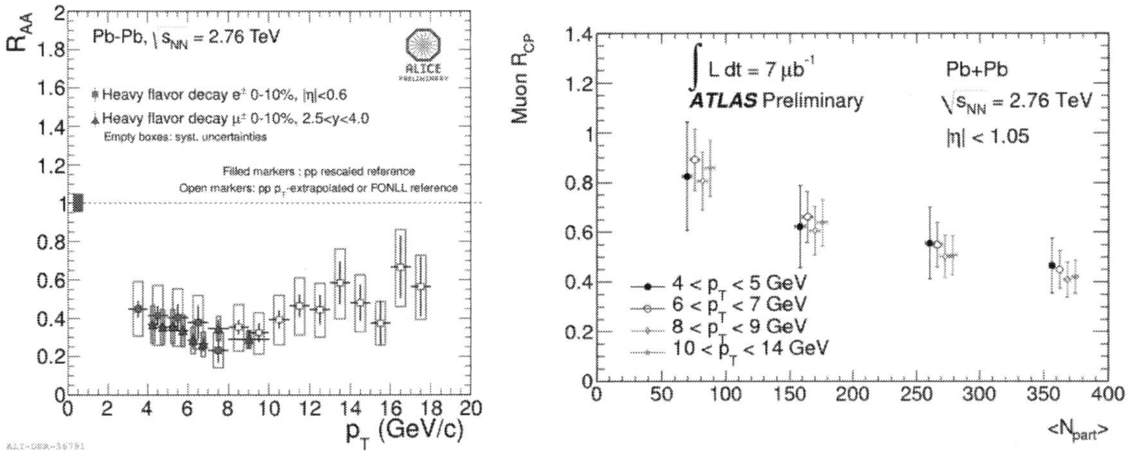

Figure 3. Suppression of heavy-flavour decay leptons at the LHC. Left: ALICE measurements of R_{AA} of heavy-flavour decay muons at forward rapidity [21] and electrons at central rapidity [23] in central (0–10%) Pb–Pb collisions, as a function of p_T. Right: ATLAS measurement of R_{CP} of heavy-flavour decay muons at central rapidity, as a function collision centrality [22].

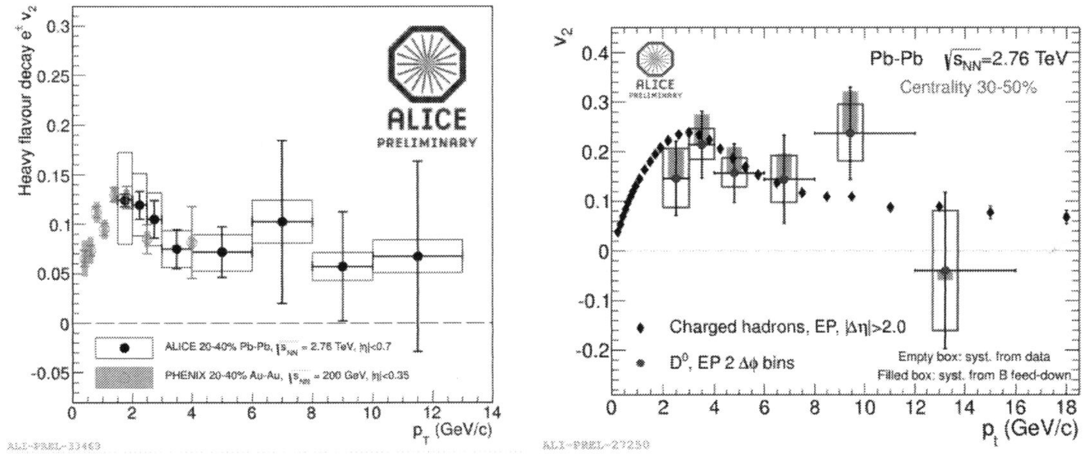

Figure 4. Elliptic flow of open heavy-flavour at RHIC and LHC. Left: v_2 of heavy-flavour decay electrons, measured by ALICE [23] and compared with the same measurement by PHENIX at RHIC [13]. Right: D^0 meson v_2 as a function of p_T, measured by ALICE and compared with charged hadrons [23].

the LHC [23, 24], though still with larger experimental uncertainties, as shown in Fig. 4 (left). The ALICE experiment has also carried out a direct measurement of v_2 for D mesons (D^0, D^+ and D^{*+}). The preliminary results, reported in Fig. 4 (right), indicate $v_2 > 0$ for D^0 mesons in the interval $2 < p_T < 6$ GeV/c [23, 32]. The observed anisotropy is comparable to that of charged particles (mostly light-flavour hadrons), although the uncertainties still prevent a precise comparison.

The v_2 measurements at RHIC and LHC are described by several models [16, 28–30] that include heavy-quark transport in the medium, with various implementations of the quark-

Journal of Physics: Conference Series **446** (2013) 012034 doi:10.1088/1742-6596/446/1/012034

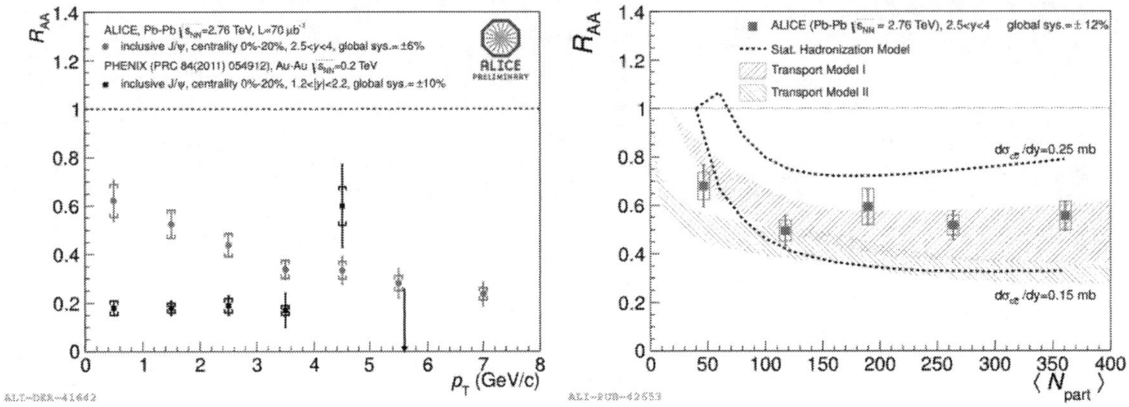

Figure 5. J/ψ suppression at RHIC and LHC. Left: J/ψ R_{AA} as a function of p_{T}, measured by the PHENIX experiment at RHIC [38] and by the ALICE [35] experiment at LHC. Right: J/ψ R_{AA} for $p_{\mathrm{T}} > 0$ at forward rapidity, as a function of centrality, measured by ALICE [36] and compared with theoretical models including a component of regeneration [39–41].

medium interaction.

The upgrades of the silicon vertex detectors at RHIC and LHC, along with the envisaged increase in heavy-ion luminosities at both machines, are expected to open a new era in terms of both high-precision measurements of the observables presented here and qualitatively new measurements, such as the production of heavy-flavour baryons and the flow of B mesons (see e.g. [33]).

4. Quarkonium production in nucleus–nucleus collisions at RHIC and LHC

The measurements of J/ψ production in Au–Au and d–Au collisions at RHIC have strengthened the conclusions, indicated by the SPS data, on the observation of a medium-induced suppression beyond the expectation from cold nuclear matter effects. However, they leave many open questions: why is the suppression the same at the SPS and at RHIC? why is the suppression at RHIC larger at forward rapidity, where the medium density should be smaller? See e.g. [34] for a review. Model calculations including cold nuclear matter effects, colour screening and a small component of regeneration can describe the data, but it is difficult to disentangle the relative weight of these effects, from data alone.

A clear indication of the relevance of the various effects is provided by the measurements at the LHC, where the ALICE experiment observed a J/ψ suppression much smaller than at RHIC in the low p_{T} region, as shown in Fig. 5 (left) [35, 37]. Since the medium temperature, thus its screening power, is expected to be larger at the LHC, this observation is an indication of the presumably important contribution of charmonium regeneration from initially-uncorrelated c and $\bar{\mathrm{c}}$ quarks that are freely roaming in the deconfined medium. Indeed, models that include this contribution can describe the data [39–41], see e.g. the comparison with R_{AA} at forward rapidity and low p_{T} in Fig. 5 (right) [36]. A further hint on the relevance of regeneration is provided by the possible observation of positive v_2 for J/ψ at the LHC. This would be consistent with the $v_2 > 0$ that is measured for D mesons. The ALICE measurement [35, 37] is compared in Fig. 6 with models and with the measurement by the STAR experiment at RHIC [42], which shows a v_2 consistent with zero.

Another important break-through at the LHC, carried out by the CMS experiment, is the first measurement of the suppression for all three states of the Υ family. Figure 7 (left) shows the dimuon invariant mass distribution in Pb–Pb collisions compared with the expected distribution obtained from the a binary (N_{coll}) scaling of the pp data [44]. This results gives a first indication of the expected sequential suppression pattern for charmonium and bottomonium

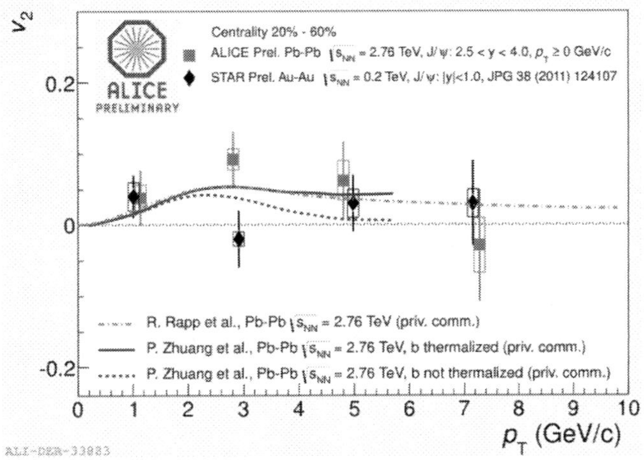

Figure 6. J/ψ elliptic flow coefficient v_2, measured by the STAR experiment [42] at RHIC and by the ALICE experiment [35] at LHC and compared with model calculations.

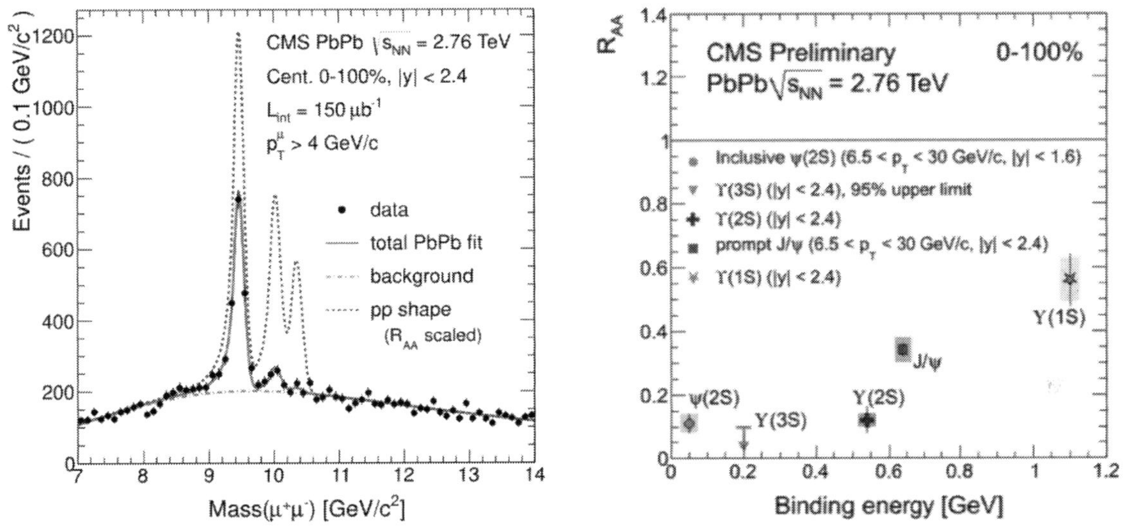

Figure 7. Left: CMS measurement of the dimuon invariant mass distribution in the Υ region in Pb–Pb collisions compared with the expected distribution obtained from the a binary ($N_{\rm coll}$) scaling of the pp data [44]. Right: CMS measurements of the suppression of various quarkonium states as a function of their estimated binding energy [20, 43, 44].

states according to their binding energy. Figure 7 (right) shows $R_{\rm AA}$ in minimum-bias Pb–Pb collisions for the five states J/ψ, ψ', Υ, Υ' and Υ'' [20, 43, 44]. Υ suppression is also observed also at RHIC by the STAR experiment [45]. This "quarkonium thermometer" of the medium will be refined with future data, using the same kinematic ranges for all particles, applying feed-down corrections and disentangling cold nuclear matter effects using p–Pb measurements. At the same time, the upgraded PHENIX and STAR detectors should allow for the same measurements to be carried out also at RHIC energy.

5. Summary

We have discussed how heavy quarks, abundantly produced at RHIC and LHC energies, allow us to address several physics issues in heavy-ion collisions. In particular, they provide tools to probe the density, via parton energy loss and its predicted mass dependence, and temperature, via quarkonia successive dissociation patterns, of the high-density QCD medium formed in ultra-relativistic heavy-ion collisions.

From the experimental point of view, we have just entered the "golden age" of heavy-flavour observables, thanks to the LHC detectors and to the upgrades at RHIC. Heavy quarks promise to be a hot topic for many Hot Quarks workshops to come.

Acknowledgement. The author warmly thanks the Hot Quarks 2012 Organizers for their kind invitation to such a lively and interesting workshop.

References

[1] Dokshitzer Y L, Kharzeev D E 2001 *Phys. Lett.* **B519** 199-206
[2] Armesto N, Salgado C A, Wiedemann U A 2004 *Phys. Rev.* **D69** 114003
[3] Matsui T and Satz H 1986 *Phys. Lett.* **B178** 416
[4] Cacciari M *et al.* 2012 *JHEP* **1210** 137
[5] Butenschoen M and Kniehl B A 2012 *Nucl. Phys. Proc. Suppl.* **222-224** 151
[6] Armesto N, Dainese A, Salgado C A, Wiedemann U A 2005 *Phys. Rev.* **D71** 054027
[7] Greco V, Ko C M and Rapp R 2004 *Phys. Lett.* **B595** 202
[8] Andronic A, Braun-Munzinger P, Redlich K and Stachel J 2003 *Phys. Lett.* **B571** 36
[9] Alessandro B *et al.* [NA50 Collaboration] 2005 *Eur. Phys. J.* **C39** 335
[10] Braun-Munzinger P and Stachel J 2000 *Phys. Lett.* **B490** 196
[11] Digal S, Petreczky P and Satz H 2001 *Phys. Rev.* **D64** 094015
[12] Mocsy A and Petreczky P 2007 *Phys. Rev. Lett.* **99** 211602
[13] Adler S S *et al.* [PHENIX Collaboration] 2006 *Phys. Rev. Lett.* **96** 032301
Adare A *et al.* [PHENIX Collaboration] 2011 *Phys. Rev.* **C84** 044905
[14] Abelev B I *et al.* [STAR Collaboration] 2007 *Phys. Rev. Lett.* **98** 192301
[15] Tlusty D [STAR Collaboration] 2012 arXiv:1211.5995 [hep-ex]
[16] He M, Fries R J and Rapp R 2012 *Phys. Rev.* **C86** 014903
[17] Abelev B *et al.* [ALICE Collaboration] 2012 *JHEP* **1209** 112
[18] Abelev B *et al.* [ALICE Collaboration] 2012 arXiv:1208.2711 [hep-ex]
[19] CMS Collaboration 2012 *JHEP* **1205** 063
[20] CMS Collaboration 2012 CMS-PAS-HIN-12-014
[21] Abelev B *et al.* [ALICE Collaboration] 2012 *Phys. Rev. Lett.* **109** 112301
[22] ATLAS Collaboration 2012 ATLAS-CONF-2012-050
[23] Conesa del Valle Z [ALICE Collaboration], to appear in the proceedings of the Quark Matter 2012 conference
[24] Fasel M [ALICE Collaboration], these proceedings
[25] Sharma R, Vitev I and Zhang B-W 2009 *Phys. Rev.* **C80** 054902
[26] He Y, Vitev I and Zhang B-W 2012 *Phys. Lett.* **B713** 224
[27] Horowitz W A and Gyulassy M 2011 *J. Phys.* **G38** 124114
[28] Alberico W M *et al.* 2011 *Eur. Phys. J.* **C71** 1666; 2011 *J. Phys.* **G38** 124144
[29] Gossiaux P B, Aichelin J, Gousset T and Guiho V 2010 *J. Phys.* **G37** 094019
[30] Fochler O, Uphoff J, Xu Z and Greiner C 2011 *J. Phys.* **G38** 124152
[31] Buzzatti A and Gyulassy M 2012 *Phys. Rev. Lett.* **108** 022301
[32] Luparello G [ALICE Collaboration], these proceedings
[33] Stiller J H [ALICE Collaboration], these proceedings
[34] Granier de Cassagnac R 2008 *J. Phys.* **G35** 104023
[35] Scomparin E [ALICE Collaboration] 2012 arXiv:1211.1623 [nucl-ex]
[36] Abelev A *et al.* [ALICE Collaboration], 2012 *Phys. Rev. Lett.* **109** 072301
[37] Lardeux A [ALICE Collaboration], these proceedings
[38] Adare A *et al.* [PHENIX Collaboration] 2011 *Phys.Rev.* **C84** 054912
[39] Andronic A, Braun-Munzinger P, Redlich K and Stachel J 2011 *J. Phys.* **G38** 124081
[40] Zhao X and Rapp R 2011 *Nucl. Phys.* **A859** 114
[41] Liu Y-P, Qu Z, Xu N, and Zhuang P-F 2009 *Phys. Lett.* **B678** 72
[42] Adamczyk L *et al.* [STAR Collaboration] 2012 arXiv:1212.3304 [nucl-ex]
[43] CMS Collaboration 2012 CMS-PAS-HIN-12-007
[44] Chatrchyan S *et al.* [CMS Collaboration] 2012 *Phys. Rev. Lett.* **109** 222301
[45] Kesich A [STAR Collaboration] 2012 *J. Phys. Conf. Ser.* **389** 012025

Journal of Physics: Conference Series **446** (2013) 012035 doi:10.1088/1742-6596/446/1/012035

Heavy quark energy loss and angular de-correlation in a quark-gluon plasma matter

Shanshan Cao, Guang-You Qin, Steffen A. Bass and Berndt Müller
Department of Physics, Duke University, Durham, North Carolina 27708, USA
E-mail: shanshan.cao@duke.edu

Abstract. We study heavy quark energy loss in a hot and dense nuclear matter in the framework of Langevin equation coupled to a (2+1)-dimensional hydrodynamic model. The classical Langevin framework is modified such that both quasi-elastic scattering and gluon radiation are incorporated. We provide good description of D meson suppression measured by the ALICE collaboration. We further investigate the angular correlation function of $c\bar{c}$ pairs, and find that it can be potentially employed to distinguish different energy loss mechanisms of heavy quarks inside QGP.

1. Introduction

It is now generally accepted that a deconfined state of QCD matter, the *strongly interacting quark-gluon plasma* (sQGP), is created during relativistic heavy-ion collisions at RHIC and LHC. Such hot and dense matter displays properties similar to a nearly perfect fluid and has been successfully described by hydrodynamic models. Among various probes of QGP properties, heavy quarks are of great interest because their high-p_T suppression and elliptic flow are revealed as significant as the light flavors in spite of their large mass [1, 2, 3]. Therefore, it is crucial to explore how heavy quarks interact with and lose energy inside such QGP matter.

Between the two energy loss mechanisms of heavy quark – gluon radiation and quasi-elastic scattering with background particles, the latter is usually considered to be dominant at low energies because the radiation phase space is restricted by the "dead-cone effect" [4]. In the limit of multiple scatterings where the momentum transfer during each interaction is small, such scatterings inside a thermalized medium can be described by the Langevin equation [5, 6]. This framework has provided good descriptions of heavy flavor suppression and elliptic flow measured at RHIC. However, when extending to the LHC energies, our previous study [7] indicated that the radiative energy loss should no longer be ignored since heavy quarks also become ultra-relativistic. We modify the classical Langevin approach such that gluon radiation is also incorporated. The momentum distribution of the radiated gluons is simulated using the Higher-Twist calculation [8, 9]. Within this improved approach, we find our calculation of D meson suppression consistent with the LHC data [2].

Another interesting quantity for exploring the energy loss of hard probes is the two particle correlations [10, 11]. In this work, we study the angular correlation function of $c\bar{c}$ pairs after they traverse the medium and find that it is sensitive to different energy loss mechanisms. If future measurements are able to provide related observations, we may acquire better understanding about how heavy quarks lose energy inside hot QGP.

2. Methodology

The heavy quark motion inside QGP is governed by the following modified Langevin equation:

$$\frac{d\vec{p}}{dt} = -\eta_D(p)\vec{p} + \vec{\xi} + \vec{f}_g. \tag{1}$$

The first two terms on the right are the drag force and the thermal random force from the classical Langevin equation for Brownian motion, and the third term $\vec{f}_g = -d\vec{p}_g/dt$ is introduced for the force exerted on the heavy quark due to gluon radiation. The probability of gluon radiation during each time interval Δt is determined by the average number of gluons:

$$P_{\rm rad}(t, \Delta t) = \langle N_g(t, \Delta t)\rangle = \Delta t \int dx dk_\perp^2 \frac{dN_g}{dx dk_\perp^2 dt}, \tag{2}$$

where $k_\perp$ is the transverse momentum of the radiated gluon, and x is the ratio between the gluon energy and the heavy quark energy. If a gluon is emitted, its energy and momentum is sampled according to the distribution from the Higher-Twist calculation [8, 9]:

$$\frac{dN_g}{dx dk_\perp^2 dt} = \frac{2\alpha_s(k_\perp)}{\pi} P(x) \frac{\hat{q}}{k_\perp^4} \sin^2\left(\frac{t - t_i}{2\tau_f}\right)\left(\frac{k_\perp^2}{k_\perp^2 + x^2 M^2}\right)^4. \tag{3}$$

With the assumption that the interaction during each scattering is small, the fluctuation-dissipation relation between the drag and the thermal force still holds – $\eta_D(p) = \kappa/(2TE)$, where κ is the momentum space diffusion coefficient defined in $\langle \xi^i(t)\xi^j(t')\rangle = \kappa\delta^{ij}\delta(t-t')$. Note that such Einstein relation between gluon radiation and absorption has not been included yet due to the lack of the latter process. However, we set a lower limit of the gluon energy at $\omega_0 = \pi T$ – the balance point between gluon emission and absorption, above which the classical Langevin equation is modified by gluon radiation. Below ω_0, the heavy quark motion is dominated by quasi-elastic scatterings whose detail balance is well defined. These assumptions guarantee the equilibrium of heavy quarks after a sufficiently long time of evolution. Different transport coefficients are related via $D = 2T^2/\kappa$ and $\hat{q} = 2\kappa C_A/C_F$, where D is the spatial diffusion coefficient of heavy quark and $\hat{q}$ is the gluon transport coefficient.

We use this modified Langevin equation to simulate the heavy quark evolution. The QGP is generated with a (2+1)D viscous hydrodynamic model, which was developed by Song [12, 13] and has recently been modified by Qiu and Shen for increased numerical stability [14]. We here employ the code version and parameter tunings for Pb+Pb collisions at LHC energies that were previously used in Ref. [14]. The MC-Glauber initialization is adopted for the hydrodynamic calculation if not otherwise emphasized. Our heavy quarks are initialized with the MC-Glauber model for the position space and a leading-order pQCD calculation for the momentum space, and are fragmented to heavy mesons via Pythia 6.4 after they traverse the medium.

3. Results

In Fig.1, we compare the D meson suppression between different energy loss mechanisms and the recent LHC data [2]. The heavy quark transport coefficient is set at $D = 6/(2\pi T)$. One observes that collisional energy loss dominates the low p_T region while gluon radiation dominates the high p_T region. The combination of the two mechanisms provides a good description of the D meson R_{AA}. The slight deviation at low p_T (below 2 GeV/c) may originate from lacking the corrections of cold nuclear matter effect and the recombination mechanism of heavy flavor hadronization in our current calculation, which will be incorporated in the future effort.

Figure 2 compares v_2 with experimental observation. Our calculation seems to underestimate the D meson v_2, but various uncertainties still exist. For instance, as shown by Fig.3, if the

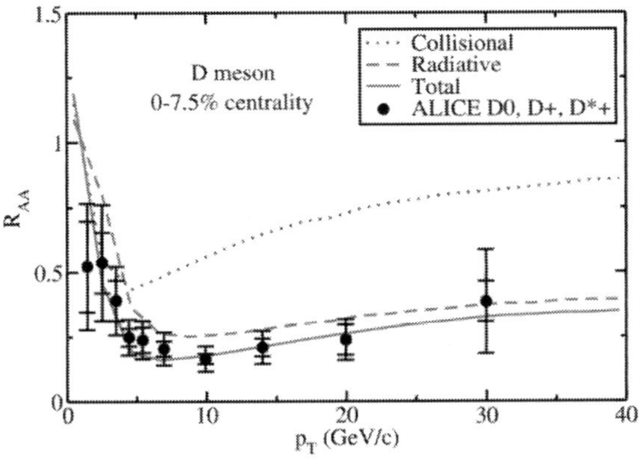

Figure 1. (Color online) Comparison of D meson R_{AA} between different heavy quark energy loss mechanisms and ALICE data.

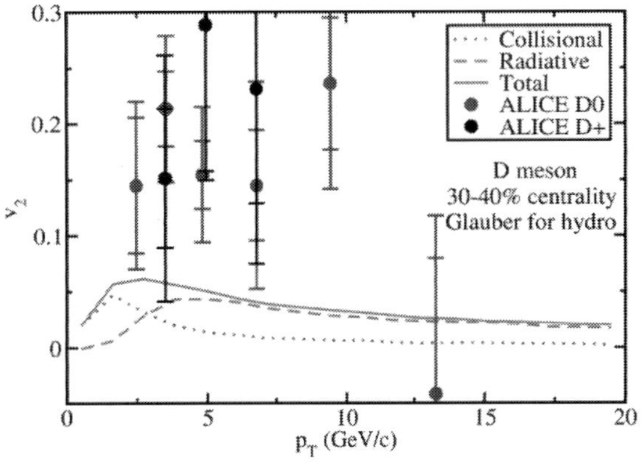

Figure 2. (Color online) Comparison of D meson v_2 between different heavy quark energy loss mechanisms and ALICE data.

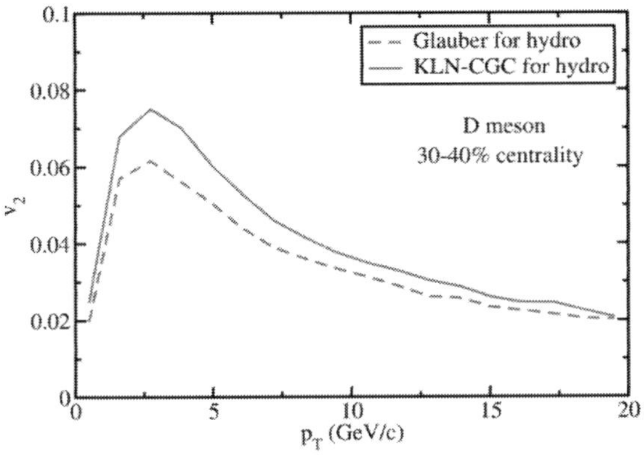

Figure 3. (Color online) Influence of the initial conditions of the hydrodynamic model on D meson v_2.

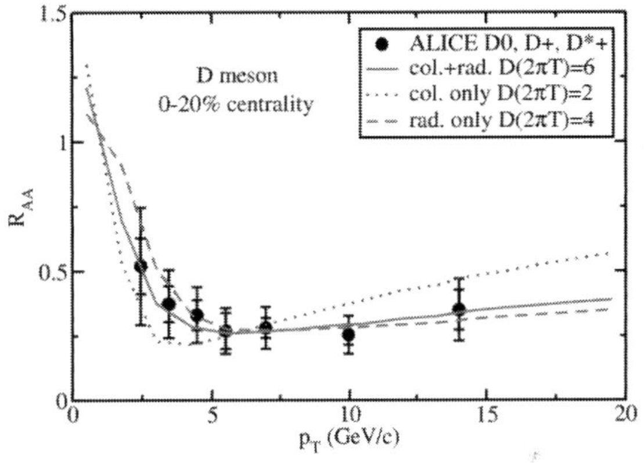

Figure 4. (Color online) Fitting D meson R_{AA} with different energy loss mechanisms by tuning the transport coefficient.

initial condition for the hydrodynamic calculation is altered from MC-Glauber to KLN-CGC, we obtain larger eccentricity of the QGP profile and therefore a larger D meson v_2 by around 30%, while the overall suppression is not significantly influenced.

Figures 4 and 5 demonstrate how the $c\bar{c}$ de-correlation behavior may help distinguish between different energy loss mechanisms. For simplicity, in this initial study, we assume back-to-back initial production of $c\bar{c}$ pairs [1]. By tuning the values of transport coefficient for each energy loss mechanism alone, we are able to obtain a reasonable fit to the data (Fig.4), however, different mechanisms lead to significantly different final state $c\bar{c}$ angular correlation (Fig.5(a)). With only collisional energy loss, the correlation function is almost flat between 0 and π, indicating the thermalization of charm quarks after they traverse QGP. On the contrary, the angular correlation distribution still peaks around π for the case of only gluon radiation, implying $c\bar{c}$ pairs strongly correlated after traversing the medium. This can be understood from the small angle dominance for the gluon radiation. Though such correlation functions are model dependent, they may

[1] Higher order contributions to heavy quark production include gluon splitting into a $c\bar{c}$ pair, contributing to small angle correlation. We shall study these in a forthcoming work.

provide deeper insight of the energy loss mechanism if a comparison can be made between theory and experiment. Last but not least, the thermalization behavior strongly depends on the momentum scale [15]. Figure 5(b) shows the result when a 2 GeV/c low p_T cut is applied.

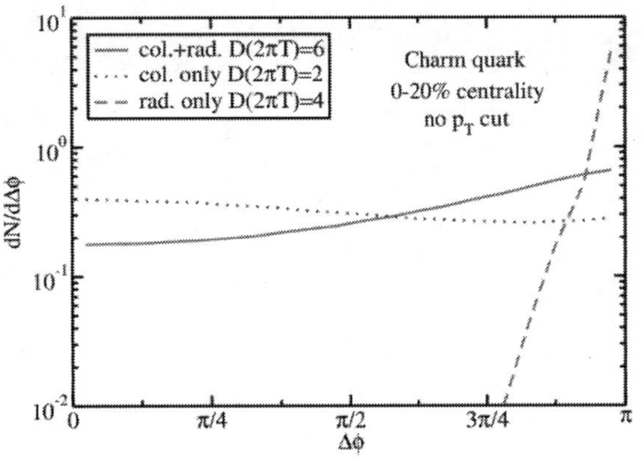

(a)

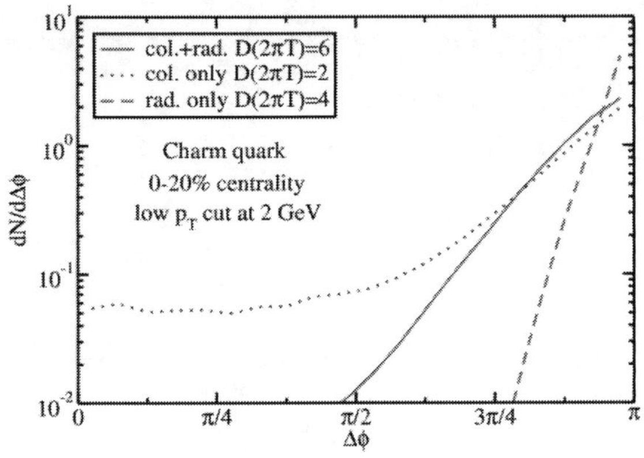

(b)

Figure 5. (Color online) Comparison of final state $c\bar{c}$ pair angular correlation functions between different energy loss mechanisms, (a) without momentum cut and (b) with 2 GeV/c low p_T cut.

4. Summary

We have studied heavy quark evolution inside a QGP medium in the framework of a Langevin equation. The classical Langevin approach has been modified such that both collisional and radiative energy loss are incorporated. Our calculation reveals a significant effect of the medium-induced gluon radiation on the heavy quark energy loss at LHC energies, and provides a good description of the D meson suppression measured by the ALICE collaboration. We have briefly discussed the $c\bar{c}$ pair angular correlation function, which may serve as an additional measurable quantity to help distinguish different energy loss mechanisms.

Acknowledgments

We thank the Ohio State University group (Z. Qiu, C. Shen, H. Song and U. Heinz) for providing the corresponding initialization and hydrodynamic evolution codes. This work was supported by the U.S. Department of Energy Grant No. DE-FG02-05ER41367.

References

[1] Adare A *et al.* (PHENIX Collaboration) 2011 *Phys. Rev.* **C84** 044905 (*Preprint* 1005.1627)
[2] Grelli A (ALICE Collaboration) 2012 (*Preprint* 1210.7332)
[3] Caffarri D (ALICE Collaboration) 2012 (*Preprint* 1212.0786)
[4] Dokshitzer Y L and Kharzeev D E 2001 *Phys. Lett.* **B519** 199–206 (*Preprint* hep-ph/0106202)
[5] Moore G D and Teaney D 2005 *Phys. Rev.* **C71** 064904 (*Preprint* hep-ph/0412346)
[6] He M, Fries R J and Rapp R 2012 *Phys. Rev.* **C86** 014903 (*Preprint* 1106.6006)
[7] Cao S, Qin G Y, Bass S A and Muller B 2012 (*Preprint* 1209.5410)
[8] Guo X F and Wang X N 2000 *Phys. Rev. Lett.* **85** 3591–3594 (*Preprint* hep-ph/0005044)
[9] Zhang B W, Wang E and Wang X N 2004 *Phys. Rev. Lett.* **93** 072301 (*Preprint* nucl-th/0309040)
[10] Qin G Y, Ruppert J, Gale C, Jeon S and Moore G D 2009 *Phys. Rev.* **C80** 054909 (*Preprint* 0906.3280)
[11] Zhang H, Owens J, Wang E and Wang X N 2009 *Phys.Rev.Lett.* **103** 032302 (*Preprint* 0902.4000)
[12] Song H and Heinz U W 2008 *Phys. Lett.* **B658** 279–283 (*Preprint* 0709.0742)
[13] Song H and Heinz U W 2008 *Phys. Rev.* **C77** 064901 (*Preprint* 0712.3715)
[14] Qiu Z, Shen C and Heinz U 2012 *Phys. Lett.* **B707** 151–155 (*Preprint* 1110.3033)
[15] Cao S and Bass S A 2011 *Phys. Rev.* **C84** 064902 (*Preprint* 1108.5101)

Journal of Physics: Conference Series **446** (2013) 012036 doi:10.1088/1742-6596/446/1/012036

Λ_c analysis with the ALICE detector at the LHC

Rosa Romita for the ALICE collaboration
University of Liverpool and Daresbury Laboratory, UK
E-mail: r.romita@liv.ac.uk

Abstract. ALICE is the LHC experiment dedicated to the study of the Quark Gluon Plasma (QGP) in high-energy nuclear collisions. Due to their large mass and their production at the early collision stage, heavy quarks (charm and beauty) are well-suited probes to study the properties of the QGP. The open charm baryon Λ_c is reconstructed in ALICE in the decay channels $\Lambda_c{}^+ \rightarrow \pi^+ + K^- + p$ and $\Lambda_c{}^+ \rightarrow K^0_S + p$, exploiting both topological selections and particle identification. We report about the analysis strategy to observe the signal in pp collisions for both decay channels and then show the invariant mass plots where the Λ_c peak is clearly visible.

1. Introduction

Heavy quarks are sensitive probes of the QGP formed in heavy-ion collisions, since they are produced in the initial hard scattering processes and experience the full collision evolution. In particular, measurements of open charm and beauty production in Pb–Pb collisions, using the pp measurements as a reference, probe the energy loss experienced by partons traversing the hot and dense medium formed in high-energy nuclear collisions.

Measuring Λ_c production in Pb–Pb collisions would provide an answer to the question whether the baryon-over-meson enhancement at intermediate momentum, observed in the light-flavor sector at RHIC and LHC, also holds in the heavy-quark sector, suggesting charm hadron production via coalescence. The measurement of the Λ_c production cross section in pp collisions, not only provides the necessary baseline to understand the heavy ions results, but has also an interest per se, since it is needed to measure the total charm cross section. The golden decay channels are: $\Lambda_c^+ \rightarrow pK^-\pi^+$, with a branching ratio of BR=$5.0 \pm 1.3\%$, and $\Lambda_c{}^+ \rightarrow K^0_S + p$, with an overall branching ratio of approximately $0.8 \pm 0.2\%$. The Λ_c cross section has been measured previously in γ-p , e^+e^- and e-p collisions [1]. The fragmentation fraction $c \rightarrow \Lambda_c$ measured by different experiments in the different collision systems has large uncertainties and varies from 8 to 14%. In addition, the open charm baryon cross section has never been measured in hadronic collisions. LHCb has also measured Λ_c, but only the ones coming from Λ_b decays [2].

ALICE (A Large Ion Collider Experiment) [3] is the LHC experiment dedicated to the study of the properties of the QGP formed in heavy-ion collisions. Its subdetectors provide excellent tracking performance, secondary vertex reconstruction, and particle identification capability, making the measurement of Λ_c in pp collisions possible down to low momenta.

2. ALICE detector: setup and data sample

For the Λ_c analysis, the main detectors involved in the central barrel ($-0.9 < \eta < 0.9$) are: the Inner Tracking System (ITS), for vertices reconstruction and tracking), the Time Projection

Journal of Physics: Conference Series **446** (2013) 012036 doi:10.1088/1742-6596/446/1/012036

Chamber (TPC), for tracking, Particle Identification (PID) and recontruction of K^0_S decays and the Time Of Flight (TOF), for PID. The ITS plays a central role in the reconstruction of primary and secondary vertices. The primary interaction vertex is reconstructed event by event with a resolution better than 100 μm. Good primary vertex and impact parameter resolution are both important for discrimination of tracks from secondary vertices close to the interaction vertex. The main ALICE tracking device is the TPC [4]. It is the largest TPC ever built and it has been designed to reconstruct efficiently up to 15000 primary tracks in a single collision, with a momentum resolution better than 1% for tracks with $p_T < 1$ GeV/c. The TPC is also the tracking device used to reconstruct the $V0$ secondary vertices: they are defined as the weak decay of a neutral particle decaying into two charged tracks after typically few cm. In particular, the $K^0_S \to \pi^+\pi^-$ decay has a $c\tau$ of 2.68 cm and is reconstructed with approximately 40% efficiency.

The results shown here are obtained analysing 3×10^8 pp minimum bias events at 7 TeV (corresponding to an integrated luminosity of 5 nb^{-1}). The detectors used for issuing the minimum bias trigger are the Silicon Pixel Detector (SPD) and the VZERO scintillator arrays, covering the rapidity ranges 2.8< η <5.1 and -3.7< η <-1.7 and placed at 3.4 m and 0.9 m from the interaction point, respectively. The trigger requires at least one hit in the SPD, combined in logical or with one signal in the VZERO counters, triggering in this way on at least one track anywhere in the eight units of pseudorapidity covered by the two trigger detectors.

3. The Λ_c analysis

For the decay channel $\Lambda_c^+ \to pK^-\pi^+$, the candidates are triplets of tracks with correct charge combination. Single tracks are first selected for their quality and transverse momentum. A second track is attached to form a pair of opposite charged tracks. A third track is attached to the selected pairs and the secondary vertex is reconstructed.

For the decay channel $\Lambda_c^+ \to K^0_S p$, the $V0$s are selected in the invariant mass window of the K^0_S. Then a track is selected for its global reconstruction quality, identified and attached to the K^0_S to form a Λ_c.

The Λ_c has a $c\tau$ of only 60 μm, so it is not easy to distinguish the secondary displaced vertices of the low momentum charmed baryon candidates. Therefore, the topological selection is less effective than for D meson decays ($c\tau >= 120$ μm).

The PID is very important for the analysis, for both decay channels under study. For this analysis the TPC and TOF detectors are used for PID. The TPC can be used to identify particles with intermediate momenta (up to $p = 1$ GeV/c for the separation of proton from K and π) thanks to the information on the specific energy deposit dE/dx in its volume. The TOF can be used to separate, for example, protons from pions up to 3 GeV/c and provide p/K separation up to 1.5 GeV/c. The two detectors cover the p_T range relevant for the identification of tracks coming from charmed baryon decays.

For the decay channel $\Lambda_c^+ \to pK^-\pi^+$, the Bayesian approach is used to combine the response of the two detectors. The probability W for a track to be of a specie i is the combination of the detectors response (s_{TPC}, s_{TOF}) and the initial value of the composition of the sample (the so called "priors", C):

$$W(i) = \frac{C_i \prod_{j=1}^{N} r(s_j|i)}{\sum_{k=e,\mu,\pi,..} C_k \prod_{j=1}^{N} r(s_j|k)}$$

To identify a track uniquely, the maximum probability criterion is applied: a track is identified as specie i if the probability $W(i)$ is the highest.

In the upper side of Fig. 1, the effect of the PID on the pKπ invariant mass distribution is shown for four different p_T ranges: the background is reduced by a factor hundred when PID is applied. The proton of the $\Lambda_c^+ \to K^0_S p$ is identified by the specific ionization energy loss in the TPC, combined in logical OR with the time of flight measured by TOF. In addition, the track is rejected if it is compatible with the kaon or pion hypothesis. In the lower side of Fig. 1, the effect of the PID on the decay channel $K^0_S p$ invariant mass distribution, p_T integrated, is shown: also in this case, the background is dramatically reduced.

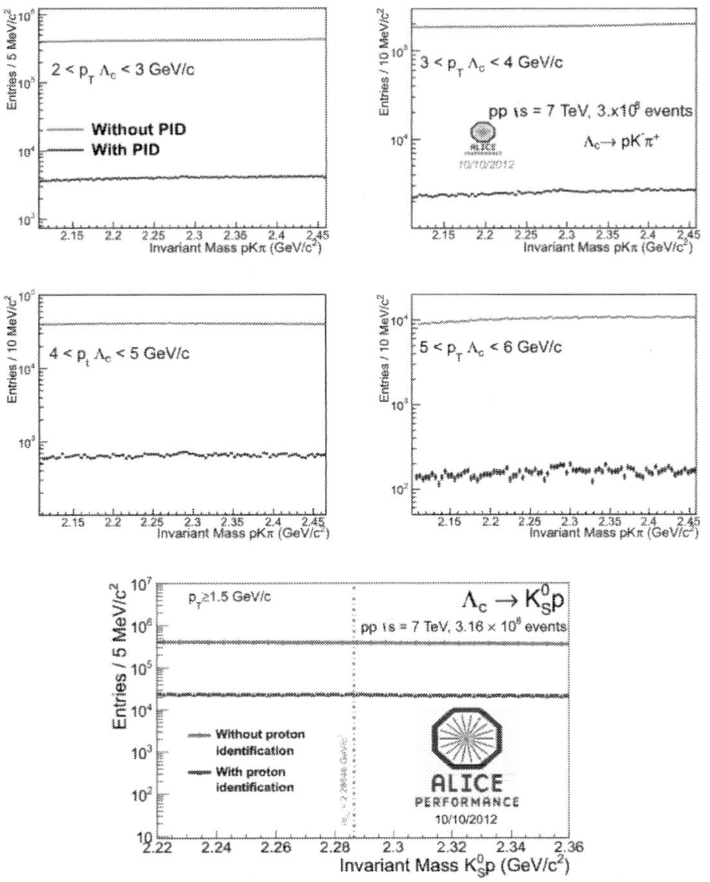

Figure 1. Invariant mass distributions for Λ_c candidates with PID (in blue) and without PID (in red), for the decay channels $pK\pi$ in the upper side and $K^0_S p$ in the lower side.

4. The Λ_c signal

The Λ_c signal can then be observed in both decay channels, in four p_T bins, as shown in Fig. 2, with a significance larger than 3. The signal is observed in the $\Lambda_c^+ \rightarrow pK^-\pi^+$ decay channel (zoom in of the case "With PID" in Fig.1) in the p_T range $2 < p_T < 6$ GeV/c, while the decay channel $\Lambda_c^+ \rightarrow K^0_S p$ covers the range $1 < p_T < 5$ GeV/c.

5. Conclusions

We have presented the analysis strategy that leads to the observation of the Λ_c signal in 3×10^8 minimum bias pp collisions at $\sqrt{s} = 7$ TeV, in the decay channels $\Lambda_c^+ \rightarrow pK^-\pi^+$ and $\Lambda_c \rightarrow K^0_S p$. The two decay channels allow us to cover a broad p_T range, with a wide overlap: not only they provide a complementary measurement, but also a good consistency check. The next step in the

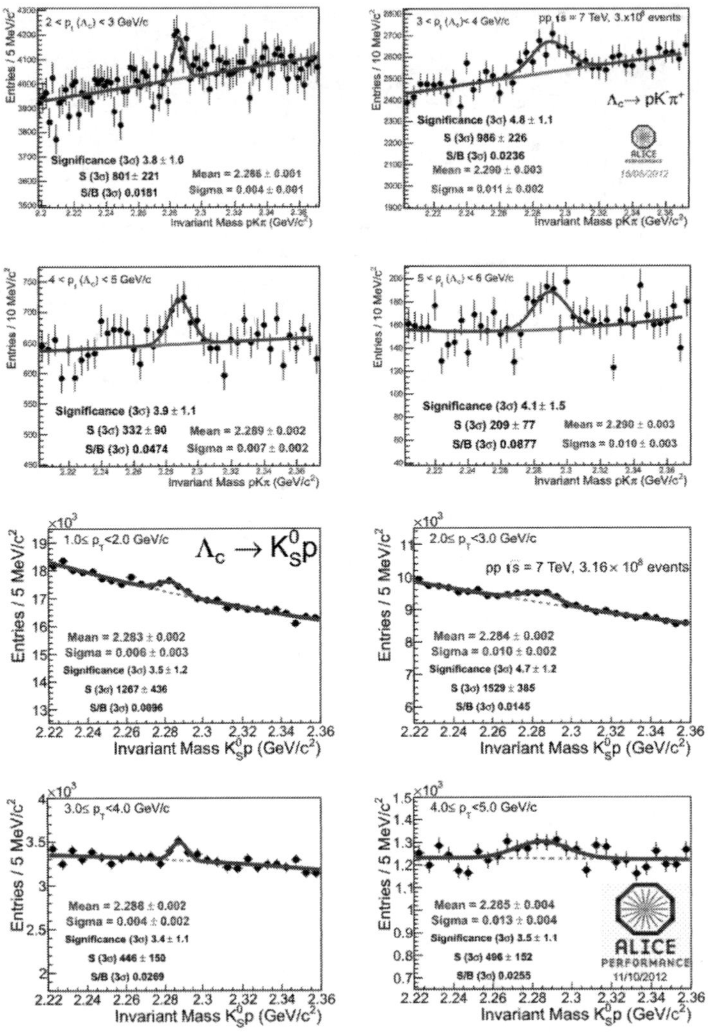

Figure 2. Invariant mass distributions for Λ_c candidates, for the decay channel $pK\pi$ in the upper side, for the decay channel $K^0_S p$ in the lower side. The background is fitted with a second order polynomial function, while the signal is fitted with a Gaussian term. The resolution in the invariant mass of the Λ_c is extracted from the fit and is determined from the width of the gaussian. It varies with the p_T of the Λ_c candidate because the momenta of the daughter particles increases with increasing momentum of the Λ_c. This leads to an increase in mass resolution.

analysis chain is to compute the efficiencies from dedicated MC simulations and estimate the feed down from beauty. Finally, the systematics errors will be evaluated.

[1] ZEUS Collaboration, H. Abramowicz et al., JHEP 11 (2010) 1-27, DOI: 10.1007/JHEP11(2010)009
[2] LHCb Collaboration, Phys. Rev. D84 (2011) 092001
[3] ALICE Collaboration, K. Aamodt et al., J. Instrum 3, S08002 (2008).
[4] ALICE Collaboration, J. Alme et al., NIMA 622 (2010) 316

Journal of Physics: Conference Series **446** (2013) 012037 doi:10.1088/1742-6596/446/1/012037

Azimuthal angular correlations between heavy-flavour decay particles and charged hadrons in pp collisions in ALICE

Deepa Thomas for the ALICE Collaboration
ERC-Research Group QGP-ALICE, Utrecht University, Princetonplein 5, 3584 CC Utrecht, The Netherlands
E-mail: deepa.thomas@cern.ch

Abstract. Heavy quarks produced in pp and heavy ion collisions are studied using heavy-flavour decay electrons and heavy-flavour mesons. Detailed understanding of the production processes and fragmentation of heavy quarks can be obtained by studying the azimuthal angular correlation of heavy-flavour hadrons. The azimuthal angular correlations of heavy-flavour mesons and charged hadrons can be used to disentangle charm and beauty-hadrons in pp collisions. In this contribution the fraction of electrons from beauty-hadron decays in the heavy-flavour decay electron yield is shown as well as the beauty production cross section in pp collisions at $\sqrt{s} = 2.76$ TeV. The measurements are compared to the predictions from next-to leading order perturbative QCD calculations. We also show the results from correlation analysis of charged D^* mesons and hadrons performed using pp collision data at $\sqrt{s} = 7$ TeV.

1. Introduction

In heavy-ion collisions heavy quarks are produced by the initial hard scattering processes and probe the high energy density medium produced in such collisions. Heavy flavour mesons can be studied via their hadronic or semi-leptonic decay channels. Detailed understanding on the interaction of heavy quarks in the medium can be achieved by studying the azimuthal angular correlations of heavy-flavour mesons. Correlation studies can be used to obtain information on in-medium partonic energy loss and modification of fragmentation function. In pp collisions, heavy-flavour correlations can be used to disentangle charm and beauty and to test the prediction of perturbative Quantum ChromoDynamic (pQCD) calculations. The measurement in pp is also a necessary baseline for Pb-Pb studies.

The azimuthal angular correlation between heavy-flavour decay electrons and charged hadrons and between D^* mesons and charged hadrons are presented here. The shape of the azimuthal angular correlations of heavy-flavour decay electrons and charged hadrons are used to determine the relative beauty contribution. Due to different decay kinematics of B and D mesons, the width of the near side correlation distribution is larger for B mesons compared to D mesons. Using the heavy-flavour electron cross section and relative beauty contribution to the heavy-flavour electron yield, we can derive the charm and beauty production cross sections separately.

2. Azimuthal angular correlation between heavy-flavour decay electrons and charged hadrons

2.1. Data sample and trigger selection

The analysis is performed using pp collisions at 2.76 TeV centre-of-mass energy collected in March 2011 with the ALICE experiment [1]. For the analysis, the detectors used are the Inner Tracking System (ITS) ($|\eta| < 0.9$, $0 < \phi < 360°$), the Time Projection Chamber (TPC) ($|\eta| < 0.9$, $0 < \phi < 360°$) and the Electromagnetic Calorimeter (EMCal) ($|\eta| < 0.7$, $80 < \phi < 180°$). The events which pass EMCal L0 trigger conditions are used. The L0 trigger is a 2×2 tower patch with a trigger cluster energy threshold of 3 GeV.

2.2. Data Analysis

Electrons are identified using information from the TPC and EMCal detectors. Particle identification in the TPC is based on the measurement of the specific ionization energy loss in the detector gas. In the EMCal, electron candidates are required to have E/p between 0.8 and 1.2. The non-heavy-flavour electrons (Non-HFE) are mainly produced by internal and in-material conversions and are identified using the invariant mass method, where the partner electron is selected with loose electron identification criteria. Electron pairs which have an invariant mass of less than 50 MeV/c^2 are tagged as Non-HFE. Above an electron p_T of 2 GeV/c, the non-HFE finding efficiency is $\approx 50\%$, as estimated from Monte Carlo (MC) simulations. The remaining Non-HFE contamination in the heavy flavour electron (HFE) sample is corrected for using this efficiency.

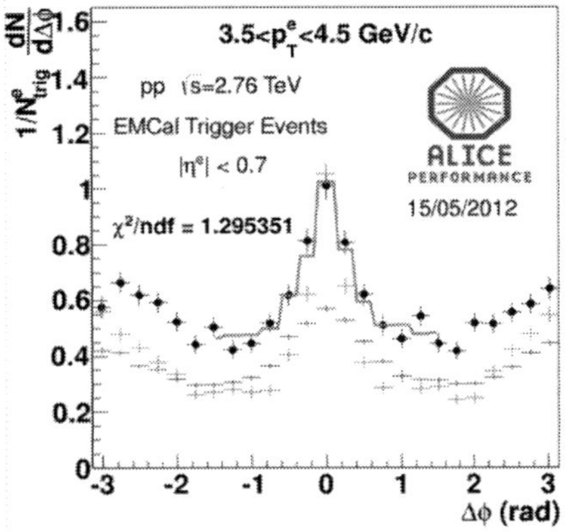

Figure 1. Azimuthal angular correlation between heavy-flavour decay electrons and charged hadrons in pp collisions at 2.76 TeV. The MC distribution for electrons from charm decay is shown in red, while the MC distribution for electrons from beauty decay is shown in blue and the full green curve is the fit to the data points.

The azimuthal angular correlations between heavy-flavour decay electron and charged hadrons is constructed using high quality tracks. In order to determine the fraction of electrons from beauty decays r_B, the measured correlation distribution is fit with the function

$$\Delta\phi^{HF}_{e-h} = const + r_B \Delta\phi^{B}_{e-h} + (1 - r_B)\Delta\phi^{D}_{e-h}, \quad (1)$$

where $r_B = \frac{e_B}{e_B + e_D}$ is the ratio of electron yield from B-meson decays to that of the heavy-flavour electron yield, $\Delta\phi^{D}_{e-h}$ ($\Delta\phi^{B}_{e-h}$) is the azimuthal angular correlation between electrons from D (B) meson decay and charged hadrons from MC simulations (PYTHIA 6.4 with Perugia-0 tune [2]) including detector response. *Const* term describes the uncorrelated background.

The fitting range used is $-1.5 < \Delta\phi < 1.5$ rad. The correlation distribution and the fit are shown in Figure 1. The beauty fraction extracted from the fit as a function of p_T is shown in Figure 2. The r_B increases with p_T and reaches ≈ 0.5 ($\frac{e_B}{e_D} \approx 1$) at around 5 GeV/$c$. This measurement is consistent, within uncertainities, with perturbative Quantum ChromoDynamics

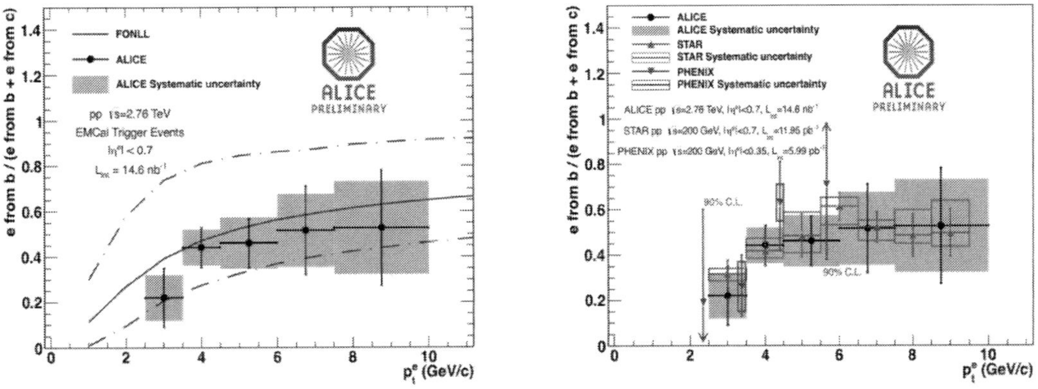

Figure 2. Relative beauty contribution to the heavy-flavour decay electron yield in pp collisions at $\sqrt{s} = 2.76$ TeV compared with FONLL calculations [3] (left) and previous RHIC measurements [4, 5] (right).

Fixed Order plus Next-to-Leading Logarithms (pQCD FONLL) calculations [3] and with results from RHIC in pp collisions at $\sqrt{s} = 200$ GeV [4, 5].

3. Beauty and charm decay electron cross section

The beauty and charm decay electron cross section are computed using the heavy-flavour decay electron cross section measured in ALICE and the relative beauty contribution. The heavy-flavour decay electron cross section is measured using the same data sample as the correlation analysis and the procedure described in [6]. Electrons are measured using the TPC and EMCal detectors. The cross section is measured after applying various corrections like tracking efficiency and unfolding, particle identification efficiency and purity and EMCal trigger efficiency. The Non-HFE background is identified and subtracted using a cocktail method where the measured η and π^0 cross section are used as an input and all the known sources of background electrons are taken into account [6].

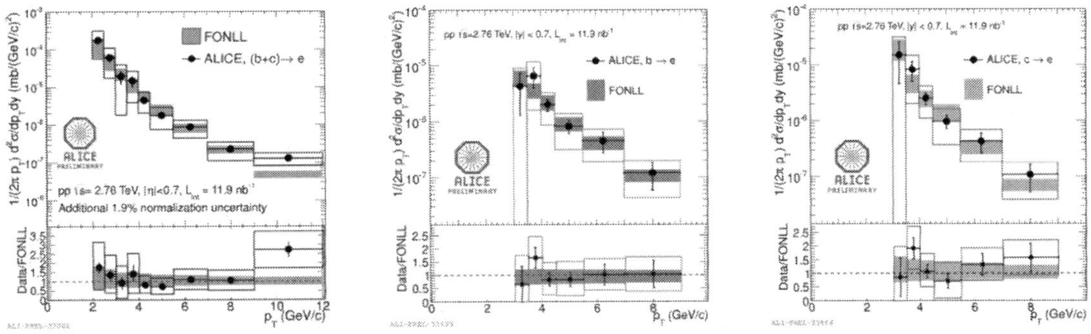

Figure 3. Heavy flavour decay electron (left), beauty (middle) and charm (right) decay electron cross section in pp collisions at $\sqrt{s} = 2.76$ TeV compared to FONLL calculations.

The beauty and charm decay electron cross section is computed as

$$\left(\frac{\mathrm{d}\sigma}{\mathrm{d}p_{\mathrm{T}}}\right)_{\mathrm{b}\to\mathrm{e}} = r_{\mathrm{B}} \times \left(\frac{\mathrm{d}\sigma}{\mathrm{d}p_{\mathrm{T}}}\right)_{\mathrm{b+c}\to\mathrm{e}} \tag{2}$$

$$\left(\frac{\mathrm{d}\sigma}{\mathrm{d}p_{\mathrm{T}}}\right)_{\mathrm{c}\to\mathrm{e}} = \left(\frac{\mathrm{d}\sigma}{\mathrm{d}p_{\mathrm{T}}}\right)_{\mathrm{b+c}\to\mathrm{e}} - \left(\frac{\mathrm{d}\sigma}{\mathrm{d}p_{\mathrm{T}}}\right)_{\mathrm{b}\to\mathrm{e}} \tag{3}$$

since r_{B} is not measured in the full p_{T} range of the HFE cross section, the beauty and charm decay electron cross section is measured from 3 to 9 GeV/c. The measured heavy-flavour electron and beauty and charm decay electron cross section is shown in Figure 3 together with FONLL pQCD calculations [3]. The cross section are consistent with FONLL calculations.

4. Azimuthal angular correlation between D* mesons and charged hadrons

The correlation between charged D* and charged hadrons is performed in pp collisions at $\sqrt{\mathrm{s}} =$ 7 TeV collected in 2010. D* mesons are reconstructed via the decay channel $\mathrm{D}^* \to \mathrm{D}^0(K\pi)\pi$ using the invariant mass method, see Figure 4 (left). The D* background is estimated using D^0 background candidates selected in the invariant mass side bands $4\sigma < |\mathrm{M}(K\pi) - \mathrm{M}(\mathrm{D}^0)| < 10\sigma$ and combining this fake D^0 with pion tracks. The D* mesons are correlated in azimuth with hadrons which pass quality track selection criteria. The reconstructed D* mesons and the correlation distribution of D* mesons with charged hadrons are shown in Figure 4 (right). The red distribution gives the azimuthal angular correction for all D* candidates and the blue distribution corresponds to sideband background candidates. The distribution can be fitted with gaussian distribution to extract the correlation parameters (yield and width).

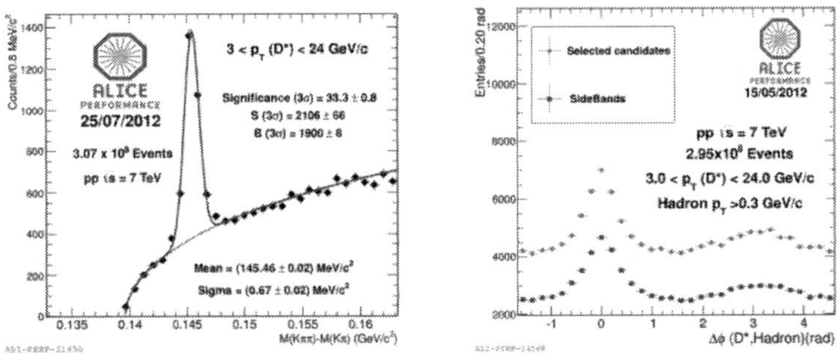

Figure 4. (Left) Invariant mass distribution of D* mesons. (Right) Azimuthal angular correlation between D* and charged hadrons in pp collisions at $\sqrt{\mathrm{s}} = 7$ TeV.

5. Results

The relative beauty contribution to the heavy-flavour decay electron yield was measured in pp collisions at $\sqrt{\mathrm{s}} = 2.76$ TeV with the ALICE detector and compared with pQCD calculations and RHIC measurements. Using this beauty fraction, the beauty-decay and charm-decay electron cross sections are derived and they are found to be consistent with FONLL pQCD calculations.

6. References

[1] K. Aamodt et al. [ALICE Collaboration], JINST 3, S08002 (2008).
[2] P. Skands, arXiv:0905.3418v1 [hep-ph] (2009).
[3] M. Cacciari et al. JHEP 0103, 006 (2001) and private communication (2012).
[4] M. M. Aggarwal et al. [STAR Collaboration], Phys. Rev. Lett. 105, 202301 (2010).
[5] A.Adare et al. [PHENIX Collaboration], Phys. Rev. Lett. 103, 082002 (2009).
[6] K. Aamodt et al. [ALICE Collaboration], arXiv:1205.5423 (2012).

Measurement of electrons from semi-leptonic heavy-flavour hadron decays in proton-proton and Pb–Pb collisions with ALICE at the LHC

Markus Fasel (for the ALICE Collaboration)
Physikalisches Institut, Ruprecht-Karls Universität Heidelberg, Im Neuenheimer Feld 226, 69120 Heidelberg

E-mail: mfasel@physi.uni-heidelberg.de

Abstract. Heavy quarks are produced in early partonic hard scatterings and experience the full history of a heavy-ion collision. Consequently heavy quarks are a good tool to study properties of the hot and dense medium. In proton-proton collisions, the study of the production of heavy quarks provides an important test for perturbative QCD and serves as a reference for heavy ion collisions. Heavy-flavour measurements in the semi-electronic channel are performed with ALICE at midrapidity. We present the measurement of the p_{T}-differential cross section in proton-proton collisions at $\sqrt{s} = 7$ TeV, and of the nuclear modification factor and elliptic flow in Pb-Pb collisions at $\sqrt{s_{\mathrm{NN}}} = 2.76$ TeV.

1. Introduction

The study of heavy-flavour production in proton-proton collisions serves as a good test for perturbative QCD calculations. In addition, it provides a reference for studies of heavy-ion collisions. Heavy quarks are produced in initial hard parton collisions and consequently are a good tool to study energy loss in the hot and dense medium. Due to the dead-cone effect [1], a mass ordering of the energy loss for different quark masses is expected, with the largest energy loss for the lightest quarks. The energy loss can be studied using the nuclear modification factor

$$R_{AA} = \frac{1}{\langle T_{AA} \rangle} \frac{\mathrm{d}N^{\mathrm{PbPb}}/\mathrm{d}p_{\mathrm{T}}}{\mathrm{d}\sigma^{\mathrm{pp}}/\mathrm{d}p_{\mathrm{T}}} \tag{1}$$

where $\langle T_{AA} \rangle$ is the nuclear overlap for a given centrality interval obtained via the Glauber model, $\mathrm{d}N^{\mathrm{PbPb}}/\mathrm{d}p_{\mathrm{T}}$ is the yield measured in Pb–Pb collisions, and $\mathrm{d}\sigma^{\mathrm{pp}}/\mathrm{d}p_{\mathrm{T}}$ is the cross section in proton-proton collisions. A study of the elliptic flow parameter v_2, defined as the second harmonic coefficient in the Fourier expansion of the azimuthal distribution of observed particles with respect to the reaction plane, could provide information on the thermalisation of heavy quarks in the hot and dense medium.

Measurements in the semi-electronic channels are well suited for heavy-flavour studies. Besides the large branching ratios of hadrons with heavy quarks into electrons ($\approx 10\%$), the electron channel allows for a measurement down to low transverse momentum p_{T}. Both charm and beauty-hadron decays contribute to the measured electron yield. These contributions can

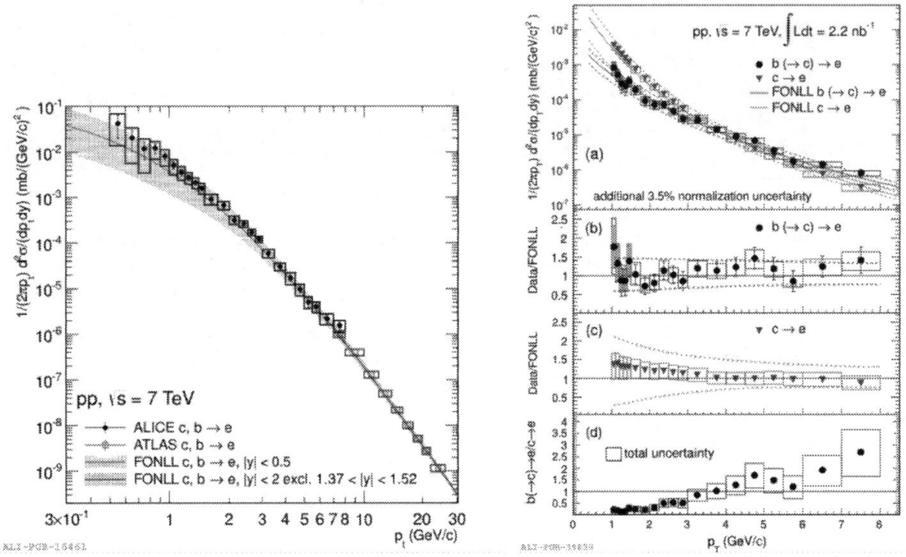

Figure 1. (Color online) Left: p_T-differential cross section of electrons from heavy-flavour hadron decays [2] from ALICE (dots) and ATLAS (open squares) [3]. Bands indicate a FONLL pQCD calculation. Right: p_T-differential cross section of electrons from beauty (black dots) and charm (blue triangles) hadron decays compared to a FONLL pQCD calculation [4]. The ratios of data over FONLL are shown for beauty in panel (b) and charm in panel (c). The ratio beauty to charm is shown in panel (d).

be disentangled even though the exact knowledge of the parent hadron kinematics is lost in the inclusive electron measurement.

The measurements are performed with ALICE, the dedicated heavy-ion experiment at the LHC, offering excellent particle identification capabilities. Various detectors are involved in the electron selection. A major device is the Time Projection Chamber (TPC), where the specific energy deposition ($\mathrm{d}E/\mathrm{d}x$) is used to distinguish electrons from other particles. At low p_T ($p_T \leq 4$ GeV/c) the Time-of-Flight detector (TOF) is used to reject protons and kaons at momenta where they are indistinguishable from electrons in the TPC based on $\mathrm{d}E/\mathrm{d}x$ alone. At high p_T hadrons are suppressed further using additional rejection power from the Transition Radiation Detector (TRD) or the Electromagnetic Calorimeter (EMCal).

Apart from electrons from heavy-flavour hadron decays, the measured electron sample contains background from various sources: photon conversions including direct photons, Dalitz-decays of light mesons, and dielectron decays of vector mesons. Two methods are used to estimate the background. One method is to calculate the background from a cocktail of various sources, based on measured π^0 and η p_T spectra, and subtract this statistically from the electron sample. As a second method, the background from internal and in-material photon conversions can be reconstructed by combining the electron with other electron candidates and selecting pairs with a small invariant mass.

2. Measurements in proton-proton collisions

The p_T-differential cross section of electrons from heavy-flavour hadron decays, measured in proton-proton collisions at $\sqrt{s} = 7$ TeV at midrapidity ($|y| < 0.5$), is shown in Fig. 1 (left) [2]. The systematic uncertainty also includes the uncertainty of the background. The comparison with FONLL pQCD calculations [5, 6] shows a good agreement between theory and data. The

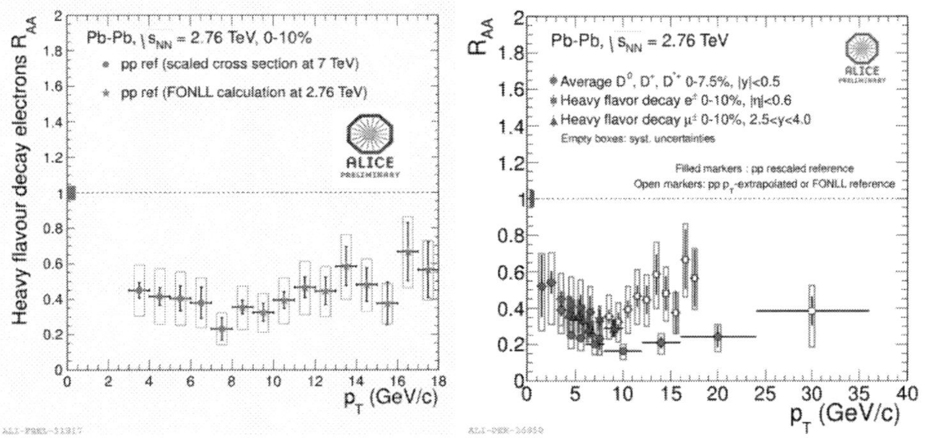

Figure 2. (Color online) Left: Nuclear modification factor of electrons from heavy-flavour hadron decays in the 0-10% most central Pb–Pb collisions at $\sqrt{s_{\rm NN}} = 2.76$ TeV [8]. Right: Comparison of the nuclear modification factors of electrons and muons, respectively, from semi-leptonic heavy-flavour hadron decays with the nuclear modification factor of D mesons [9].

$p_{\rm T}$ range covered by this measurement contains approximately 50% of the midrapidity charm cross section and approximately 90% of the midrapidity beauty cross section. The measurement, in combination with the results from the ATLAS Collaboration in the range 7 GeV/c < $p_{\rm T}$ < 26 GeV/c [3], can be used to test pQCD over a wide $p_{\rm T}$ range.

By selecting tracks with a large distance to the primary vertex in the transverse plane, the beauty component can be separated. The method exploits the longer lifetime of beauty hadrons ($c\tau \approx 500$ μm) compared to charm hadrons ($c\tau \approx 100-300$ μm). This method is complementary to the separation of electrons from beauty and charm hadron decays using angular correlations of electrons and hadrons, presented in [7]. Figure 1 (right) shows the $p_{\rm T}$-differential cross section of electrons from beauty-hadron decays [4]. The measurement is well described by FONLL pQCD calculations. Close to $p_{\rm T} = 4$ GeV/c the beauty component reaches the same cross section as the charm component.

3. Measurements in Pb–Pb collisions

The nuclear modification factor of electrons from heavy-flavour hadron decays is shown for 0-10% central collisions in Fig. 2 (left). The centrality class is determined using the VZERO detector. Below $p_{\rm T} < 8$ GeV/c the proton-proton reference was obtained from the measured $p_{\rm T}$-differential cross section at $\sqrt{s} = 7$ TeV using a FONLL-based scaling [10], above 8 GeV/c the reference is the FONLL calculation. The FONLL scaling has been verified using the measurement in proton-proton collisions at $\sqrt{s} = 2.76$ TeV. Since the FONLL scaled $p_{\rm T}$-differential cross section currently has a smaller uncertainty than the measurement at the same energy, the scaled spectrum is used as a reference. A strong suppression of electrons from heavy-flavour hadron decays is seen in central Pb–Pb collisions. At low-$p_{\rm T}$ charm hadron decays are the main source of heavy-flavour decay electrons, while at high-$p_{\rm T}$ the beauty contribution becomes dominant. A similar suppression is seen in the semi-muonic channel at forward rapidity (Fig. 2 (right)) [11]. Considering the decay kinematics the suppression in the semi-electronic channels is also similar to the suppression of D mesons.

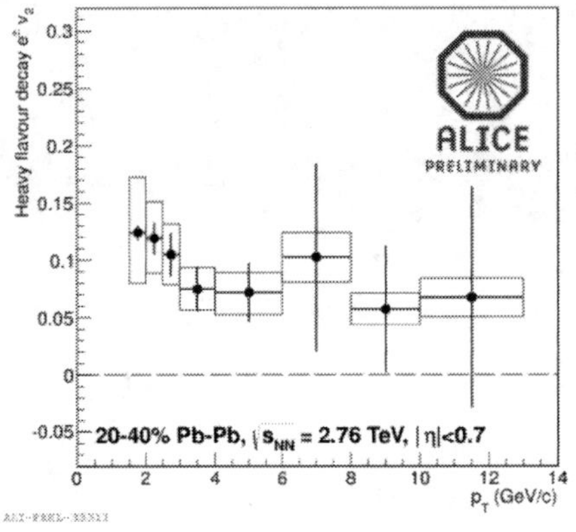

Figure 3. Elliptic flow parameter v_2 of electrons from heavy-flavour hadron decays in 20-40% semi-central collisions.

The elliptic flow parameter v_2 of electrons from heavy-flavour hadron decays is calculated via

$$v_2^{hfe} = \frac{(1+R)v_2^{incl} - v_2^{background}}{R} \tag{2}$$

where v_2^{incl} is the elliptic flow parameter of inclusive electrons, $v_2^{background}$ is the elliptic flow parameter of background electrons, and R is the signal-to-background ratio. The event plane is reconstructed using the VZERO detector. The elliptic flow parameter of background electrons is obtained using a cocktail based on the charged pion elliptic flow measurement and subtracted from the elliptic flow parameter of inclusive electrons.

Figure 3 shows the elliptic flow parameter of electrons from heavy-flavour hadron decays in 20-40% central Pb–Pb collisions. The systematic uncertainty contains both the uncertainty of the inclusive measurement and of the background. A non-zero elliptic flow parameter is observed.

4. Conclusions and Outlook

The measurement of the cross section of electrons from heavy-flavour hadron decays in proton-proton collisions at $\sqrt{s} = 7$ TeV was presented. Perturbative QCD calculations reproduce the data. Using electrons with a large distance to the primary vertex the cross section of electrons from beauty-hadron decays was extracted. In Pb–Pb collisions, a suppression was observed for electrons from heavy-flavour hadrons at high p_T in 0-10% central collisions. A non-zero elliptic flow might indicate that the heavy quarks thermalise with the hot and dense medium.

5. References

[1] Dokshitzer Y L and Kharzeev D 2001 *Phys.Lett.* **B519** 199–206 arXiv:0106202 [hep-ph]
[2] Abelev B *et al.* (ALICE Collaboration) 2012 *Phys.Rev.* **D86** 112007 (*Preprint* 1205.5423)
[3] Aad G *et al.* (ATLAS Collaboration) 2012 *Phys.Lett.* **B707** 438–458 arXiv:1109.0525 [hep-ex]
[4] Abelev B *et al.* (ALICE Collaboration) 2013 *Phys.Lett.* **B721** 13–23 arXiv:1208.1902 [hep-ex]
[5] Cacciari M, Greco M and Nason P 1998 *JHEP* **9805** 007 arXiv:9803400 [hep-ph]
[6] Cacciari M, Frixione S and Nason P 2001 *JHEP* **0103** 006 arXiv:0102134 [hep-ph]
[7] Thomas D *et al.* (ALICE Collaboration) these proceedings
[8] Sakai S (ALICE Collaboration) Proceedings of Quark Matter, to appear in Nucl. Phys. A
[9] del Valle Z C (ALICE Collaboration) Proceedings of Quark Matter, to appear in Nucl. Phys. A
[10] Averbeck R, Bastid N, del Valle Z C, Crochet P, Dainese A *et al.* 2011 arXiv:1107.3243 [hep-ph]
[11] Abelev B *et al.* (ALICE Collaboration) 2012 *Phys.Rev.Lett* **109** 112301 arXiv:1205.6443 [hep-ex]

Hot Quarks 2012 IOP Publishing
Journal of Physics: Conference Series **446** (2013) 012039 doi:10.1088/1742-6596/446/1/012039

D meson elliptic flow in Pb-Pb collisions at $\sqrt{s_{\rm NN}} = 2.76$ TeV measured with ALICE

Grazia Luparello for the ALICE Collaboration
NIKHEF, Utrecht University
E-mail: grazia.luparello@cern.ch

Abstract. These proceedings present the D^0, D^+ and D^{*+} meson v_2 measured by ALICE in Pb-Pb collisions at $\sqrt{s_{\rm NN}} = 2.76$ TeV with various methods (the standard event plane, the scalar product and the Q-cumulant) and in different centrality classes.

1. Introduction

According to Quantum Chromo Dynamics (QCD) calculations on the lattice, under the conditions of high energy density and temperature reached in high energy nucleus-nucleus collisions a phase transition to a Quark Gluon Plasma (QGP) [1] occurs: in such conditions partons are no longer confined inside hadrons. Heavy-flavour particles are good probes of the properties of the created medium. In particular, the azimuthal anisotropy of heavy-flavour hadron production can provide insight on the degree of thermalization of the heavy quarks in the expanding medium and on the path length dependence of their energy loss [2, 3, 4, 5, 6, 7]. This anisotropy, which is maximum in non-central collisions, originates from an initial geometrical anisotropy with respect to the reaction plane (the plane defined by the beam direction and the impact parameter) that is then converted to a momentum anisotropy by the different pressure gradients in the different directions during the system expansion. The anisotropic pattern can be expressed with a Fourier expansion of the particle momentum azimuthal distribution with respect to the event plane, that is an estimate of the reaction plane. The second order coefficient v_2 is called elliptic flow [8].

In Section 2 of theese proceedings the data sample used and the D meson reconstruction strategy are described, while Section 3 is about the v_2 extraction techniques. Finally in Section 4 the results are summarized.

For this analysis, the following detectors of the ALICE [9] central barrel ($|\eta| < 0.9$) were used: the Inner Traking System (ITS) composed of six layers of high resolution silicon detectors, with the two inner layers consisting of Silicon Pixel Detectors (SPD), a large volume Time Projection Chamber (TPC) and a Time of Flight Detector (TOF). They ensure good capabilities in terms of vertex reconstruction, tracking and particle identification that allow a direct charm v_2 measurement through fully reconstructed D meson hadronic decays.

2. Data sample and D meson reconstruction

The analysis was performed on a data sample of Pb-Pb collisions at a centre-of-mass energy $\sqrt{s_{\rm NN}} = 2.76$ TeV. A minimum-bias trigger was provided by the coincidence of signals in the two VZERO detectors, consisting of two arrays of scintillator counters covering the pseudo-rapidity regions $-3.7 \leq \eta \leq -1.7$ and $2.8 \leq \eta \leq 5.1$. Collisions were classified according to their centrality, defined in terms of percentiles of the hadronic Pb-Pb cross section and determined from the distribution of the signal amplitudes in the VZERO detectors. In addition, an online selection based on the VZERO signal amplitude was used to enhance the number of events with centrality < 53%. Only events with a vertex within ±10 cm from the centre of the detector along the beam line were considered for the D meson signal analysis. The number of Pb-Pb events analized was 9.5×10^6 in the centrality class 30-50%, 7.1×10^6 in 15-30% and 16×10^6 in 0-7.5%.

Open charm detection in ALICE is based on the invariant-mass analysis of fully reconstructed hadronic-decay topologies. The D meson v_2 was measured in the decay channels $D^0 \rightarrow K^-\pi^+$, $D^+ \rightarrow K^-\pi^+\pi^+$ and $D^{*+} \rightarrow D^0\pi^+$. The separation of a few hundred μm between the primary and secondary vertices, characteristic of D^0 and D^+ decays, was exploited to reduce the combinatorial background. D^{*+} candidates were obtained by combining the D^0 candidates with pion tracks. To further reduce the background, the pion and kaon identification by TPC and TOF detectors was used by applying a selection on the energy deposition dE/dx and time of flight. This PID strategy was defined with the aim of preserving most of the signal yield. Further analysis details can be found in [10].

3. v_2 extraction

Different techniques were used to measure the v_2 of the reconstructed D mesons: standard event plane, scalar product and Q-cumulants. Central for all the methods is the evaluation of the flow vector, Q, defined as $Q = (\sum_{i=0}^{N} w_i \sin 2\phi_i, \sum_{i=0}^{N} w_i \cos 2\phi_i)$ where ϕ_i is the angle of the ith track in the ALICE reference frame, w_i are weights applied to tracks depending on their angle in the ALICE reference frame to account for different efficiencies among TPC sectors.

In the standard event plane method (EP) the event plane angle is calculated as $\Psi_n = \arctan(Q_{n,y}, Q_{n,x})/n$. In this case only tracks in the pseudo-rapidity region $0 < \eta < 0.8$ were used to improve the event plane flatness. The invariant mass distribution of the reconstructed D meson candidates was split into in-plane and out-of-plane regions defined as the regions $(0 < \Delta\phi < \frac{\pi}{4}) \bigcup (\frac{3\pi}{4} < \Delta\phi < \pi)$ and $(\frac{\pi}{4} < \Delta\phi < \frac{3\pi}{4})$ respectively, where $\Delta\phi = \phi^D - \Psi_2$. An invariant mass analysis was used to extract the D meson yield in the two regions (N_{in},N_{out}). Knowing N_{in} and N_{out}, v_2 can be directly calculated from Eq. 1, where R is the event plane resolution:

$$v_2 = \frac{\pi}{4} \frac{1}{R} \frac{N_{in} - N_{out}}{N_{in} + N_{out}} \tag{1}$$

The main drawback of the standard event plane method is that the event plane determination is affected by correlations not coming from the genuine correlation of particles with the true reaction plane. This introduces a bias in the flow estimates. Therefore it is useful to compare the results with methods not relying on the event plane determination.

The scalar product method (SP) [11] is based on the evaluation of the quantity $\langle \mathrm{u} \cdot Q^* \rangle$ (where $\mathrm{u} = e^{\imath 2\varphi}$ is the unity vector of the D meson candidate and Q^* is the complex coniugate of the flow vector Q) averaged over all D meson candidates in an event and over all events as in Eq. 2.

$$v_2 = \frac{\langle\langle \mathrm{u} \cdot \frac{Q^*}{M_Q} \rangle\rangle_e}{\sqrt{\langle \frac{Q_A}{M_A} \cdot \frac{Q_B^*}{M_B} \rangle_e}} \tag{2}$$

The vectors Q_A and Q_B are two flow vectors constructed from particles coming from two subevents chosen in different pseudorapidity regions, respectively $0 < \eta < 0.8$ and $-0.8 < \eta < 0$. The D meson candidates are taken from the opposite η region. The vectors Q, Q_A and Q_B are normalized by the number M of particles in each event to reduce the effects of multiplicity fluctuations. Eq. 2 differs from the event plane method, since it uses the magnitude of the flow vector as a weight.

The Q-Cumulants method (QC) [12] is based on the unique relationship between the v_2 coefficient and the so-called Q-cumulants. The mixed correlator between the D meson candidate and reference flow particles (RFP) is calculated. It is defined as $e^{i2(\varphi-\phi)}$, where φ^D (ϕ) is the azimuthal angle of the D meson candidate (reference particle) in the ALICE reference frame. The mixed correlator is normalized by the reference flow calculated using RFP. For this analysis reference particles were TPC tracks in the pseudo-rapidity range $|\eta| < 0.8$.

For both the SP and QC method, D mesons candidates were reconstructed in bins of η and p_T and the independent particle correlators were calculated for each bin obtaining a distribution of v_2 as a function of the mass of the D meson. The v_2 of the signal (v_2^S) was disentangled from the v_2 of the background (v_2^B) with a simultaneous fit of the invariant mass yield and of the v_2 vs. mass distribution. A Gaussian shape was assumed for the signal peak while the v_2 distribution was fitted with $v_2(x) = S(x)/(S(x)+B(x)) \cdot v_2^S(x) + B(x)/(B(x)+S(x)) \cdot v_2^B(x)$ where x is the mass of the D meson. $v_2^B(x)$ was parametrized with a first order polynomial.

Several sources of systematic uncertainties were considered in the analysis. The main contributions to the total systematic uncertainty come from topological cut selection and from uncertainties on the yield extraction. The analysis was repeated using three different sets of cuts, while the systematic uncertainty on yield extraction was estimated using different background functions in the invariant mass fit and repeating the fit in different mass ranges. For the EP analysis the yields were extracted using also a bin counting method. For the SP and QC methods also the parametrization of the v_2 of the background was varied. The absolute value of the systematic uncertainty from cut variation and yield extraction is ~ 0.05 for D^0 mesons. For the EP method, the variation of the event plane resolution in the centrality range 30-50% introduces a systematic uncertainty of $^{+7}_{-3}\%$

The signal sample considered contains also a fraction of D mesons coming from B decays, thus the measured elliptic flow is a combination of prompt and secondary D meson anisotropy. The v_2 of prompt D mesons was obtained assuming that these two contributions have the same elliptic flow and a systematic uncertainty was estimated using as input pQCD calculations of B production, as explained in [10] and including a conservative variation of the unknown R_{AA} (the ratio between the yield measured in AA collisions and the cross section measured in pp

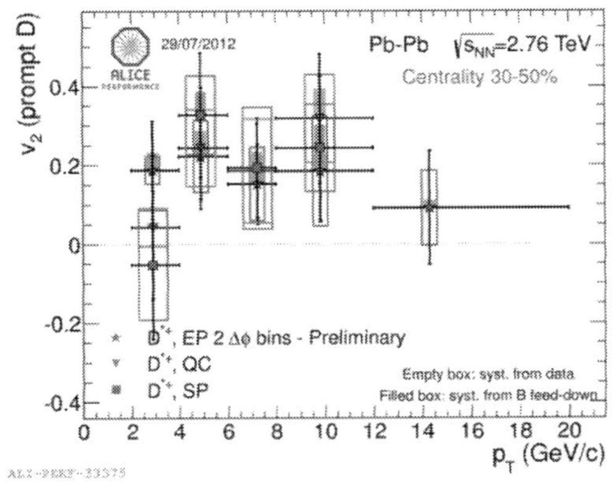

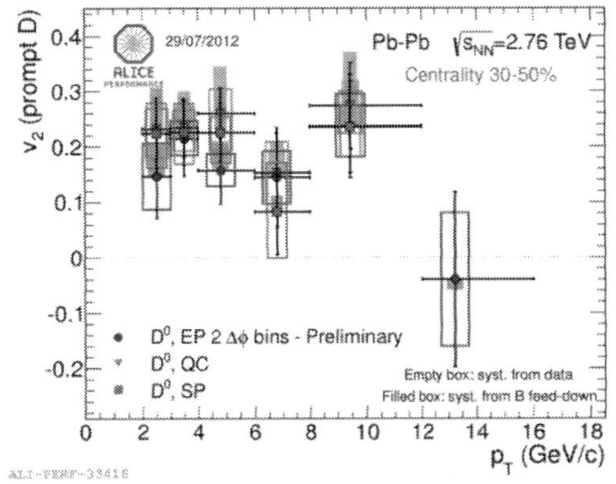

Figure 1. D^{*+} v_2 (left) and D^0 v_2 (right) measured by ALICE in the 30-50% centrality class with EP, SP and QC methods. Empty boxes show systematic uncertainty from data, shaded areas the uncertainty from B feed-down.

collisions, scaled by the overlap nuclear function) and v_2 of D mesons from B decays. This uncertainty extends only towards higher v_2 values and it amounts to up to +23%.

4. Results

The D meson anisotropy was measured in the 30-50% centrality class for D^0 in the range $2 < p_T < 16$ GeV/c, D^+ in $3 < p_T < 8$ GeV/c and D^{*+} in $2 < p_T < 20$ GeV/c. For D^0 and D^{*+} the measurement was carried out using different techniques and the results are in agreement within uncertainties, as shown in Figure 1. The measurement of the v_2 of the D^+ with SP and QC methods is ongoing. Figure 2 (left) shows that the v_2 of D^0, D^+ and D^{*+} is compatible within uncertainties and it is also of similar magnitude as the ALICE measurement of charged hadrons in the same rapidity interval. In Figure 2 (right) the D^0 v_2 measurement in three centrality classes is reported.

In conclusion, the D^0 v_2 is above 0 in the p_T range 2-6 GeV/c with 3σ significance and there is a hint of increasing v_2 from central to semi-peripheral collisions.

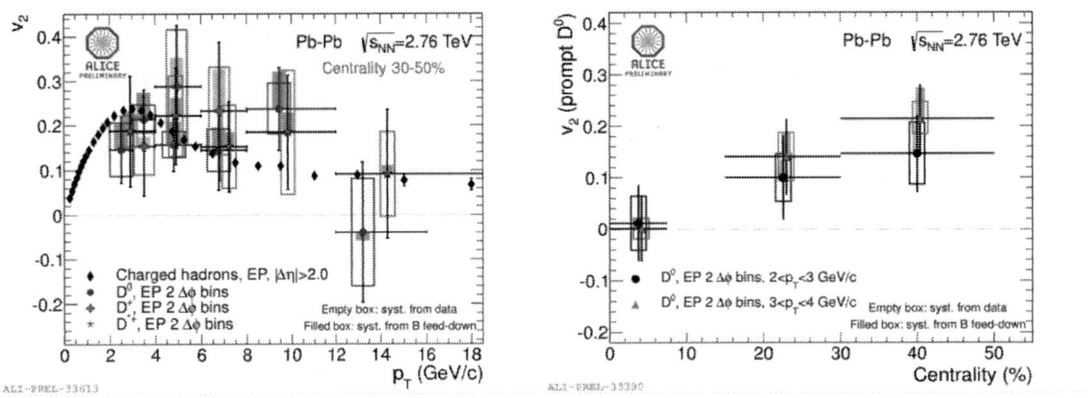

Figure 2. *Left* : D^0, D^+, D^{*+} v_2 measured by ALICE in the 30-50% centrality class with the EP method. Empty boxes show systematic uncertainty from data, shaded areas the uncertainty from B feed-down. Black points show charged hadron v_2[13]. Right: D^0 v_2 mesured by ALICE in the centrality classes 0-7.5%, 15-30% and 30-50%. Empty boxes show systematic uncertainty from data, shaded areas the uncertainty from B feed-down.

Reference

[1] F. Karsch, J. Phys. Conf. Ser. **46**, 122 (2006) [hep-lat/0609008].
[2] H. van Hees, V. Greco, R. Rapp, Phys. Rev. C **73** 034913 (2006).
[3] D. Molnar, J. Phys. G **31** S421S428 (2005).
[4] W. M. Alberico *et al.* Eur. Phys. J. C **71**, 1666 (2011) [arXiv:1101:6008 [hep-ph]].
[5] P. B. Gossiaux, R. Bierkandt and J. Aichelin, Phys. Rev. C **79**, 044906 (2009) [arXiv:0901.0946 [hep-ph]].
[6] J. Uphoff, O. Fochler, Z. Xu and C. Greiner, arXiv:1112.1559 [hep-ph].
[7] M. He, R. J. Fries and R. Rapp, arXiv:1204.4442 [nucl-th].
[8] J. Y. Ollitrault, Eur. J. Phys. C **29**, 275302 (2008).
[9] K. Aamodt *et al.* ALICE Collaboration, JINST 3 S08002 (2008)
[10] B. Abelev *et al.* [ALICE Collaboration], JHEP **1209** 112 (2012) [arXiv:1203.2160 [nucl-ex]]
[11] C. Adler *et al.* [STAR Collaboration], Phys. Rev. C **66** 034904 (2002)
[12] A. Bilandzic, R. Snellings and S. Valoshin, Phys. Rev. C **83** 044913 (2011)
[13] B. Abelev *et al.* [ALICE Collaboration], arXiv:1205.5761 [nucl-ex]

Journal of Physics: Conference Series **446** (2013) 012040 doi:10.1088/1742-6596/446/1/012040

D_s^+ production at central rapidity in pp collisions at $\sqrt{s} = 7$ TeV and in Pb–Pb collisions at $\sqrt{s_{NN}} = 2.76$ TeV with the ALICE detector

Gian Michele Innocenti for the ALICE Collaboration
Università degli Studi di Torino and INFN, Torino, Italy
E-mail: ginnocen@cern.ch

Abstract. We present the measurements of the D_s^+ production in pp collisions at $\sqrt{s} = 7$ TeV and in Pb–Pb collisions at $\sqrt{s_{NN}} = 2.76$ TeV performed in the central rapidity region with the ALICE detector through the reconstruction of the hadronic decay channel $D_s^+ \rightarrow \phi\pi^+ \rightarrow K^+K^-\pi^+$. The ratios of the yields of D_s^+ and non-strange D mesons as a function of the transverse momentum are also shown for both pp and Pb-Pb collisions.

1. Introduction

The study of charm production in proton-proton and in heavy-ion collisions allows one to investigate properties of the high-density QCD medium created in heavy-ion collisions and to study the mechanism of in-medium energy loss of heavy quarks [1]. In addition, since strange quarks are abundant in the medium [2], the relative yield of D_s^+ mesons with respect to non-strange charm mesons (D^0, D^+ and D^{*+}) is predicted to be largely enhanced if in-medium hadronization is the dominant mechanism for charm hadron formation in the low momentum region [3, 4]. An effective observable to study these processes is the nuclear modification factor which is defined for a given particle species h as $R^h_{AA}(p_T) = \frac{dN^h_{AA}/dp_T}{\langle T_{AA}\rangle \times d\sigma^h_{pp}/dp_T}$, where N^h_{AA} is the yield measured in heavy-ion collisions, $\langle T_{AA}\rangle$ is the average nuclear overlap function calculated using the Glauber model and σ^h_{pp} is the production cross section in pp collisions at the same energy. We present the measurement of the nuclear modification factor of the D_s^+ meson in Pb–Pb collisions performed at central rapidity ($|y| < 0.5$) with the ALICE detector [5] at 2.76 TeV. The proton-proton measurement at $\sqrt{s} = 7$ TeV used as a reference for this analysis is described in details in [6].

2. Detector layout and data sample

Charmed mesons are reconstructed in ALICE using the tracking detectors and particle identification systems of the central barrel, which cover the pseudo-rapidity region $-0.9 < \eta < 0.9$ and is embedded in a magnetic field of B = 0.5 T. A brief description of the detectors utilized in these analyses will be given in this section (more details in [5]). The Time Projection Chamber (TPC) is the main tracking detector, which provides track reconstruction and particle identification via the measurement of the specific energy deposit dE/dx. The Inner Tracking System (ITS) is the central barrel detector closest to the beam axis and it is composed of six

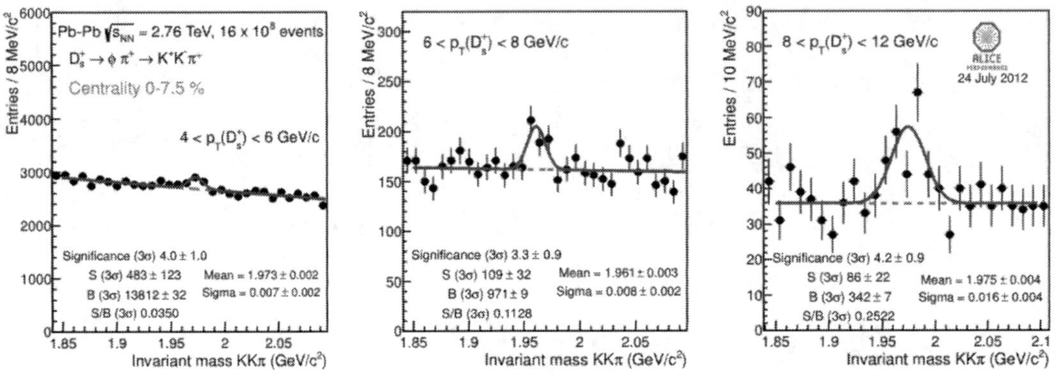

Figure 1. Invariant mass distributions for D_s^+ candidates and charge conjugates in the three considered p_T intervals in the range $4 < p_T < 12$ GeV/c obtained from the analysis of 16 million Pb–Pb events in the centrality range 0–7.5%.

cylindrical layers of silicon detectors. The ITS is essential to reconstruct secondary vertices originating from open charm decays, because it allows for a precise measurement of the track impact parameter to the primary vertex. Finally, the Time-of-Flight (TOF) detector provides pion, kaon and proton identification on the basis of their time of flight. All three detectors have full azimuthal coverage. The data sample used for this analysis was collected during the 2011 LHC run (November-December 2011) with Pb–Pb collisions at $\sqrt{s_{NN}} = 2.76$ TeV and it consists of 16 million events in the centrality range 0–7.5%. The centrality selection was performed with the amplitude of the VZERO detectors, which are two scintillator hodoscopes positioned in the forward and backward regions of the experiment.

3. D_s^+ meson reconstruction in ALICE

D_s^+ mesons and their antiparticles ($c\tau = 150$ μm) were reconstructed in the decay channel $D_s^+ \to \phi\pi^+ \to K^+K^-\pi^+$ (and its charge conjugate) which has a branching ratio (BR) of $2.28 \pm 0.12\%$ [7]. The analysis strategy is based on the reconstruction and selection of secondary vertex topologies with significant separation from the primary vertex. The D_s^+-meson candidates were defined starting from combinations of tracks selected with respect to their pseudorapidity ($|\eta| < 0.8$), transverse momentum and number of associated space points. In order to improve the statistical significance of the signal, the D_s^+ candidates were filtered by applying kinematical and topological cuts together with particle identification criteria. The selection was based on the decay length, the cosine of the pointing angle and on the sum of the distances of the decay tracks to the reconstructed decay vertex. Moreover, the D_s^+ candidates were required to have one of the two pairs of opposite-charge tracks with an invariant mass compatible with the PDG value for the ϕ meson mass [7]. The selection strategy adopted in this analysis (similar to the one performed in the pp measurement) aims at reducing the combinatorial background while keeping the selection efficiency as large as possible. In Fig. 1 the invariant mass distributions of D_s^+ candidates in the centrality interval 0–7.5% in three p_T intervals from 4 to 12 GeV/c are shown. In order to evaluate the signal yield, a fit to the distributions with a Gaussian function for the signal and an exponential shape for the background was used. The raw yields were then corrected for acceptance and selection efficiency of prompt D_s^+ mesons using Monte Carlo simulations based on PYTHIA (Perugia-0 tuning) [8] and HIJING [9]. The correction factor that accounts for the feed-down contribution from B meson decays was evaluated using FONLL

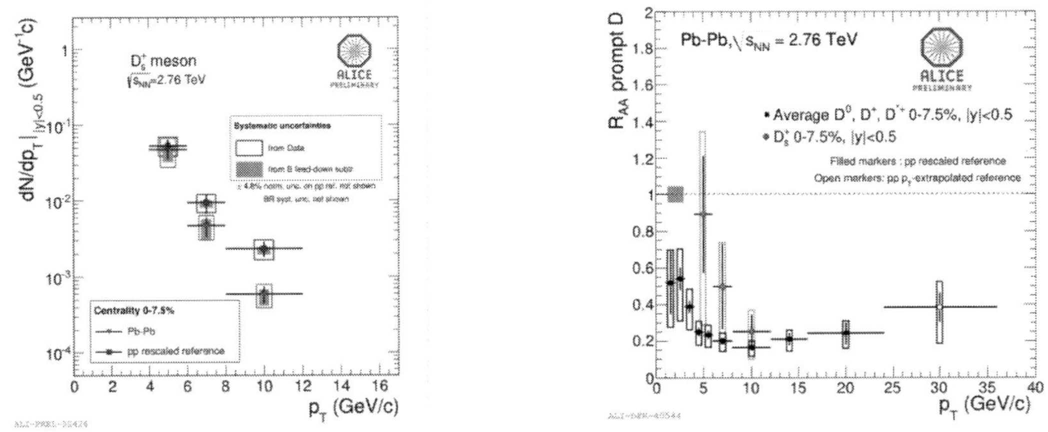

Figure 2. (Left) D_s^+ yield for central Pb–Pb collisions (0–7.5%) at $\sqrt{s_{NN}}$=2.76 TeV compared to the reference p_T spectra. (Right) D_s^+ nuclear modification factor measured in central (0–7.5%) Pb–Pb collisions compared to the average R_{AA} of D^0, D^+ and D^{*+} also measured in ALICE.

predictions for the p_T-differential cross section of B mesons, the $B \to D_s^+$ decay kinematics and the Monte Carlo efficiency for feed-down D_s^+ mesons. Indeed, FONLL calculation describe well bottom production at Tevatron [10] and at the LHC [11, 12]. The contribution estimated with FONLL was multiplied in the Pb–Pb analysis by the average nuclear overlap factor estimated via the Glauber Model $\langle T_{AA} \rangle$ and by a hypothesis on the suppression of D_s^+ from B decays ($R_{AA}^{\text{feed-down}}$) (more details in [13]). The estimated contribution of D_s^+ from B meson decays is $\approx$ 20% in the three p_T intervals considered. To estimate the systematic uncertainties on the B feed-down subtraction the full spread of the results obtained by varying the FONLL parameters and the values of $R_{AA}^{\text{feed-down}}$ was considered. The latter was varied in the range $1/3 < R_{AA}^{\text{feed-down}}/R_{AA}^{D} < 3$. The reference for the nuclear modification factor was obtained by scaling the pp cross section at 7 TeV to $\sqrt{s}$ = 2.76 TeV using the ratio of FONLL cross sections for meson production at the two energies. To compute the systematic uncertainties on the scaling (about 15% in the three p_T intervals), the parameters of the FONLL calculation (factorization and renormalization scales, charm quark mass) were varied. The validity of the energy scaling was tested for D^0, D^+ and D^{*+} mesons through the comparison with the measured p_T-differential cross section in pp collisions at 2.76 TeV [14].

4. Results

The D_s^+ yield measured in Pb–Pb collisions at $\sqrt{s_{NN}}$ =2.76 TeV in the centrality range 0–7.5% is shown in the left panel of Fig. 2 together with the pp reference which is obtained by scaling with FONLL the cross-section measured at $\sqrt{s}$ = 7 TeV at the Pb–Pb center of mass energy and by multiplying it by the corresponding $\langle T_{AA} \rangle$ value. In the right panel of Fig. 2 the D_s^+ R_{AA} in the same centrality class is shown and compared to the average of D^0, D^+ and D^{*+} [15]. The statistical uncertainties are shown as vertical bars while the boxes around the data points represent the systematic uncertainties, which include the contribution of signal extraction, track reconstruction efficiency, topological and PID selection and of the B feed-down subtraction. The D_s^+ nuclear modification factor in the highest p_T interval is $\approx$ 0.25 and it is compatible within uncertainties with the value measured for non-strange D mesons. At lower transverse momenta, the D_s^+ R_{AA} presents an increase which is however not significant within present uncertainties if compared to the average value of D^0, D^+ and D^{*+}. In Fig. 3, the ratios between the D_s^+ yield and

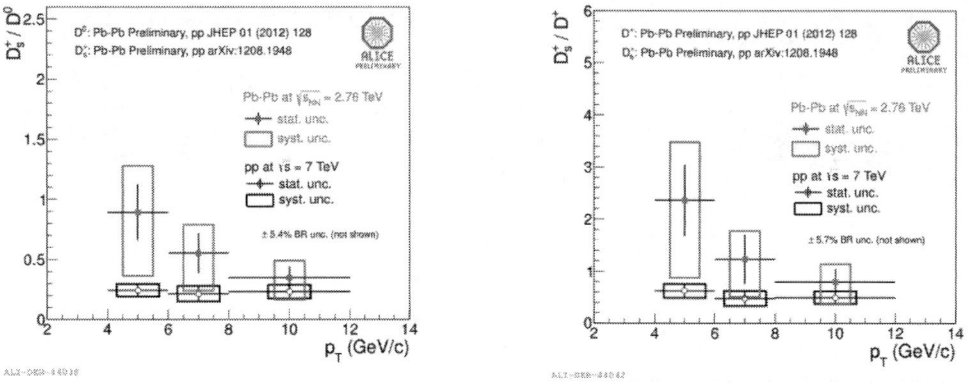

Figure 3. Ratios of the yield D_s^+/D^0 (left) and D_s^+/D^+ (right) measured in Pb–Pb collisions in the centrality range 0–7.5% as a function of p_T compared with the ratios of the production cross sections measured in pp collisions at 7 TeV.

the D^0 and D^+ ones measured in Pb–Pb collisions in the centrality range 0–7.5% are presented as a function of p_T together with the ratios of the production cross sections measured in pp collisions at 7 TeV. In the evaluation of the systematic uncertainties, the sources of correlated and uncorrelated systematic effects were treated separately. In particular, the contributions of the yield extraction, cut efficiency and PID selection were considered as uncorrelated and summed in quadrature, while the uncertainty on the tracking efficiency, being completely correlated, cancels in the ratio. The FONLL contribution to the B feed-down systematic uncertainty was considered as completely correlated both in pp and in Pb–Pb and estimated as the spread of the ratios obtained by varying the factorization and renormalization scales and the heavy quark mass in FONLL coherently for all mesons. In the Pb–Pb analysis, the contribution from the hypothesis on the $R_{AA}^{\text{feed-down}}$ was considered completely uncorrelated between the considered D meson species and summed in quadrature due to the possible different nuclear modification factor of B mesons which decay in D_s^+ (mainly B_s^+ mesons) and D^0, D^+ and D^{*+} (mainly B^0 and B^+ mesons). Larger values of the ratios are observed in Pb–Pb collisions in the lowest p_T interval but they are still compatible with the pp results within uncertainties. Therefore, larger statistics samples from the future LHC runs are needed to draw firm conclusions on a possible enhancement of D_s^+ production in Pb–Pb collisions at low-intermediate p_T.

References

[1] Y.L. Dokshitzer and D.E. Kharzeev, Phys. Lett B519 (2001), 199.
[2] M. Nicassio for the ALICE Collaboration, Acta Physica Polonica B Proc. Suppl. 5 (2012), 237.
[3] M. He, R. J. Fries and R. Rapp, arXiv:1204.4442 [nucl-th].
[4] I. Kuznetsova and J. Rafelski, Eur. Phys. J. C 51 (2007) 113.
[5] The ALICE experiment at CERN LHC, JINST 0803 (2008) S08002.
[6] B. Abelev et al. [ALICE Collaboration] Phys. Lett. B 718 (2012), 233.
[7] J. Beringer et al. [Particle Data Group], Phys. Rev. D 86 (2012) 010001.
[8] T. Sjostrand, S. Mrenna, P. Skands, JHEP 05 (2006) 026.
[9] X.N.Wang and M. Gyulassy, Phys. Rev. D 44 (1991) 3501.
[10] M. Cacciari et al., JHEP 0407 (2004) 033; private communication.
[11] R. Aaij et al. [LHCb Coll.], Phys. Lett. B694 (2010) 209-216.
[12] V. Khachatryan et al. [CMS Coll.], EPJC C 71 (2011) 1575.
[13] B. Abelev et al. [ALICE Collaboration] JHEP 09 (2012) 112.
[14] B. Abelev et al. [ALICE Collaboration], JHEP 07 (2012) 191.
[15] A. Grelli for the ALICE collaboration, arXiv:1210.7332 [hep-ex].

Journal of Physics: Conference Series **446** (2013) 012041 doi:10.1088/1742-6596/446/1/012041

New Measurement of Heavy-Flavor Electron v_2 and Separation of Bottom and Charm Using VTX in PHENIX

Lei Ding for the PHENIX Collaboration
Iowa State University
E-mail: `lding@iastate.edu`

Abstract.
Heavy quarks are important probes to understand the properties of the Quark Gluon Plasma (QGP) created in the relativistic heavy-ion collision. The PHENIX measurement of electrons from heavy-flavor decays shows large flow and strong suppression in Au+Au collisions at 200 GeV [1][2], which is not well understood. In these proceedings, we will present both the new PHENIX heavy-flavor electron v_2 results in Au+Au collisions at 62.4 GeV collisions using the Hadron Blind Detector (HBD) and the latest PHENIX measurement of charm and bottom separately in p+p collisions at 200 GeV with the recently installed Silicon Vertex Track (VTX).

1. Introduction

Heavy quarks are hard probes to study the properties of the hot dense matter created in relativistic heavy ion collisions at RHIC because they are produced in parton-parton collisions at the early stage of the reaction, then propagate through and interact with the Quark Gluon Plasma (QGP). Because of their heavy mass, small angle gluon radiation is suppressed when heavy quarks interact with the medium: this is the dead cone effect [3, 4] . Therefore, heavy quarks are predicted to lose less energy by gluon radiation compared to light quarks. However, unexpected large flow and strong suppression of electrons from heavy-flavor meson semi-leptonic decays are observed at PHENIX in Au+Au collisions at $\sqrt{s_{NN}} = 200$ GeV [1] [2]. These results indicate that heavy quarks are suppressed in a similar way in the QGP compared to light quarks. The energy loss mechanism of heavy quarks is still a puzzle for us to solve.

In 2010, PHENIX took high quality data in Au+Au collision at $\sqrt{s_{NN}} = 62.4$ GeV. In order to understand the density dependence and the geometrical dependence of the partonic energy loss, we measured heavy-flavor electron azimuthal anisotropy (v_2) at this lower beam energy of 62.4 GeV, where the medium formed in collisions is expected to have a lower energy-density than at 200 GeV.

Additionally, all the previous PHENIX measurements are a mixture of electrons from charm and bottom quarks. Because bottom quarks have much larger mass than charm quarks, they may interact with the medium differently. To separate charm and bottom, an accurate measurement of secondary decay vertex or the distance of closest approach (DCA) is required. The Silicon Vertex Track (VTX) installed in PHENIX in December 2010 makes the separation of charm and bottom possible in the PHENIX experiment.

Journal of Physics: Conference Series **446** (2013) 012041 doi:10.1088/1742-6596/446/1/012041

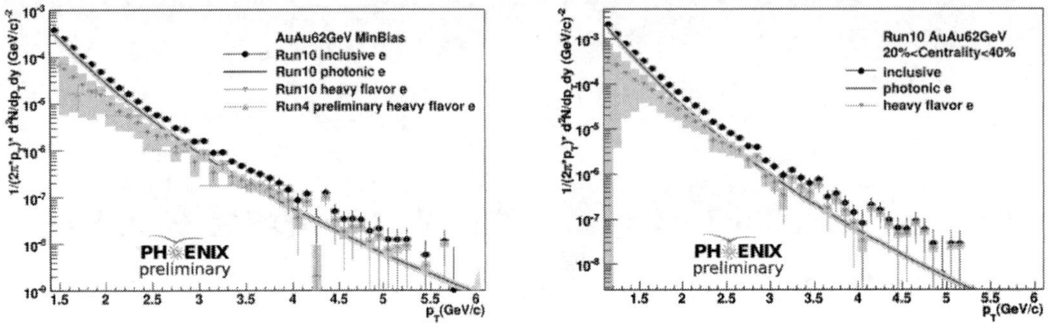

(a) (b)

Figure 1. Inclusive, photonic and heavy-flavor electron spectra for (a) MinBias and (b) $20\% < centrality < 40\%$ in Au+Au collisions at $\sqrt{s_{NN}} = 62.4$ GeV.

2. Heavy-Flavor Electrons v_2 Measurement in Au+Au Collisions at 62.4 GeV

The PHENIX Central Arm detectors, which cover $|\eta| < 0.35$ and $|\Delta\phi| = \pi$, provide track reconstruction and electron identification for this analysis. In Run-10, the Hadron Blind Detector (HBD) was also installed in PHENIX. The HBD is a windowless Cherenkov detector which is blind for hadrons up to around 4.5 GeV/c. Typically, the photoelectron associated with a single electron is around 20. By carefully tuning the cut on the HBD charge, we can achieve around 90% efficiency for single electrons, while rejecting more than 90% of the background from random matching and conversions that happened at or after the HBD backplane.

After applying all the cuts, the inclusive electrons contain heavy-flavor electrons and photonic electron background. The photonic electron sources include light meson Dalitz decays, photon conversions in the material (e.g. beam pipe, HBD entrance, HBD gas), vector meson decays, Ke3 decays, direct photon decays. The dominant backgrounds come from the π^0 and η Dalitz decay and photon conversions. The ingredients for all relevant background sources are measured and used to produce a cocktail estimate of the photonic background.

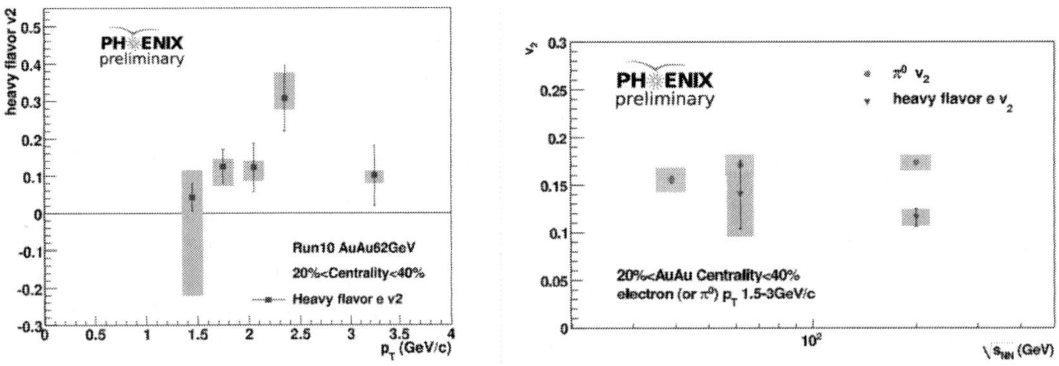

(a) (b)

Figure 2. Left panel is the heavy-flavor electron v_2 as a function of p_T for 20%<centrality<40% data set in Au+Au collisions at $\sqrt{s_{NN}} = 62.4$ GeV. Right panel shows the v_2 for heavy-flavor electrons and π^0 as a function of beam energy at $1.5 < p_T < 3$ GeV/c, $20\% <$ centrality $< 40\%$.

Heavy-flavor electron spectra for MinBias (Figure 1(a)) and centrality 20% – 40% (Figure 1(b)) data sets are obtained by subtracting the photonic cocktail from the inclusive electron spectra. Figure 1(a) shows a good agreement with PHENIX run4 preliminary results which gives us the confidence that the HBD detector is well understood in the analysis. The heavy-flavor electron v_2 is calculated from the inclusive electron v_2 by subtracting photonic electron v_2 at centrality 20% – 40% (Figure 2(a)). For p_T >1.5GeV/c, the heavy-flavor electron has a nonzero v_2 at 62.4 GeV given the statistical and systematic errors. Comparing the heavy-flavor electron v_2 and π^0 v_2 at 1.5< p_T <3GeV/c, 20% < *centrality* < 40% as a function of beam energy at RHIC (Figure 2(b)), we observe the heavy-flavor v_2 at 62.4 GeV is consistent with 200 GeV, and π^0 flow is saturated from 39GeV to 200 GeV. The error bars for heavy-flavor can be further reduced in the future analysis by tuning the HBD cuts to reject pairs of photonic e^+ and e^- that produce a double signal in the HBD.

3. Charm and Bottom Separation Using VTX Detector

The VTX detector was ready to take data in Run2011. The VTX detector has large coverage: $|\eta|$ <1.2, ∼2π in azimuth [5]. Due to the fine spatial resolution (∼ 80μm in Au+Au MB events), the most important capability VTX provides is to measure distance of closest approach (DCA) to separate charm and bottom components of heavy-flavor spectra. DCA is the shortest distance from the track to the primary vertex. Since B mesons have longer life time than D mesons, they have a wider width compared to the DCA distribution of D mesons. The bottom to charm ratio can be obtained by fitting the electron DCA.

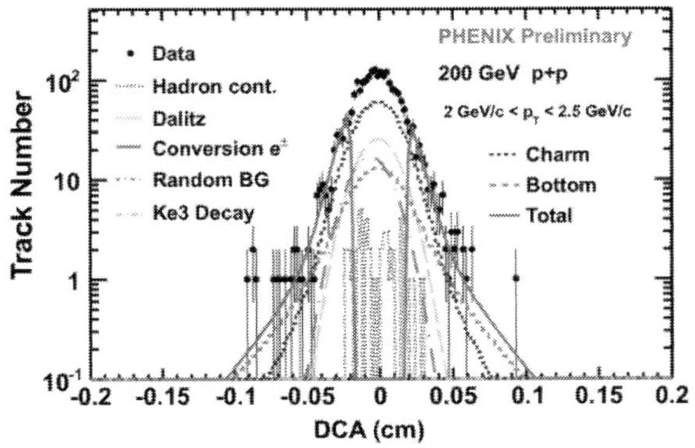

Figure 3. DCA decomposition at electron p_T range 2 to 2.5GeV/c in p+p collisions at $\sqrt{s_{NN}}$ = 200 GeV. The black dots are raw DCA distribution from data. The color curves represent the decomposed contribution from different sources. The total curve in red agrees with the raw DCA shape in the fitting range 0.2 < $|DCA|$ < 1.5 mm.

Similar to the previous analysis, firstly, the Central Arm electron identification cuts are applied to data to select the inclusive electron candidates. We also require electron candidates to associate with VTX clusters, especially a cluster in the most inner pixel layer to maximally reduce the photon conversions that happen in the VTX. To further reduce the photonic contribution, isolation cuts, i.e. conversion tagging cuts, are applied to each VTX layer to suppress the photonic electrons from Dalitz decay and photon conversions that happen at the beam pipe and the most inner VTX layer. The efficiency of Central Arm cuts and VTX cuts are studied through GEANT simulation.

After surviving all these cuts, the raw DCA is decomposed to different sources to obtain the bottom to charm ratio. Backgrounds in the raw DCA distribution are from hadron contamination, random matching with a VTX cluster, remaining photonic electrons. They can be estimated and subtracted from the raw DCA distribution. In p+p collisions at 200 GeV, the DCA shape of c→e and b→e are studied by GEANT simulation assuming the PYTHIA parent (e.g. D, B) p_T distribution and decay kinematics. Then the DCA distributions are fitted by the background components and the signal components to obtain b→e to c→e ratio at each p_T bin

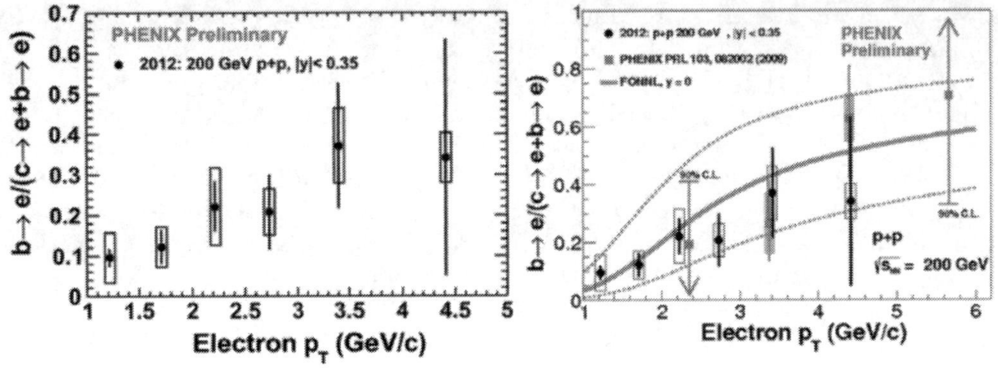

(a) (b)

Figure 4. $b \to e/(b \to e + c \to e)$ ratio as a function of electron p_T in p+p collisions at $\sqrt{s_{\rm NN}}$ = 200 GeV. The fraction is extracted from the DCA decomposition at each p_T bin assuming PYTHIA D and B parents p_T spectra. The right panel shows a comparison between the $b \to e/(b \to e + c \to e)$ ratio in p+p at $\sqrt{s_{\rm NN}} = 200$ GeV and the FONLL calculation [7].

[5]. Figure 3 shows the raw DCA distribution of electrons and all the signal and background contributions in the raw DCA in the range $2< p_T^e <2.5$GeV/c in p+p collisions at 200 GeV.

Figure 4(a) shows the resulting $b \to e/(b \to e + c \to e)$ ratio as a function of p_T^e in p+p $\sqrt{s_{\rm NN}}$ = 200 GeV collisions. Figure 4(b) shows that the latest results are consistent with the previous PHENIX measurement [6]. The fixed order plus next to leading log (FONLL) perturbative QCD calculation [7] also agrees with the PHENIX measurements. The first direct measurement of charm and bottom separately by DCA decomposition using VTX detector in p+p collision at RHIC has been achieved.

4. Summary

Heavy-flavor electron flow at 20% $<$ *centrality* $<$ 40% is measured in Au+Au collisions at $\sqrt{s_{\rm NN}}$ = 62.4 GeV in PHENIX. The HBD detector is working well on rejecting background including photon conversions happened at HBD backplane. The 62.4 GeV heavy-flavor electrons have non-zero flow at p_T >1.5GeV/c and v_2 is consistent with the 200 GeV result at $1.5< p_T <3$GeV/c, given the stated statistical and systematic uncertainties.

Using the VTX detector, the first direct measurement of charm and bottom separately in p+p collision at RHIC has been achieved by an accurate DCA measurement assuming PYTHIA parent meson (D and B) p_T distribution. The FONLL perturbative QCD calculation of $b \to e/(b \to e + c \to e)$ agrees with the p+p data. For the charm and bottom separation in Au+Au collisions, the PHENIX Collaboration has identified different DCA unfolding procedures to properly account for the modification of parent p_T spectra and is actively working towards robust, self-consistent results.

References

[1] A. Adare et al. [PHENIX] 2007 Phys. Rev. Lett. **98** 172301
[2] A. Adare et al. [PHENIX] 2011 Phys. Rev. **C 84** 044905
[3] Y L Dokshitzer and D E Kharzeev 2001 Phys. Lett. **B 519** 199
[4] N Armesto, A Dainese, C A Salgado and U A Wiedemann 2005 Phys. Rev. **D 71** 054027
[5] R. Nouicer et al. [PHENIX] 2011 Nuclear Physics A 862 64 and references therein
[6] A. Adare et al. [PHENIX] 2009 Phys. Rev. Lett. **103** 082002
[7] M. Cacciari et al. 2005 Phys. Rev. Lett **95** 122001

J/ψ production in Pb-Pb collisions at $\sqrt{s_{\rm NN}}$=2.76 TeV in the ALICE experiment

Antoine Lardeux

SUBATECH, CNRS/IN2P3, École des Mines de Nantes, Université de Nantes, 4 rue Alfred Kastler, 44307 Nantes, France

E-mail: `Antoine.Lardeux@subatech.in2p3.fr`

Abstract. Quarkonium states are expected to provide essential information on the properties of the high-density strongly-interacting system formed in the early stages of heavy-ion collisions. In particular the J/ψ suppression via color screening is a direct consequence of deconfinement. ALICE is the LHC experiment specifically designed to study nucleus-nucleus collisions. The production of heavy quarkonium states is measured by ALICE down to zero transverse momentum via the $\mu^+\mu^-$ decay channel in the Forward Muon Spectrometer ($2.5 < y < 4$) and via the e^+e^- decay channel in the central barrel at mid rapidity ($|y| < 0.9$). The analysis of the inclusive J/ψ production in Pb-Pb collisions at a center of mass energy per nucleon pair $\sqrt{s_{\rm NN}}$=2.76 TeV is presented. The inclusive J/ψ nuclear modification factor as a function of centrality, transverse momentum and rapidity is shown and compared to similar measurements by other experiments and to theoretical predictions. Finally, a hint of a low transverse momentum J/ψ yield excess, with respect to the expected hadronic channel production, is discussed.

For more than 25 years, J/ψ production in heavy-ion collisions has been a promising probe of Quark Gluon Plasma (QGP), motivated by the prediction of Matsui and Satz in 1986 [1] of J/ψ dissociation due to color-screening in the QGP. J/ψ suppression has been extensively studied experimentally in heavy-ion collisions, first at the SPS and then at RHIC and it was found to be significantly larger than the suppression expected from cold nuclear matter effects, such as shadowing and nuclear absorption. However, in spite of the different center of mass energy of the two accelerators, the observed suppression patterns present similar features. This could be explained by an additional J/ψ production mechanism in heavy-ion collisions via the combination of initially uncorrelated charm and anti-charm quarks [2–5]. Another possibility is that directly produced J/ψ are not suppressed at both SPS and RHIC, and that the ~40% suppression due to hot matter is due to the suppression of the J/ψ coming from the decay of ψ(2S) and χ_c quarkonia [6]. The measurement of J/ψ suppression is of special interest at the Large Hadron Collider (LHC), where the high energy density of the medium and the large number of charm quark pairs produced in central Pb-Pb collisions should help to disentangle the suppression and recombination scenarios. In the following we describe the measurement by the ALICE experiment at the LHC of charmonium production at forward and mid-rapidity. The dependence of the J/ψ nuclear modification factor ($R_{\rm AA}$) on centrality, transverse momentum and rapidity are compared to previous measurements and theoretical predictions. Finally, an intriguing observation of a low transverse momentum J/ψ yield excess with respect to the expected hadronic channel production is discussed.

The ALICE experiment, described in detail in [7], is specifically designed to study nucleus-nucleus collisions. The dielectron analysis is based on a data sample of 12.8 million Pb-Pb collisions (0-80% most central) from the 2010 runs and 17.5 million Pb-Pb collisions (10 million 0-10% and 7.5 million 10-40%) from the 2011 runs, both collected with a minimum bias (MB) trigger. This corresponds to an integrated luminosity $\mathcal{L}_{\rm int}\sim 15\mu b^{-1}$. Electrons are reconstructed in the central barrel using the high precision tracking and particle identification of the Inner Tracking System (ITS) and of the Time Projection Chamber (TPC). Particle identification (PID) is achieved via the measurement of the specific energy loss of particles in the TPC, with dE/dx resolution of $\sim$ 5%. For the dimuon analysis at the same center of mass energy, 17.7 million Pb-Pb events were collected in 2011 with a dimuon trigger. This corresponds to an integrated luminosity $\mathcal{L}_{\rm int}\sim 70\mu b^{-1}$. Muons are reconstructed in the forward muon spectrometer, which consists of a front absorber to remove hadrons, followed by five tracking stations, the third one being embedded in a 3 Tm dipole magnet. The muon spectrometer also includes a triggering system located downstream of a 1.2 m iron wall, which absorbs secondary hadrons and low momentum muons, mainly from light hadron decays. Two scintillator arrays (VZERO), located on either side of the interaction vertex, are used for triggering, centrality determination and background removal. To further reject electromagnetic interactions and satellite collisions, the two Zero Degree Calorimeters, placed at $\pm$114 m from the collision, are used. The analysis was limited to 90% most central collisions, for which the MB trigger is fully efficient.

In both channels, the J/ψ yield is extracted from the invariant mass distribution of opposite sign (OS) lepton pairs. In case of the dielectron channel [8], this distribution is dominated by the contribution of uncorrelated pairs, which is subtracted using an event mixing technique. The mixed event background distribution is scaled to match the integral of the signal distribution in the invariant mass range between 3.2 and 4.0 GeV/c^2. The remaining contribution to the J/ψ signal from the correlated e^+e^- continuum is small and is included in the systematic uncertainty. In the dimuon channel, an extended Crystal-Ball function (CB2), which consists of a gaussian core with two power law tails, is fit to the J/ψ signal. These CB2 tails are tuned on a Monte-Carlo (MC) production of J/ψ embedded into real events. The background contribution is described by a gaussian function with a mass-dependent width. Alternatively, an event mixing procedure has also been applied to subtract the background contribution, before fitting the signal. The residual background is then described by an exponential or a first order polynomial function. The raw J/ψ yield is then determined as the average of the results obtained with the two approaches, as detailled in [9]. In both cases, the extracted raw yield is corrected for the acceptance $\times$ efficiency ($A\epsilon$) of the apparatus. This correction is evaluated, for the dielectron analysis using simulated Pb-Pb collisions generated by the HIJING event generator enriched with primary J/ψ, while for the dimuon analysis the correction is determined by embedding generated J/ψ particles into real events. The kinematic distributions of the simulated J/ψ are obtained from an interpolation of RHIC, Fermilab and LHC data [10]. For the dielectron channel, $A\epsilon$ is 8%, while for the dimuon channel it is about 14%. A weak dependence (not more than 12%) of the efficiency on the collision centrality is observed in both cases.

The in-medium modification of J/ψ production can be quantified through measurement of the nuclear modification factor, defined as the ratio of the inclusive J/ψ yield measured in Pb-Pb collisions to the expected yield, where the latter is obtained by scaling the pp J/ψ production cross section at the same center of mass energy per nucleon pair by the nuclear overlap function ($T_{\rm AA}$). Details of pp reference in both rapidity intervals are found in [11]. These $R_{\rm AA}$ is measured separately for each kinematic bin under study. The main contribution to the systematic uncertainty for the dielectron channel results from the description of the combinatorial background. The main contribution to the global systematic uncertainty comes from the pp reference cross section. The centrality dependence of the inclusive J/ψ $R_{\rm AA}$, integrated over $p_{\rm T}$ and y, is shown in Fig.1 (left). For the most central 0-10% collisions we obtain at mid-rapidity

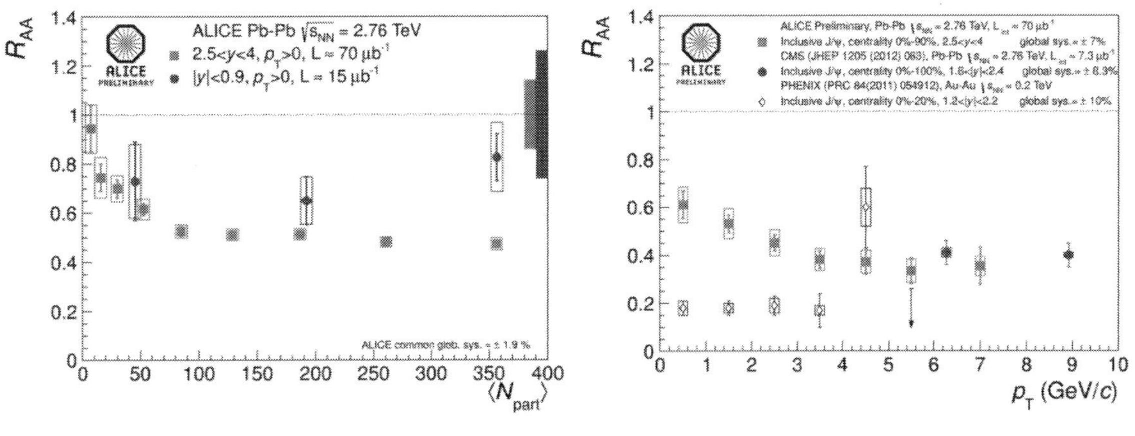

Figure 1: Left: J/ψ $R_{\rm AA}$ as a function of the number of participants at forward and mid-rapidity. Statistical errors are shown as lines, while systematic uncertainties are shown as boxes around the points. Right: J/ψ $R_{\rm AA}$ as a function of $p_{\rm T}$ at forward rapidity. Comparisons to the CMS and PHENIX results are shown.

an $R_{\rm AA}$ value of $0.83 \pm 0.09(stat) \pm 0.26(syst)$. At forward rapidity, the inclusive $R_{\rm AA}$ value averaged over centrality, $p_{\rm T}$ and y is $0.497 \pm 0.006(stat) \pm 0.078(syst)$, exhibiting less suppresion than at RHIC energies. In addition, the observed pattern has weaker centrality dependence and a smaller suppression for central collisions than the PHENIX measurements [12]. The ALICE data suggest that the $R_{\rm AA}$ decreases with increasing rapidity (more details in [8]) and suggest a different interplay of suppression and regeneration mechanisms compared to PHENIX data. Fig.1 (right) shows $R_{\rm AA}$ as a function of $p_{\rm T}$ as measured by ALICE, compared to CMS data [13] as well as to results from PHENIX [12]. $R_{\rm AA}$ is decreasing from 0.6 at low $p_{\rm T}$ to about 0.4 at higher $p_{\rm T}$. The CMS results are in agreement with the ALICE measurements in their common $p_{\rm T}$ range. The lower energy results from PHENIX show a significantly smaller $R_{\rm AA}$. The $p_{\rm T}$ dependence of J/ψ $R_{\rm AA}$ for different centrality classes is shown in the left pannel of Fig.2. The data are fairly well reproduced by theoretical model predictions that include a large fraction of J/ψ produced from regenerated pairs [14]. This contribution is expected to be dominant at low $p_{\rm T}$, especially in central collisions. Additional results on the $<p_{\rm T}>$ of the J/ψ as a function of the centrality [15] and on the J/ψ elliptic flow [16] are in agreement with such a scenario.

The $p_{\rm T}$ yield of OS dimuon pairs in the mass range $2.95 < {\rm M}_{\mu^+\mu^-} < 3.3$ GeV$/c^2$, integrated over y and for peripheral collisions (70-90%), is shown in Fig.2 (right). In this mass range, the J/ψ contribution dominates, with a signal over background ratio of 3.96 ± 0.52. No $A\epsilon$ correction has been applied. The red line[1] represents the expected contribution from hadronic J/ψ production [10]. A clear excess is observed at very low $p_{\rm T}$, below 300 MeV$/c$. This excess is also seen in the semi-peripheral collisions 50-70%. J/ψ particles are also be generated via a coherent photo-production mechanism [17], as is measured at forward rapidity in Ultra Peripheral Pb-Pb Collision (UPC) at $\sqrt{s_{\rm NN}}$=2.76 TeV [18]. The unique feature of this mechanism is the production of J/ψ at very low $p_{\rm T}$ ($<p_{\rm T}>\approx 60$ MeV$/c$). Theoretical photo-production predictions can be found in [19] although the calculations are done for UPC, *i.e.* for an impact parameter larger than twice the nuclear radius. In Pb-Pb peripheral collisions, J/ψ photonic and hadronic production could exhibit similar production cross-sections [20]. The low $p_{\rm T}$ excess shown in Fig.2 (right) could then be due to the contribution of photo-produced J/ψ. Since this production only occurs in the initial stage of the collision, it may be a novel way to study the medium induced dissociation of the J/ψ.

[1] Also called "red wine" for Hot Quarks participants.

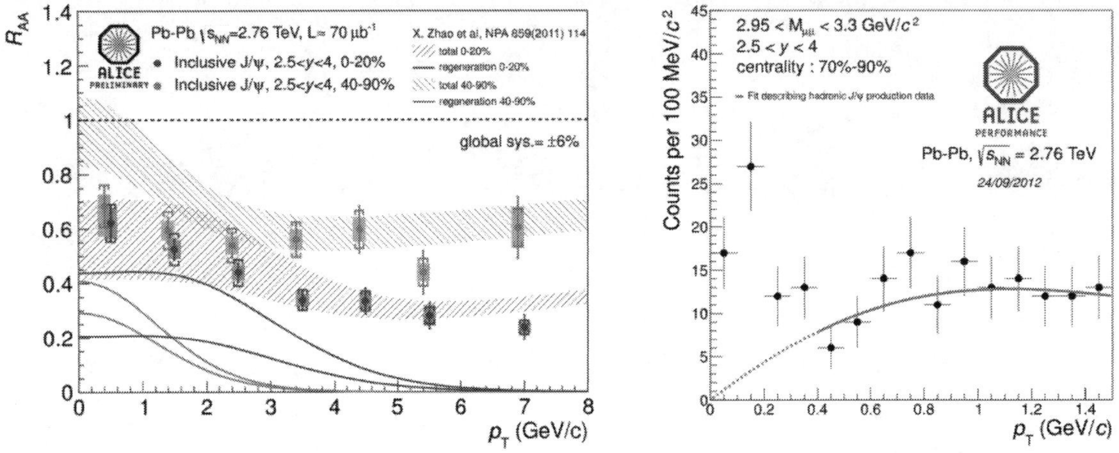

Figure 2: Left: J/ψ R_{AA} p_{T} dependence at forward rapidity for two centrality classes. Systematic uncertainties are split between uncorrelated and partially correlated components, shown, respectively, as boxes and brackets around the points. Right: Dimuon p_{T} distribution from opposite sign invariant mass spectrum around J/ψ mass at forward rapidity.

In summary, ALICE has measured J/ψ R_{AA} as a function of centrality, p_{T} and rapidity, in Pb-Pb collisions at $\sqrt{s_{\mathrm{NN}}}$=2.76 TeV. At mid-rapidity, the R_{AA} indicates a moderate J/ψ suppression, with no significant centrality dependence. At forward rapidity, the R_{AA} shows a clear reduction of the J/ψ yield, with negligible centrality dependence and clear p_{T} dependence, especially in central collisions. These features can be qualitatively described by theoretical models which include regeneration as an additional mechanism for J/ψ production in heavy-ion collisions. Further insight into J/ψ production and suppression still needs a precise knowledge of the cold nuclear matter effects, which will be studied with p-Pb collisions in early 2013. Finally, a low transverse momentum J/ψ yield excess in hadronic collision has been observed, and may arise from J/ψ photo-production mechanism.

References

[1] Matsui T and Satz H 1986 *Phys.Lett.* **B178** 416
[2] Braun-Munzinger P and Stachel J 2000 *Phys.Lett.* **B490** 196–202
[3] Thews R L, Schroedter M and Rafelski J 2001 *Phys.Rev.* **C63** 054905
[4] Andronic A, Braun-Munzinger P, Redlich K and Stachel J 2007 *Phys.Lett.* **B652** 259–261
[5] Zhao X and Rapp R 2008 *Phys.Lett.* **B664** 253–257
[6] Karsch F, Kharzeev D and Satz H 2006 *Phys.Lett.* **B637** 75–80
[7] Aamodt K *et al.* (ALICE Collaboration) 2008 *JINST* **3** S08002
[8] Arsene I C (for the ALICE Collaboration) 2012 (*Preprint* 1210.5818)
[9] Abelev B *et al.* (ALICE Collaboration) 2012 *Phys.Rev.Lett.* **109** 072301
[10] Bossu F, del Valle Z C, de Falco A, Gagliardi M, Grigoryan S *et al.* 2011 (*Preprint* 1103.2394)
[11] Abelev B *et al.* (ALICE Collaboration) 2012 *Phys.Lett.* **B718** 295
[12] Adare A *et al.* (PHENIX Collaboration) 2011 *Phys.Rev.* **C84** 054912
[13] Chatrchyan S *et al.* (CMS Collaboration) 2012 *JHEP* **1205** 063
[14] Zhao X and Rapp R 2011 *Nucl.Phys.* **A859** 114–125
[15] Arnaldi R (for the ALICE Collaboration) 2012 (*Preprint* 1211.2578)
[16] Massacrier L (for the ALICE Collaboration) 2012 (*Preprint* 1208.5401)
[17] Frankfurt L, Strikman M and Zhalov M 2002 *Phys.Lett.* **B540** 220–226
[18] Abelev B *et al.* (ALICE Collaboration) 2012 (*Preprint* 1209.3715)
[19] Rebyakova V, Strikman M and Zhalov M 2012 *Phys.Lett.* **B710** 647–653
[20] Nystrand J 2010 ReteQuarkonii workshop Nantes

Journal of Physics: Conference Series **446** (2013) 012043 doi:10.1088/1742-6596/446/1/012043

Observation of sequential Υ suppression in PbPb collisions

Zhen Hu, on the behalf of the CMS collaboration
Purdue University, West Lafayette, IN 47906, USA
E-mail: zhen.hu@cern.ch

Abstract. The sequential suppression of the individual Υ states in PbPb collisions with respect to their yields in *pp* data has been measured. The PbPb and *pp* datasets used in these proceedings correspond to integrated luminosities of $150/\mu$b and 230/nb, respectively, collected by the CMS experiment at the LHC at 2.76 TeV in 2011. The relative suppression of the excited Υ states has been measured with respect to the Υ(1S) ground state, expressed as a double ratio $[\Upsilon(\mathrm{nS})/\Upsilon(1\mathrm{S})]_{\mathrm{PbPb}}/[\Upsilon(\mathrm{nS})/\Upsilon(1\mathrm{S})]_{pp}$. The absolute suppression of the Υ(nS) yields in PbPb relative to the yields in *pp* scaled by the number of nucleon-nucleon collisions, R_{AA}, is measured as a function of the collision centrality. Integrated over centrality, the R_{AA} values for Υ(1S), Υ(2S), and the upper limit of Υ(3S) R_{AA} are reported in this proceeding, which demonstrate the sequential suppression of the Υ(nS) states in PbPb collisions at LHC energies.

1. Introduction

The LHC allows for the detailed studies of the heavy quarkonium states in ultra-relativistic heavy-ion collisions. Suppression of heavy quarkonium states has been proposed as a probe of the properties of the hot and dense medium created in high-energy heavy-ion collisions [1]. Quantum chromodynamics (QCD) predicts that strongly interacting matter undergoes a phase transition to a deconfined state, often referred to as the quark-gluon plasma (QGP), in which quarks and gluons are no longer bound within hadrons. If the QGP is formed in heavy-ion collisions, it is expected to screen the confining potential of heavy quark-antiquark pairs [1], leading to the melting of charmonium and bottomonium states: J/ψ, $\psi(2S)$, χ_c, Υ(1S), Υ(2S), Υ(3S), χ_b, etc. The melting temperature depends on the binding energy of the quarkonium state. The ground states, J/ψ and Υ(1S), are expected to dissolve at significantly higher temperatures than the more loosely bound excited states. Quenched lattice QCD calculations [2][3] originally predicted that the Υ states melt at 1.2 T_c (3S), 1.6 T_c (2S), and above 4 T_c (1S), while modern spectral-function approaches with complex potentials [4] favor somewhat lower dissolution temperatures. This sequential melting pattern is generally considered a smoking-gun signature of the QCD deconfinement transition. Given the mass resolution achieved, the CMS detector suits these measurements well.

The indication of the Υ excited states suppression at the LHC was published in [5] in 2011 with the first $7.28/\mu$b PbPb data at $\sqrt{s_{NN}} = 2.76$ TeV. In these proceedings, we report a follow up study revealing the observation of sequential Υ suppression with full $150/\mu$b PbPb data collected in 2011 by the CMS experiment [6].

Journal of Physics: Conference Series **446** (2013) 012043 doi:10.1088/1742-6596/446/1/012043

2. Signal extraction

A detailed description of the CMS detector can be found in [7]. The trigger conditions, offline event selection, and muon reconstruction criteria are discussed in [6]. The same offline reconstruction algorithm and selection criteria are applied to the PbPb and pp data samples. In order to reduce the background in the Υ mass region, only muons with a transverse momentum (p_T^{μ}) higher than 4 GeV/c are considered [5].

An extended unbinned maximum likelihood fit is performed to extract the signal yields, following the method described in [6][8]. The measured mass line shape of each Υ state is parametrized by a Crystal Ball (CB) function, i.e. a Gaussian resolution function with the low-side tail replaced by a power law describing final state radiation (FSR). The mass differences between the states are fixed to their world average values [9] and the mass resolution is forced to scale with the resonance mass. The signal shape parameters are treated as common for both PbPb and pp data sets via a simultaneous fit. The background model for the pp data set consists of a second-order polynomial, as was used in Ref. [5]. The larger PbPb data set requires a more detailed background model, so an exponential function multiplied by an error function is used to describe its low-mass turn-on effect due to the p_T^{μ} selection threshold.

The dimuon invariant mass spectra are shown in Fig. 1 for PbPb and in Fig. 2 for pp. The solid (signal + background) and dashed (background-only) curves show the simultaneous fit to the two data sets. The three Υ peaks are clearly observed in the pp case, but the Υ(3S) peak is not visible over the residual background in PbPb collisions.

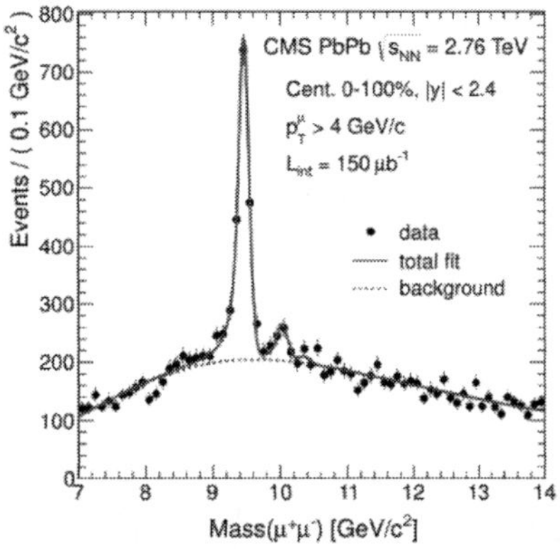

Figure 1. Dimuon invariant-mass distributions in PbPb data at $\sqrt{s_{NN}} = 2.76$ TeV.

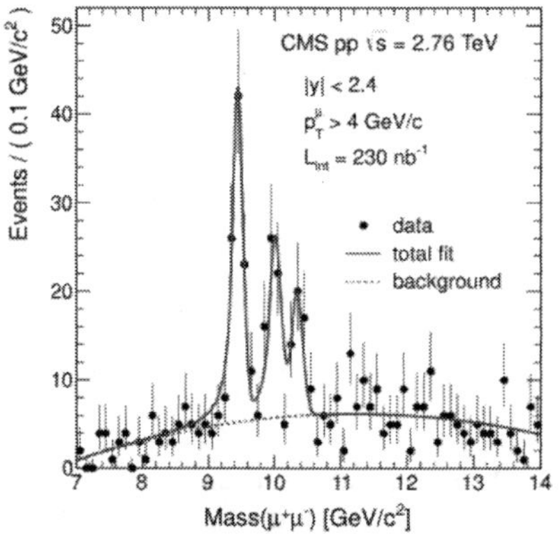

Figure 2. Dimuon invariant-mass distributions in pp data at $\sqrt{s} = 2.76$ TeV.

3. Relative suppression of excited Υ states in PbPb collisions

The simultaneous fit to the PbPb and pp mass spectra gives the ratio of the Υ(nS)/Υ(1S) ratios in PbPb and pp collisions:

$$\begin{aligned}
[\Upsilon(2S)/\Upsilon(1S)]_{\mathrm{PbPb}}\,/\,[\Upsilon(2S)/\Upsilon(1S)]_{pp} &= 0.21 \pm 0.07\,(\mathrm{stat.}) \pm 0.02\,(\mathrm{syst.})\,, \\
[\Upsilon(3S)/\Upsilon(1S)]_{\mathrm{PbPb}}\,/\,[\Upsilon(3S)/\Upsilon(1S)]_{pp} &= 0.06 \pm 0.06\,(\mathrm{stat.}) \pm 0.06\,(\mathrm{syst.}) \\
&< 0.17\ (95\%\ \mathrm{CL})\,. \qquad (1)
\end{aligned}$$

This double ratio measurement benefits from an almost complete cancellation of possible efficiency and acceptance differences among the reconstructed resonances. The systematic uncertainties from the fitting procedure are evaluated by varying the fit function. An additional systematic uncertainty (1%), estimated from MC simulation, is included to account for possible imperfect cancellations of efficiency and acceptance.

The double ratios, defined in Eq. (1), are expected to be compatible with unity if there is no suppression of the excited Υ states relative to the $\Upsilon(1S)$ state. Instead, the measured values are considerably smaller than unity. The significance of the observed suppression exceeds $5\,\sigma$.

In order to investigate the dependence of the suppression on the centrality, the double ratio is displayed as a function of $N_{\rm part}$ in Fig. 3. The results are constructed from the single ratio $[\Upsilon(2S)/\Upsilon(1S)]_{\rm PbPb}$ measured in bins of PbPb centrality, using the *pp* ratio as normalization. The dependence on centrality is not pronounced. More data, in particular more *pp* collisions, are needed to further establish possible dependences on centrality and kinematic variables.

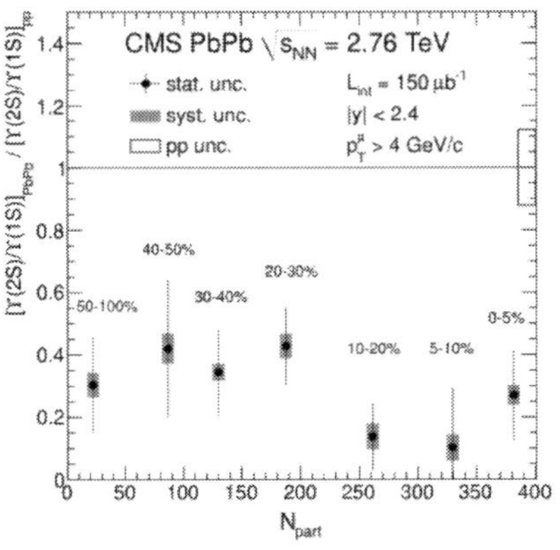

Figure 3. Centrality dependence of the double ratio. The relative uncertainties from $N_{\rm part}$-independent quantities are represented by the box at unity, and are not included in the data points as these uncertainties do not affect the point-to-point trend. The event centrality bins used are indicated by percentage intervals.

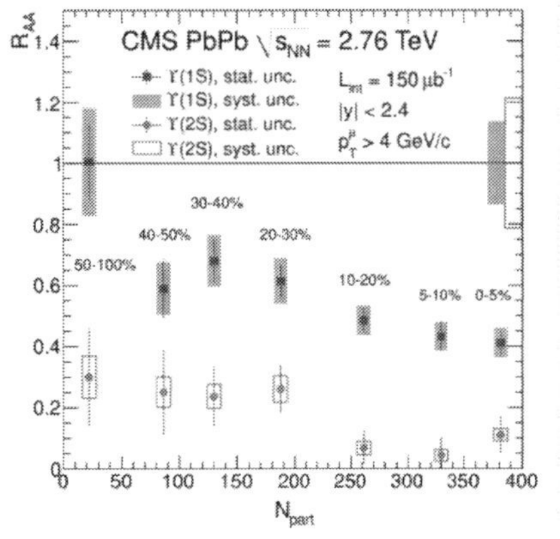

Figure 4. Centrality dependence of the nuclear modification factors for the $\Upsilon(1S)$ and $\Upsilon(2S)$ states. The two relative uncertainties from $N_{\rm part}$-independent quantities (*pp* yields, efficiency and integrated luminosity) are represented by the boxes at unity. These common uncertainties are not included in the data points.

4. Absolute suppression of the Υ(nS) states in PbPb collisions

Absolute suppressions of the individual Υ states and their dependence on the collision centrality are studied using the nuclear modification factor, R_{AA}, defined as the yield per nucleon-nucleon collision in PbPb relative to that in *pp* [6].

The centrality-integrated (0–100%) R_{AA} values for the three individual Υ states are:

$$\begin{aligned} R_{AA}(\Upsilon(1S)) &= 0.56 \pm 0.08\,(\text{stat.}) \pm 0.07\,(\text{syst.})\,, \\ R_{AA}(\Upsilon(2S)) &= 0.12 \pm 0.04\,(\text{stat.}) \pm 0.02\,(\text{syst.})\,, \\ R_{AA}(\Upsilon(3S)) &= 0.03 \pm 0.04\,(\text{stat.}) \pm 0.01\,(\text{syst.}) \\ &< 0.10\ (95\%\ \text{CL})\,. \end{aligned} \qquad (2)$$

The centrality dependences of the $\Upsilon(1S)$ and $\Upsilon(2S)$ R_{AA} are obtained by performing the measurement in ranges of centrality, as displayed in Fig. 4. The results indicate a significant suppression of the Υ(nS) states in heavy-ion collisions compared to *pp* collisions at the same per-nucleon-pair energy. The data support the hypothesis of increased suppression of less strongly bound states: the Υ(1S) is the least suppressed and the Υ(3S) is the most suppressed of the three states. The Υ(1S) and Υ(2S) suppressions are observed to increase with collision centrality. The observed Υ(nS) yields contain contributions from heavier bottomonium states decays. As a result, the measured suppression is affected by the dissociation of these states. The detailed discussion about the feed-down contribution is in [6]. These results indicate that the directly produced Υ(1S) state is not significantly suppressed, however quantitative conclusions will require precise estimations of the feed-down contribution matching the phase space of the suppression measurement. In addition to QGP formation, cold-nuclear-matter effects can also lead to the differences between quarkonium production yields in PbPb and *pp* collisions [10]. More discussion can be found in [6]. The comparisons to two theoretical models are shown in Fig. 5 and Fig. 6.

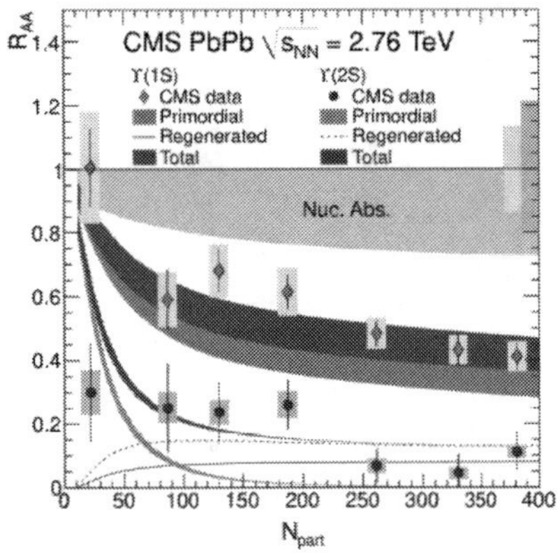

Figure 5. Strong-binding scenario (SBS) prediction, produced using the calculations performed in the paper [11], based upon the approach developed in [12].

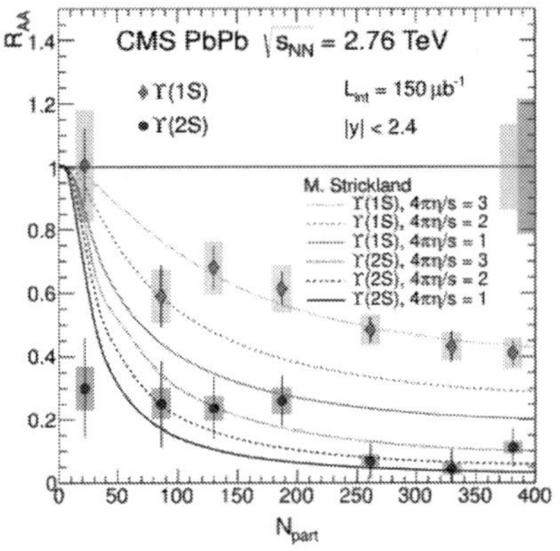

Figure 6. Comparison of the Υ(1S) and Υ(2S) nuclear modification factor R_{AA} centrality dependence result to theory prediction by M. Strickland [13][14].

In conclusion, the nuclear modification factors for the Υ(nS) states have been measured in this analysis. The observation of sequential suppression of the Υ(nS) states in heavy-ion collisions is reported in these proceedings.

References

[1] Matsui T and Satz H 1986 *Phys. Lett.* B **178** 416
[2] Satz H 2006 *J. Phys.* G **32** R25
[3] Wong C Y 2005 *Phys. Rev.* C **72** 034906
[4] Petreczky P, Miao C and Mocsy A 2011 *Nucl. Phys.* A **855** 125
[5] CMS Collaboration *et al* 2011 *Phys. Rev. Lett.* **107** 052302
[6] CMS Collaboration *et al* 2012 *Phys. Rev. Lett.* **109** 222301
[7] CMS Collaboration *et al* 2008 *JINST* **3** S08004
[8] CMS Collaboration *et al* 2011 *Phys. Rev.* D **83** 112004
[9] Nakamura K *et al* (Particle Data Group) 2010 *J. Phys. G: Nucl. Part. Phys.* **37** 075021
[10] Vogt R *et al* 2010 *Phys. Rev.* C **81** 044903
[11] Emerick A, Zhao X and Rapp R 2012 *Eur. Phys. J.* A **48** 72
[12] Grandchamp L *et al* 2006 *Phys.Rev.* C **73** 064906
[13] Strickland M 2011 *Phys. Rev. Lett.* **107** 132301
[14] Strickland M 2012 Thermal Bottomonium Suppression *Preprint* arXiv:1207.5327

Hot Quarks 2012 IOP Publishing
Journal of Physics: Conference Series **446** (2013) 012044 doi:10.1088/1742-6596/446/1/012044

Open heavy flavor and J/psi at RHIC and LHC within a transport model

Jan Uphoff,[1] Oliver Fochler,[1] Zhe Xu,[2] and Carsten Greiner[1]

[1] Institut für Theoretische Physik, Johann Wolfgang Goethe-Universität Frankfurt, Max-von-Laue-Str. 1, D-60438 Frankfurt am Main, Germany

[2] Department of Physics, Tsinghua University, Beijing 100084, China

E-mail: `uphoff@th.physik.uni-frankfurt.de`

Abstract. The production and space-time evolution of heavy quarks and J/ψ in the quark gluon plasma is studied within the partonic transport model *Boltzmann Approach to MultiParton Scatterings* (BAMPS). This framework allows interactions among all partons: gluons, light quarks and heavy quarks. Heavy quarks, in particular, interact with the rest of the medium via binary scatterings with a running coupling and a more precise Debye screening which is derived from hard thermal loop calculations. We compare our results of the elliptic flow and nuclear modification factor not only to experimental data of heavy flavor electrons at RHIC, but also to LHC data of heavy flavor electrons, muons, D mesons, and non-prompt J/ψ. Where no data is available yet, we make predictions for those observables. Furthermore, results on prompt J/ψ elliptic flow are reported for RHIC energy within the same framework, taking the dissociation as well as regeneration of J/ψ in the quark-gluon plasma into account.

1. Introduction

In ultra-relativistic heavy ion collisions at the Relativistic Heavy Ion Collider (RHIC) [1, 2] and the Large Hadron Collider (LHC) [3] a unique state of matter is produced, in which quarks and gluons form the relevant degrees of freedom. This quark gluon plasma (QGP) has exciting properties such as collective behavior like a nearly perfect liquid or the quenching of high energy particles.

In particular heavy quarks (charm and bottom) provide an insightful way to learn more about the properties of this matter. Since they are heavy, their production time is at a very early stage of the heavy ion collision when enough energy is available [4]. In consequent interactions with other particles they are influenced by the medium without fully thermalizing [5] and, therefore, carry information about its properties. Charm and bottom quarks travel through the medium, lose energy and participate in the collective motion. The exact mechanism of the intense interaction with the medium is actively debated, most recently, for instance, in [6, 7, 8, 9, 10, 11, 12, 13, 14, 15].

Complimentary to these open heavy flavor particles are particles in which the flavor is hidden. That is, bound states of a heavy quark and an anti-heavy quark. They can survive in the QGP to some extent, but melt if the temperature of the medium is too high [16]. Therefore, they could be used as a thermometer. In addition to this melting they can also be regenerated by two heavy quarks which meet in the medium. One prominent hidden heavy flavor particle is the J/ψ, which consists of a charm and an anti-charm quark.

In this paper we employ the partonic transport model BAMPS to study the suppression and elliptic flow of both open and hidden heavy flavor. In detail, we make calculations for open heavy flavor observables such as the nuclear modification factor R_{AA} and elliptic flow v_2 of heavy flavor muons and electrons, D mesons and non-prompt J/ψ which stem from B mesons. Furthermore we present v_2 calculations of prompt J/ψ.

2. Partonic transport model BAMPS

For the simulation of the QGP we use the partonic transport model *Boltzmann Approach to MultiParton Scatterings* (BAMPS) [17, 18], which describes the full space-time evolution of the QGP by solving the Boltzmann equation,

$$\left(\frac{\partial}{\partial t} + \frac{\mathbf{p}_i}{E_i}\frac{\partial}{\partial \mathbf{r}}\right) f_i(\mathbf{r}, \mathbf{p}_i, t) = \mathcal{C}_i^{2\to 2} + \mathcal{C}_i^{2\leftrightarrow 3} + \dots \,, \tag{1}$$

for on-shell partons and pQCD interactions. On the heavy flavor sector all possible binary collisions are implemented, such as $gg \leftrightarrow Q\bar{Q}$, $q\bar{q} \leftrightarrow Q\bar{Q}$, $gQ \to gQ$, $qQ \to qQ$, and $J/\psi + g \leftrightarrow c + \bar{c}$. As a note, heavy quark production via gluon fusion within BAMPS is discussed in Ref. [4], but will not be addressed in this paper.

The divergent t channel of elastic processes is regularized with a screening mass μ which is determined by matching energy loss calculations within the hard thermal loop approach [19, 20, 21]. Furthermore, the running of the coupling is explicitly taken into account. We note that currently for light partons neither a running coupling nor an improved Debye screening is employed, which we plan, however, to do in the future. Inelastic processes $2 \to 3$ are currently being studied for heavy quarks, but have already been implemented for light partons. Details can be found in the following references: the model itself [17, 18], the open heavy flavor implementation [4, 21, 8], and the J/ψ details [22, 23].

3. Results and comparison with data

The implemented running coupling and the improved screening procedure, which reproduces the energy loss from hard thermal loop calculations, effectively enhances the heavy quark cross section with the medium. Quantitative comparisons [24, 21, 25, 7, 8] show that elastic processes contribute significantly to the energy loss of heavy quarks. However, they alone can reproduce neither the data of the nuclear modification factor nor the elliptic flow of any heavy flavor particle species. This is not too surprising since we expect that radiative $2 \to 3$ processes also play an important role and that both processes together should account for the measured suppression and flow. The quantitative contribution of radiative processes will be studied in a forthcoming investigation.

In this paper we mimic the radiative influence by effectively increasing the elastic cross section by a factor $K = 3.5$ which fits the heavy flavor electron elliptic flow data from PHENIX in the centrality class 20-40 % [8]. On the left hand side of Fig. 1 we show our prediction for 0-60 % centrality class for the same parameters, which is compared to new data from STAR [26]. Not surprisingly the agreement is very good for small and intermediate transverse momenta. Even the bump around 1.7 GeV is nicely described. At large p_T, however, the data is significantly larger than our calculations. This could also be due to jet-like correlations which are not considered in the analysis [26]. Simultaneously, the R_{AA} data at RHIC can also be described with the same parameter [8]. Whether this effective description is valid and radiative contributions really boil down to simply multiplying the elastic cross section by a constant factor will be studied with BAMPS in the future.

On the right hand side of Fig. 1 the nuclear modification factor of D mesons for very central events at LHC is depicted. Our prediction with BAMPS is slightly smaller than the experimental

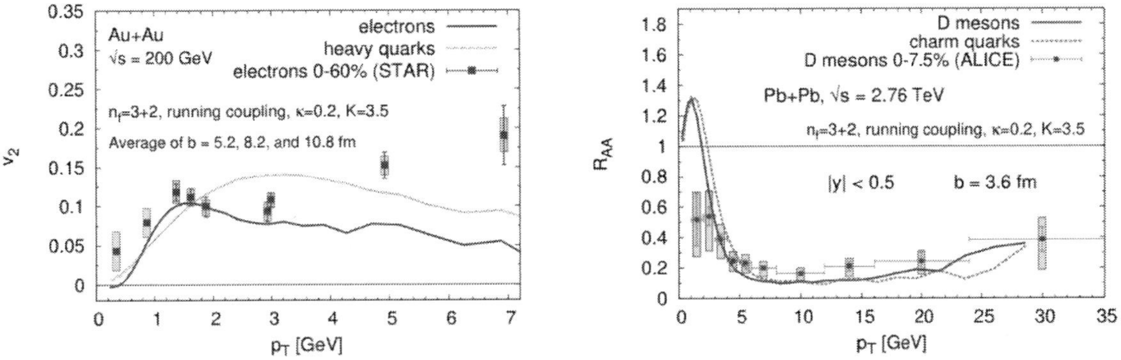

Figure 1. Left: Elliptic flow v_2 of heavy flavor electrons at RHIC for 0-60 % centrality class as a function of transverse momentum with STAR data [26]. Right: Nuclear modification factor R_{AA} of D mesons for $b = 3.6$ fm with data [27].

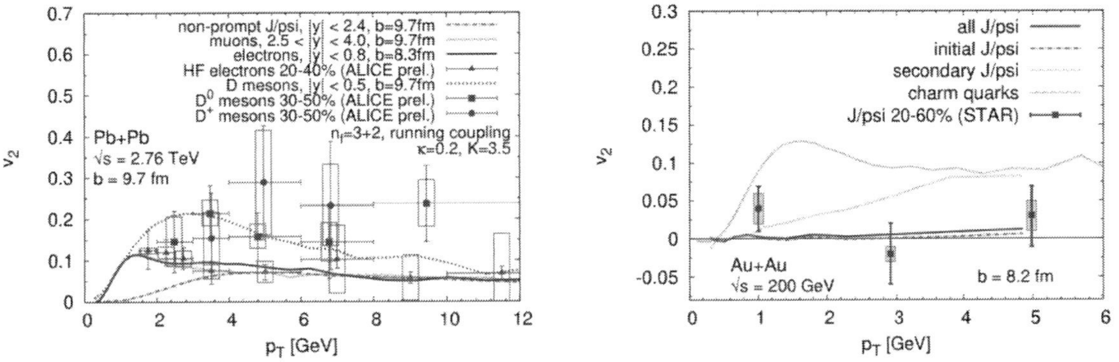

Figure 2. Left: Elliptic flow v_2 of D mesons, non-prompt J/ψ, muons, and electrons at LHC together with available data for D mesons and electrons [28, 29]. Right: Elliptic flow of J/ψ with data [30]. For comparison the charm quark v_2 is also shown.

data points. This is in agreement with the observation that we also underestimate the R_{AA} of non-prompt J/ψ, heavy flavor electrons, and muons [8].

The left hand side of Fig. 2 shows our predictions of the elliptic flow of D mesons, non-prompt J/ψ, heavy flavor electrons, and muons. For D mesons and electrons there is already data available which agrees very well with our calculations.

Experimental measurements at RHIC showed that the v_2 of prompt J/ψ is very small [30]. This is in contradiction with the regeneration picture where the flow of the charm quarks should be transferred to the J/ψ. BAMPS is an ideal framework to study this in more detail since it reproduces the open heavy flavor flow and also allows recombination of charm quarks to J/ψ. The right panel of Fig. 2 shows that in BAMPS even with regeneration the elliptic flow of all J/ψ is compatible with the data.

4. Summary

We presented heavy flavor calculations with the parton cascade BAMPS. So far only elastic interactions with a running coupling and an improved Debye screening are implemented on the heavy flavor sector. To account for missing radiative contributions we scale the elastic cross section with $K = 3.5$ which is determined from comparing to the PHENIX v_2 data at RHIC. With this parameter we find also a good agreement to recent v_2 data from STAR in another centrality class. The predictions for all open heavy flavor R_{AA} at the LHC consistently underestimate the data slightly. However, the predicted v_2 of D mesons and electrons is in very good agreement with recent data from ALICE. The radiative processes [31] are currently being implemented in BAMPS. It will be interesting to see if they can indeed account for scaling the binary interactions with a constant factor. Furthermore, we showed some first BAMPS calculations on prompt J/ψ v_2 at RHIC.

Acknowledgements

J.U. is grateful for the kind hospitality at Tsinghua University, where part of this work has been done. Furthermore, J.U. would like to thank the DAAD and the Helmholtz Research School for Quark Matter studies for financial support. O.F. acknowledges support by the BMBF.

The BAMPS simulations were performed at the Center for Scientific Computing of the Goethe University Frankfurt. This work was supported by the Helmholtz International Center for FAIR within the framework of the LOEWE program launched by the State of Hesse.

References

[1] Adams J *et al.* (STAR) 2005 *Nucl. Phys.* **A757** 102–183 (*Preprint* nucl-ex/0501009)
[2] Adcox K *et al.* (PHENIX) 2005 *Nucl. Phys.* **A757** 184–283 (*Preprint* nucl-ex/0410003)
[3] Müller B, Schukraft J and Wyslouch B 2012 (*Preprint* 1202.3233)
[4] Uphoff J, Fochler O, Xu Z and Greiner C 2010 *Phys. Rev.* **C82** 044906 (*Preprint* 1003.4200)
[5] Cao S and Bass S A 2011 *Phys.Rev.* **C84** 064902 (*Preprint* 1108.5101)
[6] Abir R, Jamil U, Mustafa M G and Srivastava D K 2012 *Phys.Lett.* **B715** 183–189 (*Preprint* 1203.5221)
[7] Meistrenko A, Peshier A, Uphoff J and Greiner C 2012 (*Preprint* 1204.2397)
[8] Uphoff J, Fochler O, Xu Z and Greiner C 2012 *Phys.Lett.* **B717** 430–435 (*Preprint* 1205.4945)
[9] He M, Fries R J and Rapp R 2012 (*Preprint* 1204.4442)
[10] Horowitz W 2012 (*Preprint* 1210.8330)
[11] Lang T, van Hees H, Steinheimer J and Bleicher M 2012 (*Preprint* 1208.1643)
[12] Buzzatti A and Gyulassy M 2012 (*Preprint* 1207.6020)
[13] Alberico W, Beraudo A, De Pace A, Monteno M, Molinari A *et al.* 2012 (*Preprint* 1208.0705)
[14] Cao S, Qin G Y, Bass S A and Muller B 2012 (*Preprint* 1209.5410)
[15] Gossiaux P B, Nahrgang M, Bluhm M, Gousset T and Aichelin J 2012 (*Preprint* 1211.2281)
[16] Mocsy A and Petreczky P 2007 *Phys. Rev. Lett.* **99** 211602 (*Preprint* 0706.2183)
[17] Xu Z and Greiner C 2005 *Phys. Rev.* **C71** 064901 (*Preprint* hep-ph/0406278)
[18] Xu Z and Greiner C 2007 *Phys. Rev.* **C76** 024911 (*Preprint* hep-ph/0703233)
[19] Gossiaux P B and Aichelin J 2008 *Phys. Rev.* **C78** 014904 (*Preprint* 0802.2525)
[20] Peshier A 2012 *Nucl.Phys.* **A888** 7–22 (*Preprint* 0801.0595)
[21] Uphoff J, Fochler O, Xu Z and Greiner C 2011 *Phys.Rev.* **C84** 024908 (*Preprint* 1104.2295)
[22] Uphoff J, Zhou K, Fochler O, Xu Z and Greiner C 2011 *PoS* **BORMIO2011** 032 (*Preprint* 1104.2437)
[23] Uphoff J, Fochler O, Xu Z and Greiner C 2012 (*Preprint* 1208.1970)
[24] Uphoff J, Fochler O, Xu Z and Greiner C 2011 *Nucl.Phys.* **A855** 444–447 (*Preprint* 1011.6183)
[25] Fochler O, Uphoff J, Xu Z and Greiner C 2011 *J.Phys.G* **G38** 124152 (*Preprint* 1107.0130)
[26] Mustafa M (STAR Collaboration) 2012 (*Preprint* 1210.5199)
[27] Grelli A (ALICE collaboration) 2012 (*Preprint* 1210.7332)
[28] Ortona G (ALICE Collaboration) 2012 (*Preprint* 1207.7239)
[29] Sakai S (ALICE collaboration) 2012 Quark Matter 2012 proceedings
[30] Powell C B (STAR Collaboration) 2011 *PoS* **EPS-HEP2011** 485 (*Preprint* 1111.6944)
[31] Abir R, Greiner C, Martinez M, Mustafa M G and Uphoff J 2012 *Phys.Rev.* **D85** 054012 (*Preprint* 1109.5539)

Journal of Physics: Conference Series **446** (2013) 012045 doi:10.1088/1742-6596/446/1/012045

Direct photon measurements in pp collisions at $\sqrt{s} = 7$ TeV with ALICE

Alexis Mas (for the ALICE collaboration)
SUBATECH, Ecole des Mines de Nantes, Université de Nantes, CNRS/IN2P3, 4 Rue A. Kastler, Nantes, France
E-mail: Alexis.Mas@subatech.in2p3.fr

Abstract. The production of direct photons in pp collisions via Compton scattering and $q\bar{q}$ annihilation leading order processes is of great interest to test pQCD predictions and also to study parton fragmentation. The fragmentation of the recoil scattered parton can be studied using correlation between the photon and the charged particles emitted in the opposite direction. We present the study of the parton fragmentation in pp collisions at $\sqrt{s} = 7$ TeV measuring the imbalance parameter $x_E = -\vec{p}_T^{\,h}.\vec{p}_T^{\,\gamma}/|\vec{p}_T^{\,\gamma}|^2$. Furthermore, a first uncorrected direct photon spectrum measured with the ALICE electromagnetic calorimeter (EMCal) is shown. These measurements open the way to a straightforward comparison between pp and Pb-Pb collisions concerning parton fragmentation and direct photon production.

1. Introduction

In pp collisions, direct photon measurements can provide strong constraints on parton distribution functions (PDF) [1, 2]. Compton scattering ($qg \rightarrow \gamma q$) is the dominant channel of direct photon production at LHC energies (in the p_T range investigated here: 8-50 GeV/c) making the cross-section QCD predictions very sensitive to the poorly known gluon PDF and then opening the opportunity to better constrain it. The γ-jet events are naturally pointed out to study the modification of the fragmentation functions that occurs in heavy-ion collisions [3]. Indeed, the photon is a colorless particle that does not interact strongly with the hot and dense medium created [4]. Therefore the direct photon is a reliable reference to study the fragmentation of the oppositely scattered parton in both pp and heavy-ion collisions. From the experimental point of view, direct photons are not discernable from higher order processes (i.e., bremsstrahlung). Nevertheless, isolation techniques provide a strong discrimination between them and other photon production channels (dominated by meson decays). We present first experimental details as well as the direct photon selection techniques used. Then we report on two different analyses: the γ-hadron correlation measurements and the direct photon analysis.

2. Experimental details and direct photon selection

Both analyses are based on pp collisions at $\sqrt{s} = 7$ TeV from data taken in year 2011. The photons are measured in the ALICE Electromagetic Calorimeter (EMCal) [5] which is a sampling calorimeter (Pb-Scintillator) with an acceptance of $\Delta\varphi = 100°$ and $|\eta| < 0.7$ and a pseudorapidity $\times$ azimuth granularity of $\delta\eta \times \delta\varphi = 0.143 \times 0.143$. The tracks of charged hadrons are measured in the Inner Tracking System and the ALICE Time Projection Chamber [6]

($\Delta\varphi = 360°$ and $|\eta| < 0.9$). The EMCal L0 trigger is used to select events with high energy deposition (> 5 GeV) in a 4x4 EMCal tower patch. It reduces the data volume as well as the detector dead time allowing to enhance the integrated luminosity recorded.

Particles depose their energy in the calorimeter creating clusters (groups of neighboring fired cells). Direct photon candidates are selected from those clusters using various requirements. The clusters produced by charged particles are rejected using a track extrapolation technique. But the main background comes from neutral meson decays (π^0, η). In the p_T range considered (8-50 GeV/c) these decay photons more likely merge into single elongated clusters. Then the selection is also based on the shape of the electromagnetic shower in the calorimeter requiring $\lambda_0^2 < 0.3$ with:

$$\lambda_0^2 = 0.5 \times (d_{\eta\eta} + d_{\varphi\varphi}) + \sqrt{0.25 \times (d_{\eta\eta} - d_{\varphi\varphi})^2 + d_{\eta\varphi}^2} \tag{1}$$

where $d_{\eta\eta}$, $d_{\varphi\varphi}$ and $d_{\eta\varphi}$ are given by: $d_{AB} = \frac{\sum_i w_i A_i B_i}{\sum_i w_i} - \frac{(\sum_i w_i A_i)(\sum_i w_i B_i)}{(\sum_i w_i)^2}$, where A_i and B_i are the coordinates of the cell i (in η or φ) and for each cell, $w_i = 4.5 - \log(\frac{E_i}{E_{cluster}})$ are weights based on their energy E_i. Isolation techniques also decrease contamination from decay photons and reject most of fragmentation photons [7]. Indeed, produced inside a jet, these photons are surrounded by hadronic activity. Different isolation criteria were used for the two analyses described below: to have no cluster or no track with $p_T > 0.5$ GeV/c in a cone radius of $R = \sqrt{\Delta^2\phi + \Delta^2\eta} = 0.4$ with $\Delta\phi = \varphi - \varphi_{photon}$ and $\Delta\phi = \eta - \eta_{photon}$ (x_E analysis, **section 3**) and that $\sum E_T < 2$ GeV in the same cone radius (direct photon spectrum, **section 4**).

3. Direct photon imbalance parameter measurement

Two-particle correlations can be used to study parton fragmentation in both pp and Pb-Pb collisions [8, 9]. In this analysis (~10 millions events) we use the imbalance parameter $x_E = -\vec{p}_T^{\,h}.\vec{p}_T^{\,\gamma}/|\vec{p}_T^{\,\gamma}|^2$ that correlates a photon candidate with its associated charged hadrons in the opposite hemisphere where $\vec{p}_T^{\,\gamma}$ and $\vec{p}_T^{\,h}$ are respectively their transverse momentum. This measurement provides a good approximation of fragmentation function [10] in $x_E \in$ [0.2-0.8] [11]. Experimentally the uncorrected γ-hadron correlation is computed using:

$$x_E^{cluster\ iso} = -\frac{p_T^h}{p_T^\gamma} cos(\Delta\Phi) - x_E^{UE} \tag{2}$$

where $\Delta\Phi = |\phi_\gamma - \phi_{hadron}|$. Figure 1 shows the measurement of $x_E^{cluster\ iso}$ selecting tracks within $2\pi/3 < \Delta\Phi < 4\pi/3$ and subtracting the underlying event [12] contribution (x_E^{UE}) estimated in $\pi/3 < \Delta\Phi < 2\pi/3$ and $4\pi/3 < \Delta\Phi < 5\pi/3$. But, even after identification (λ_0^2 cut and isolation), the direct photon candidate sample still contains a contamination that comes mainly from π^0 decays. The following formula, where p is the purity of direct photon sample, allows to extract the $x_E^{\gamma\ iso}$ distribution of direct photons by subtracting the isolated π^0 contribution ($x_E^{\pi^0\ iso}$):

$$x_E^{\gamma\ iso} \approx \frac{1}{p} x_E^{cluster\ iso} - \frac{(1-p)}{p} x_E^{\pi^0\ iso} \tag{3}$$

The $x_E^{\pi^0\ iso}$ distribution shown on figure 2 has been obtained identifying clusters as π^0 by an invariant mass analysis of distinct clusters. The purity estimation method is based on the difference of cluster shower shape profile (λ_0^2) between direct photons and background particles. The signal template is estimated using γ-jet events simulated with PYTHIA [13] tracked through the detector using GEANT3 [14]. The background template is derived from data: the shower shape of particles that failed the isolation criterion has been used. The result presented on figure 3 shows the combined fit of background plus direct photon templates (for the bin [12-16] GeV/c). The purity is extracted from these two distributions by evaluating them in the same λ_0^2 range

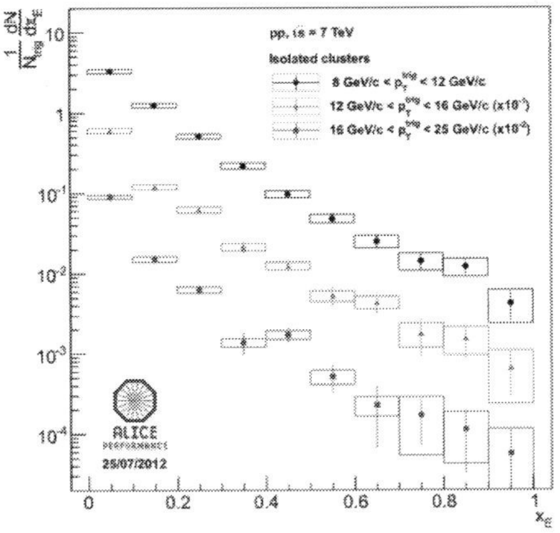

Figure 1. x_E distributions of isolated clusters for 3 different p_T bins [8-12], [12-16] and [16-25] GeV/c.

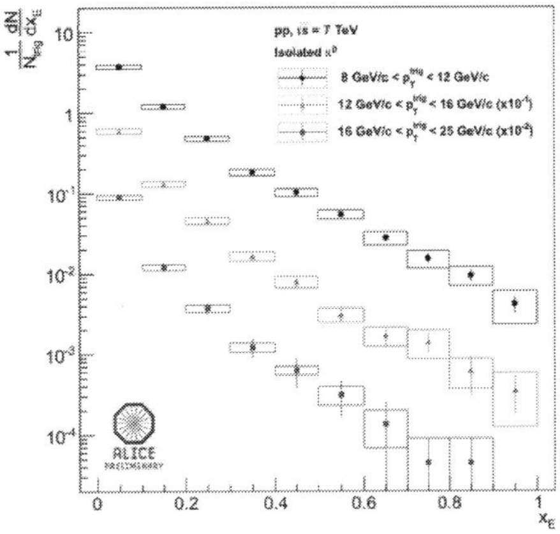

Figure 2. x_E distributions of isolated π^0 for 3 different p_T bins [8-12], [12-16] and [16-25] GeV/c.

than the one used for photon selection. Finally, combining these results with formula (3) the $x_E^{\gamma\, iso}$ distribution is obtained (figure 4). This distribution and its exponential slope will serve as baselines to study medium induced parton fragmentation modification in Pb-Pb collisions.

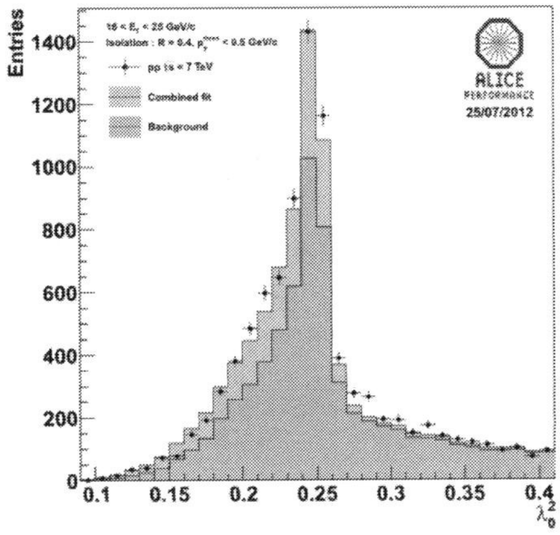

Figure 3. Fit result of the shower shape template method for a leading particle with $12 < p_T < 16$ GeV/c.

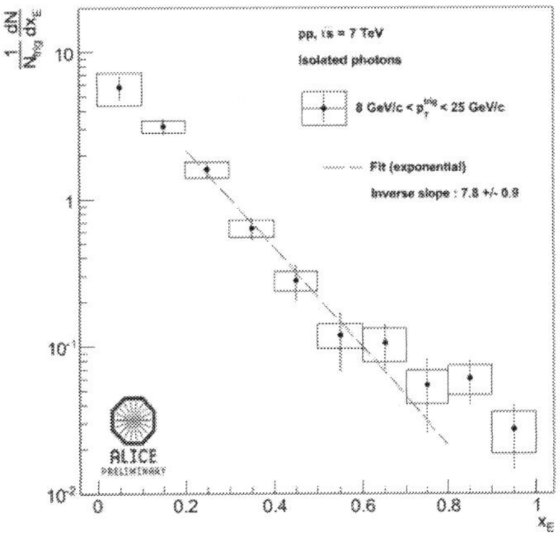

Figure 4. x_E distribution of isolated photons for p_T 8-25 GeV/c (3 bins combined) fitted by an exponential in the range $x_E \in [0.2\text{-}0.8]$.

4. Direct photon spectrum measurement

Beyond correlation studies, the measurement of direct photon cross-section in pp collisions is also an important goal. The complementarity of ALICE and other LHC experiments (ATLAS and CMS) in terms of energy range is valuable to test QCD predictions [15] and to strongly constrain the PDF determination. The pp isolated cluster spectrum at $\sqrt{s} = 7$ TeV measured with ALICE is presented here. The main differences of this analysis compared to the correlation study are firstly that the full available statistics is used (less constraints on TPC acceptance) and secondly that a strict fiducial cut ($\Delta\eta = 0.3$ and $\Delta\varphi = 54°$) is established so that the isolation cone is always contained in the EMCal acceptance. Figure 5 presents the yield extraction procedure performed with a data driven method. The background template (BG) is estimated from clusters having a large shower shape ($0.5 < \lambda_0^2 < 2$) and the signal plus background template (Signal+BG) comes from clusters with $0.1 < \lambda_0^2 < 0.3$. For both templates the $E_T^{cone} - E_T^{UE}$ (ISO) distribution has been computed where E_T^{cone} is the transverse energy in the isolation cone and E_T^{UE} is the underlying event estimated in a different η range than the cone. The yield shown on figure 6 is obtained by subtracting the BG distribution (normalized with tails: ISO > 15 GeV) from the Signal+BG one. This result shows the expected performance for direct photon cross-section measurement using 2011 data. To obtain it, the next step will be to correct and normalize the raw spectrum (purity, efficiency, systematic errors).

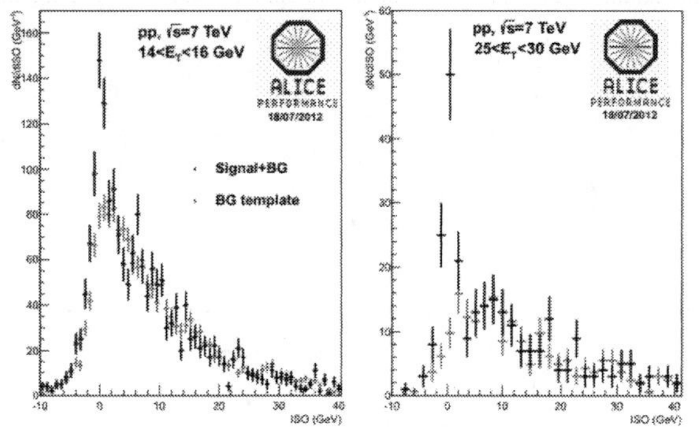

igure 5. ISO distributions ($E_T^{cone} - E_T^{UE}$) for signal and background for 2 different E_T bins [14-16] (left panel) [25-30] GeV (right panel).

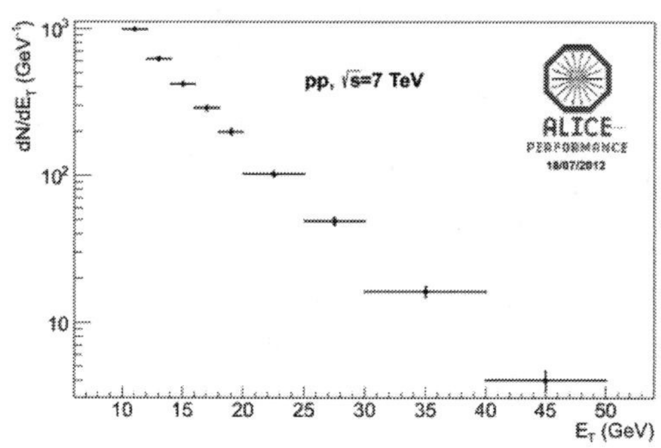

Figure 6. Isolated cluster distribution for pp collisions at 7 TeV.

5. Conclusion

The results summarized here show the potential of the ALICE-EMCal calorimeter to provide direct photon measurements. Moreover, the direct photon imbalance parameter analysis presented gives perspective to study parton fragmentation modification in Pb-Pb collisions using γ-jet events. Finally a first step toward the measurement of a direct photon cross-section in pp collisions has been shown. These results, among others, will enable to quantify the medium induced modification of PDF and fragmentation functions in Pb-Pb collisions in ALICE.

References

[1] F. Arleo *et al.* 2011 JHEP 04 055
[2] R. Ichou, D. D'Enterria, 2010 *Phys. Rev. D* **82** 014015
[3] CMS Collaboration, 2013 *Phys. Lett. B* **718** 773-794
[4] K. Werner, Iu. Karpenko, T. Pierog, M. Bleicher, K. Mikhailov, 2010 *Phys. Rev. C* **82** 044904
[5] U. Abeysekara *et al.* ALICE EMCAL Physics performance report, 2010 arXiv:1008.0413
[6] ALICE Collaboration, 2006 *J. Phys. G: Nucl. Part. Phys.* **32** 1295-2040
[7] S. Catani, M. Fontannaz, J. Guillet, 2002 JHEP 05 028
[8] A. Adare *et al.* PHENIX collaboration, 2006 *Phys. Rev. D* **74** 072002
[9] A. Adare *et al.* PHENIX collaboration, 2010 *Phys. Rev. D* **82** 072001
[10] Z. Belghobsi *et al.* 2009, *Phys. Rev. D* **79** 114024
[11] N. Arbor, Quark Matter 2012 proceedings, arXiv:1211.6620
[12] A. Affolder *et al.* CDF Collaboration, 2002 *Phys. Rev. D* **65** 092002
[13] PYTHIA 6.400 manual, 2006 JHEP 05 026
[14] R. Brun *et al.*, 1994 *CERN Program Library Long Write-up W5013*
[15] CMS Collaboration, 2012 *Phys. Lett. B* **710** 256-277

Electroweak bosons in heavy-ion collisions with the CMS detector at $\sqrt{s_{NN}}$=2.76 TeV

Alice Florent, on behalf of the CMS collaboration
LLR, Ecole Polytechnique, Palaiseau, France
E-mail: alice.florent@cern.ch

Abstract. Electroweak gauge bosons W and Z, do not interact strongly, and thus constitute clean probes of the initial state of nucleus-nucleus collisions. The comparison of their production cross-sections in pp and in nuclear collisions provides an estimate of the nuclear parton distribution functions. Despite the low production cross section of weak bosons compared to other nuclear processes, the relatively clean signal of their leptonic decay channel allows their detection. This paper reports measurements of Z and W bosons, produced in PbPb and pp collisions both at nucleon-nucleon center of mass energy $\sqrt{s_{NN}}$=2.76 TeV with the CMS detector. The Z boson yield and the nuclear modification factor (R_{AA}) corresponding to the integrated luminosity of 150 μb^{-1} for PbPb collisions are presented. The search for W bosons has been performed in the muon plus neutrino channel, using the data sample with integrated luminosity of 7.2 μb^{-1} for PbPb collisions. Event centrality an muon pseudorapidity dependencies are studied for the complete W candidate sample as well as samples separated by charge (W^+ and W^-).

1. Introduction

The unprecedented high energies reached in heavy-ion collisions at the LHC have opened the possibility of producing weakly interacting bosons, Z and W, with significant cross sections. These vector bosons are produced in the primary binary nucleon-nucleon collisions and decay before formation of the quark-gluon plasma. Their leptonic decay products do not interact with the hot and dense coloured matter and they are sensitive only to the initial state of the collision and in particular to the parton distribution functions (PDF) in the nuclei. They can be used as a reference to normalize the production mechanisms for other processes.

2. Muon reconstruction

One of the particularly strong features of the CMS detector is its muon reconstruction [1]. The muon signature is extremely clear with a very good momentum resolution (1-2% up to 100 GeV/c at mid-rapidity) due to the precise silicon tracking to which tracks in the muon chambers are matched. In addition the large rapidity coverage of the muon chambers results in large acceptance for single and di-muons. The results presented in this paper are for electroweak boson decays with muons in the final state. The dense environment produced in heavy-ion collisons requires special care in particle reconstruction. The pp reconstruction algorithms were re-optimized for PbPb using HYDJET simulations [2]. HYDJET is an event generator that simulates jet production, jet quenching and flow effects in ultra relativistic heavy-ion collisions.

3. The $Z \to \mu^+\mu^-$ analysis

From the 2011 data, 616 Z candidates have been extracted with a dimuon invariant mass between 60 and 120 GeV/c^2 [3]. The invariant mass of pairs of opposite charge muons, having a transverse momentum (p_T) larger than 20 GeV/c and passing all established quality cuts, is shown in Figure 1. No same-charge pairs (combinatorial background) are found passing the kinematic and reconstruction quality criteria. Once corrected for the efficiency and acceptance, the yield of Z's can be described as a function of the collision centrality. If the Z boson is not affected by the dense medium then the yield per binary nucleon-nucleon collisions should not depend of centrality. Figure 2 shows the Z yield per unity of rapidity, represented as a function of the number of participants in the collision. In this figure the Z yield is divided by the average integral of nuclear overlap function (T_{AA}), which is the number of collisions divided by the cross section in pp collisions. The T_{AA} scaling is the natural scale to work with and allows to factor out the dependence of the pp cross section. The results presented in Figure 2 show that the Z yield is independent of collision centrality, moreover, the POWHEG generator describes the data very well. POWHEG is an next-to-leading-order (NLO) generator that is known to reproduce the Tevatron (2 TeV) and the LHC (7 TeV) data. The POWHEG predictions have about 5% uncertainty due to next-to-next-to-leading-order (NNLO) corrections and PDF uncertainties [4].

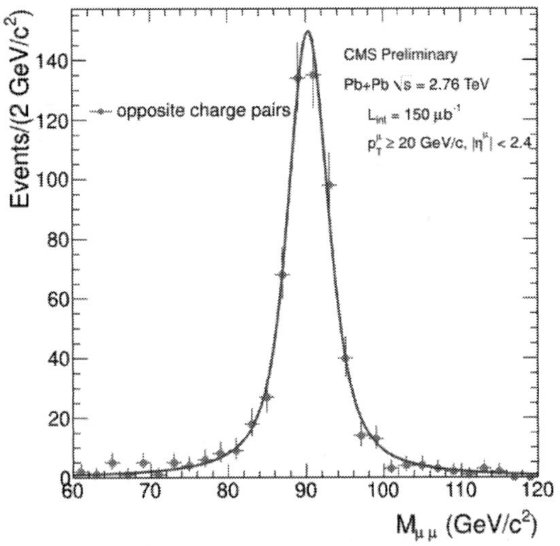

Figure 1: Dimuon invariant mass spectrum between 60 and 120 GeV/c^2.

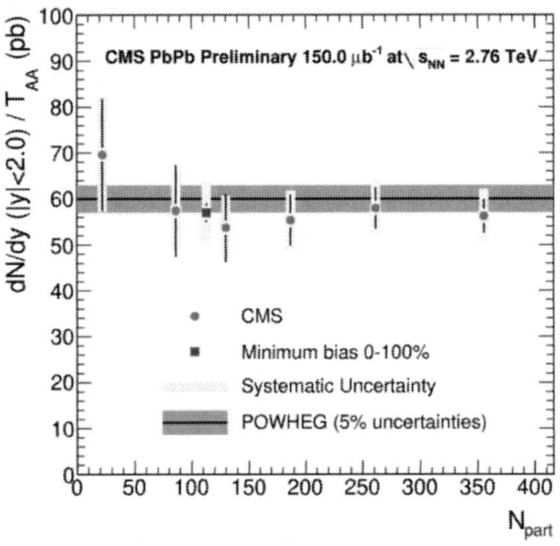

Figure 2: Collision centrality dependence of the $Z \to \mu^+\mu^-$ yield per event divided by expected average nuclear overlap function T_{AA}.

The non-dependence on centrality is a first confirmation that Z bosons are not affected by the dense medium. An absolute comparison to the pp production is provided by the nuclear modification factor (R_{AA}) which is the ratio between the Z yield in PbPb and the Z yield in pp collisions, scaled by the number of collisions. If the ratio is close to one then there are no significant changes between a pp and a PbPb collision. The pp reference for the R_{AA} measurement is calculated using theoretical calculations due to the very low available pp statistics at 2.76 TeV: $R_{AA}=\frac{dN_{AA}}{T_{AA}\times\sigma_{pp}}$ where σ_{pp} is given by POWHEG and $\frac{dN_{AA}}{T_{AA}}$ is extracted from Figure 2. With a R_{AA}=0.95 ± 0.03($stat$) ± 0.13($syst$), the Z boson through its leptonic decay is, as expected, medium-blind and confirms the previous Z results obtained from the 2010 run [5].

4. The W $\rightarrow \mu\nu$ analysis

Data collected in 2010 (7.2 μb^{-1}) was used to search for the W bosons decaying into muon and neutrino [6]. Only muons with p_T larger than 25 GeV/c were considered in this analysis. The high p_T cut significantly increases the signal to background ratio. Dimuon events were rejected to remove Z decays. A missing transverse energy (MET) is calculated following the principle of energy balance in the transverse plane. Events containing undetected high energy neutrinos are expected to have non-zero MET.

In this analysis, a missing transverse momentum is calculated by summing all the transverse momenta of all charge particle tracks in the detector (p_T). To reduce the background fluctuations only tracks with a p_T above 4 GeV/c are taken into account in the sum. From the muon energy and the obtained MET a transverse mass can be calculated. The distribution is shown in Figure 3 and compared with the W transverse mass obtained in pp collisions and Monte Carlo (MC) describing the W signal embedded in heavy-ion collisions (HYDJET). The three distributions completely match between each other.

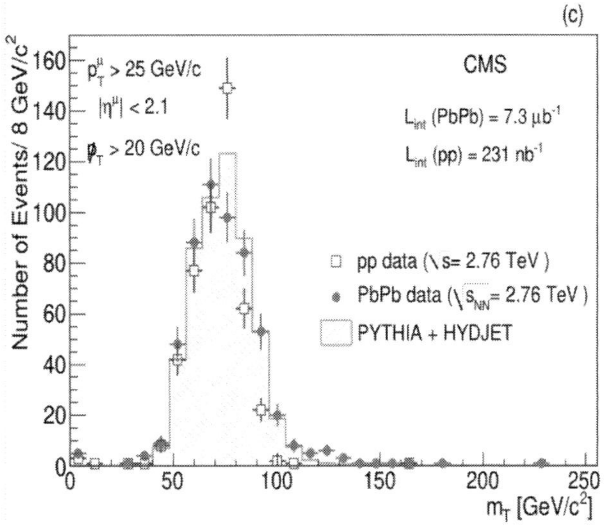

Figure 3: Distribution of transverse mass, m_T (colour online), for the selected PbPb data (red dots), the embedded W MC (green histogram) and pp data at 2.76 TeV after the same selection procedure as the one applied for PbPb (blue dots).

There are 539 W split into 275 W$^+$ and 264 W$^-$. These numbers have to be compared with the ones obtained in the pp sample at 2.76 TeV: 466 W with 301 W$^+$ and 165 W$^-$ in the same window of pseudorapidity ($|\eta|\leq 2.1$). A W$^+$ (W$^-$) is produced at first order by the fusion of a valence quark u (d) and an anti-quark d (u) from the sea. Hence a pp collisions is expected to produce more W$^+$ than W$^-$. In PbPb collisions there are more d than u quarks because of the neutrons present in Pb nuclei so one would expect many more events containing W$^-$ than W$^+$, contrary to what we observe. This can be explained by the muon charge asymmetry.

When the W is produced, it is boosted in the direction of the valence quark. The neutrino is always left-handed and the anti-neutrino right-handed. Hence the μ^+ coming from W$^+$ is by preference right-handed because of spin and momentum conservation. It is then produced backward with respect to the W$^+$, meaning mainly at central rapidity. The μ^- coming from W$^-$ is this time left-handed and then produded in the W$^-$ direction, at forward rapidity. By chosing $|\eta_{muon}|\leq 2.1$ in the analysis, the yield of W$^+$ relative to W$^-$ is enhanced. Figure 4 represents the charge asymmetry ($\frac{N(W^+)-N(W^-)}{N(W^+)+N(W^-)}$) as a function of $|\eta|$, for pp data and PbPb data.

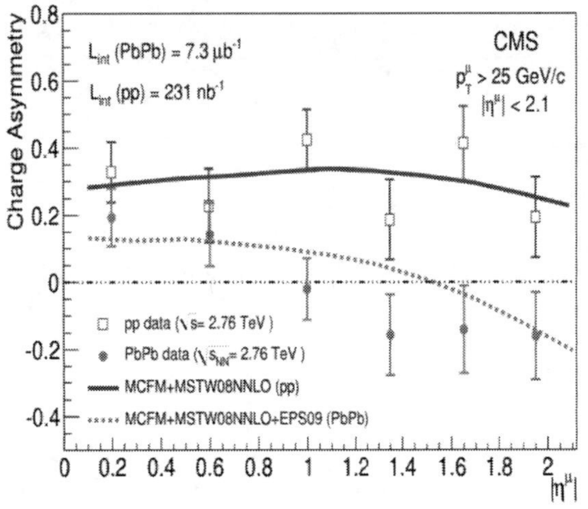

Figure 4: Muon charge asymmetry as a function of muon pseudo-rapidity for pp and PbPb collisions at 2.76 TeV.

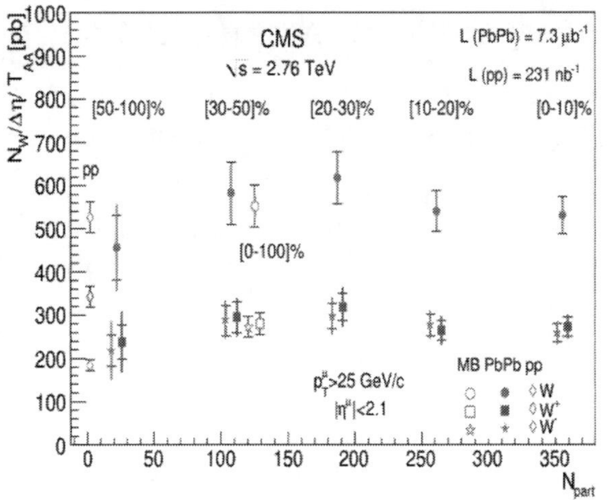

Figure 5: Collision centrality dependence of the W→ $\mu\nu$ yield per event divided by expected average nuclear function T_{AA}.

Similarly to the Z, the yield of W as a function of centrality was studied. Figure 5 shows that the binary-scaled yield of W^+ and W^- is independent of centrality. When no distinction of the W charge is made, the binary collision scaled yield in PbPb is compatible with pp. From these yields a R_{AA} can be calculated by using the pp data instead of POWHEG prediction, since the statistics coming fom the 2010 run in PbPb is close to the one in pp at 2.76 TeV. Hence $R_{AA}(W)$=1.04 ± 0.07($stat$) ± 0.12($syst$) with $R_{AA}(W^+)$=0.82 ± 0.07($stat$) ± 0.09($syst$) and $R_{AA}(W^-)$=1.46±0.14($stat$)±0.16($syst$). These observed modifications are compatible with this muon charge asymmetry, also called *isospin* effect.

5. Conclusion

The main conclusion of the above studies can be summarized this way : weak boson production iscconsistant with the binary-collision scaling hypothesis (R_{AA}=1) and can thus serve as reference to modified probes. The further analysis of the 2011 data of W boson and the Z → e^+e^- decay channels is in progress and we expect new results shortly. The improved statistics will improved significantly the precision of charge asymmetry and R_{AA} and it could reveal parton distribution functions modifications in the nucleus as compared to proton.

Speaker received funding from the ERC, grant no. 259612

References

[1] Chatrchyan S *et al.* (CMS Collaboration) 2008 *JINST* **3** S08004
[2] Lokhtin I and Snigirev A 2006 *Eur.Phys.J.* **C45** 211–217
[3] CMS Collaboration 2012 (*Preprint* CMS-PAS-HIN-12-008)
[4] Alioli S, Nason P, Oleari C and Re E 2009 *JHEP* **0909** 111
[5] Chatrchyan S *et al.* (CMS Collaboration) 2011 *Phys.Rev.Lett.* **106** 212301
[6] Chatrchyan S *et al.* (CMS Collaboration) 2012 *Phys.Lett.* **B715** 66–87

Journal of Physics: Conference Series **446** (2013) 012047 doi:10.1088/1742-6596/446/1/012047

Di-electron elliptic flow in $\sqrt{s_{NN}} = 200$ GeV Au+Au collisions at STAR

Xiangli Cui for the STAR Collaboration

University of Science and Technology of China, Hefei, 230026, China
Brookhaven National Laboratory, Upton, NY, 11973, USA

E-mail: cuixl@rcf.rhic.bnl.gov

Abstract. In minimum-bias (0-80%) Au+Au collisions at $\sqrt{s_{NN}} = 200$ GeV, the elliptic flow (v_2) measurements of di-electrons for the low mass region ($M_{ee} < 1.1$ GeV/c^2) are presented. The differential v_2 as a function of transverse momentum in different mass regions are also reported. The simulated and measured v_2 of di-electrons are compared in the low mass region.

1. Introduction

Ultra-relativistic heavy ion collisions provide a unique environment to study the properties of strongly interacting matter at high temperature and high energy density [1]. Di-leptons constitute one of the crucial probes of this strongly interacting matter because they are not affected by the strong interaction once produced. In the low mass region of produced lepton pairs ($M_{ll} < 1.1$ GeV/c^2), we can study the vector meson in-medium properties through their di-lepton decays, where any modifications observed may be related to the possibility of chiral symmetry restoration. In the intermediate mass region ($1.1 < M_{ll} < 3.0$ GeV/c^2), the di-lepton spectra are directly related to the thermal radiation of Quark-Gluon Plasma (QGP) [2, 3]. However, other sources significantly contribute in this mass region, such as $c\bar{c} \to l^+l^-X$ or $b\bar{b} \to l^+l^-X$ [4].

Elliptic flow (v_2) is generated in the early stage of heavy-ion collisions, via the transformation of the initial spatial eccentricity of the nuclear overlap region into momentum anisotropy through the action of azimuthally anisotropic pressure gradients [5]. Di-leptons, which escape from the expanding fireball with minimum interaction, will be able to probe specifically this early stage where the v_2 first develops. It has been proposed that transverse momentum (p_T) and M_{ll} dependence of v_2 of di-leptons could provide very rich information on specific stages of the fireball expansion, distinguishing partonic and hadronic radiation sources [5].

At STAR, the newly installed Time-of-Flight detector (TOF) which covers $|\eta| < 0.9$ and with complete azimuthal coverage ($\Delta\phi = 2\pi$), offers large acceptance and high efficiency for particle identification [6]. The TOF, combined with the measurements of the ionization energy loss (dE/dx) from the Time Projection Chamber (TPC) [7, 8, 9], enables the electron identification with high purity from low to intermediate p_T [9, 10, 11, 12]. In this article, we present the M_{ee}, p_T, and centrality dependence of di-electron elliptic flow in $\sqrt{s_{NN}} = 200$ GeV Au+Au collisions at STAR.

Journal of Physics: Conference Series **446** (2013) 012047 doi:10.1088/1742-6596/446/1/012047

2. Data analysis and results

The TPC is the primary tracking device at STAR which provides the dE/dx, momentum, and path-length measurements of particles produced in the collision. The dE/dx is used for particle identification [8, 9]. The TOF measures the particles' time of flight from their decay vertex to a matched hit in the TOF detector. Combined with a path-length measurement, the particle velocity (β) extends the identification of π and K up to 1.6 GeV/c and the proton to 3 GeV/c in momentum [10, 11]. We utilize 720 million minimum-bias (0-80% top centrality) Au+Au collisions taken in years 2010 and 2011 with full TOF system coverage. In these events, primary z-vertex position is required to be within $\pm$ 30 cm, which is the distance from the center of the detector. With the β and p_T dependent dE/dx cuts for $p_T > 0.2$ GeV/c and $|\eta| < 1$, electrons can be clearly identified from low to intermediate p_T. The purity of electron candidates is about 97% in minimum-bias Au+Au collisions.

The e^+ and e^- candidates from the same events are combined to reconstruct the invariant mass distributions (M_{ee}) marked as unlike-sign distributions, which contain both signal and background. The background contains random combinatorial, and other correlated pairs. We use two methods to estimate the background: same-event like-sign and mixed-event unlike-sign techniques. In the same-event like-sign technique, electron candidates from the same event with same charge sign are combined. In the mixed-event technique, unlike-sign electron candidates from different events are combined. In the mixed-event technique, the event used to mix must be in the same centrality bin, vertex z-coordinate bin and event plane angle bin. We divide the centrality into 9 bins, vertex-z into 10 bins, and event plane angle into 100 bins so that the events used to mix have a similar structure. In the mass region $M_{ee} < 0.7$ GeV/c^2, we subtract the like-sign background. In the mass region $M_{ee} > 0.7$ GeV/c^2, we subtract the unlike-sign mixed-event background: since the like-sign and mixed-event background distributions are consistent for the higher mass region ($M_{ee} > 0.7$ GeV/c^2), we use mixed-event background for better statistics.

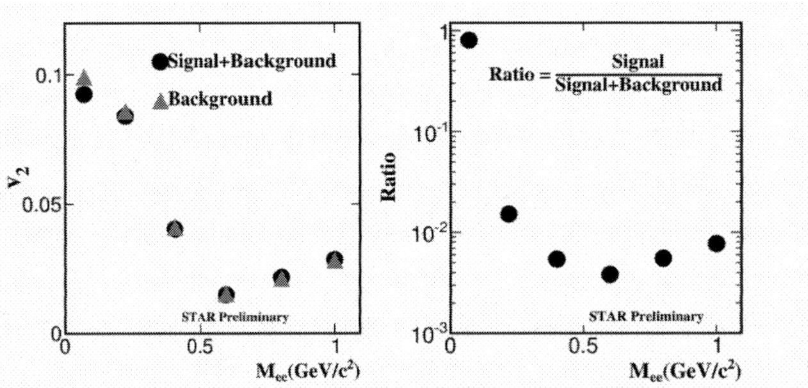

Figure 1. (left panel) The di-electron v_2 as a function of M_{ee} for Au+Au $\sqrt{s_{NN}} = 200$ GeV collisions. The circles represent the v_2 of unlike-sign pairs (v_2^T). The triangles represent the v_2 of background pairs (v_2^B). (right panel) Ratio = N_S/N_{S+B} as a function of M_{ee} for Au+Au $\sqrt{s_{NN}} = 200$ GeV collisions.

We use the event-plane method to calculate the v_2 of the di-electron signal. The event-plane is reconstructed using tracks from the TPC. The details of the method are in Refs. [13, 14]. The di-electron signal v_2 can be obtained by using the following two formulas:

$$v_2^S \times \frac{N_S}{N_{S+B}} = v_2^T - v_2^B \times (1 - \frac{N_S}{N_{S+B}}), \quad (1)$$

$$v_2 \;=\; \langle \cos(2(\phi_i - \psi_2))/r_j \rangle, \tag{2}$$

where v_2^T is the v_2 of the signal+background, v_2^S is the di-electron signal v_2, v_2^B is the background v_2, and we use the formula (2) to calculate v_2^T and v_2^B which are shown in the left panel of Fig.1. N_S/N_{S+B} is the ratio of signal over signal+background shown in the right panel of Fig. 1. We use the sub-event method to calculate r_j which is the resolution of event-plane in centrality j, and ϕ_i is the di-electron pair angle, and ψ_2 is the event-plane angle. Angle brackets denote averages over all di-electron pairs in all events. In each mass bin, with the v_2^T, v_2^B, and N_B/N_{S+B}, we calculate the v_2^S by the Eqn. (1).

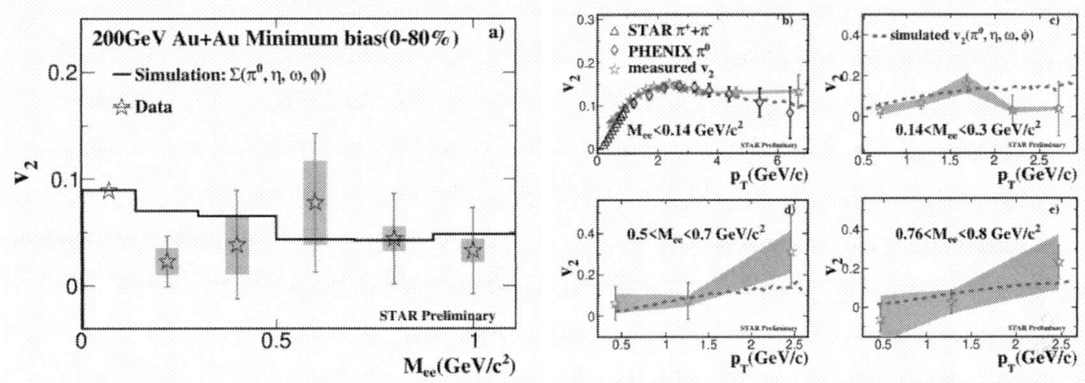

Figure 2. (left panel (a)) The di-electron v_2 as a function of M_{ee}. The star symbol is the measured di-electron v_2. The line shape is the simulated di-electron v_2. (right panels (b-e)) The v_2 of di-electron as a function of p_T for different mass regions. The star symbol is for the measured results. The dashed curve is for the simulated results. The charged pion v_2 [15] and neutral pion v_2 [16] are also shown in panel (b) for comparison. The error bars and the band represent statistical and systematic uncertainties, respectively.

Figure 2 shows the di-electron v_2 as a function of M_{ee} (left panel) and v_2 as a function of p_T in different mass regions (right panels) in minimum-bias Au+Au collisions at $\sqrt{s_{NN}} = 200$ GeV. The star symbol is the measured di-electrons v_2. The line shapes are the simulated cocktail v_2. The components of the cocktail are π^0, η, ω and ϕ. We assume that the η has the same v_2 as K_S^0 [14] because the value of their mass are nearly the same, and also assume the ω v_2 same as the ϕ v_2. For each component, we parameterize its v_2, simulate its two body or Dalitz decays, and obtain the di-electron v_2 from its decay. The cocktail v_2 is obtained by combining the di-electron v_2 from each component, which are weighted by their yields. The cocktail v_2 is consistent with the measured v_2 as a function of M_{ee} as well as p_T. The consistency between the simulation and measurements demonstrates the credibility of our method to obtain the di-electron v_2.

The left panel of Fig. 3 shows the p_T dependent v_2 of di-electrons in the mass region of $0.98 < M_{ee} < 1.06$ GeV/c^2, which is consistent with the measured v_2 of ϕ meson through the K^+K^- decays [17]. The simulated cocktail v_2 is also consistent with the measured v_2. The right panels of Fig. 3 shows the centrality dependence of di-electron v_2 for the mass region $M_{ee} < 0.14$ GeV/c^2. In this mass region, the π^0 Dalitz decay contribution dominates. We parameterize π^0 v_2 from low to high p_T [16, 18], do the π^0 Dalitz decay, and obtain the simulated di-electron v_2 from π^0 Dalitz decay shown by the dashed curve. In each centrality interval, the measured di-electron v_2 is consistent with the simulated di-electron v_2.

3. Summary

Measurements of the di-electron v_2 are presented as a function of M_{ee} and p_T for the low mass regions in minimum-bias Au+Au $\sqrt{s_{NN}} = 200$ GeV collisions. The v_2 of di-electrons in different

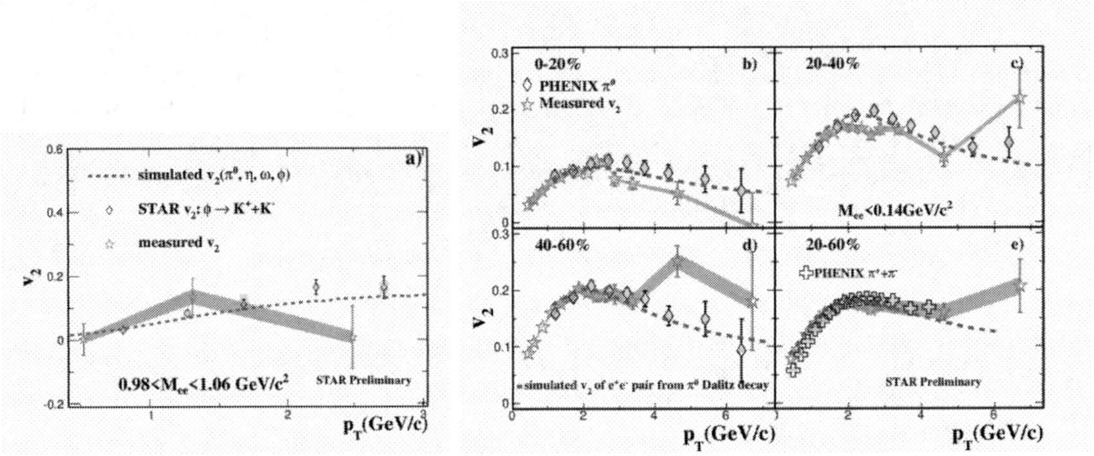

Figure 3. (left panel (a)) The di-electron v_2 as a function of p_T in minimum-bias Au+Au collisions at $\sqrt{s_{NN}} = 200$ GeV for $0.98 < M_{ee} < 1.06$ GeV/c^2. The measured v_2 of ϕ meson through the K^+K^- decay is also shown [17]. (right panels (b-e)) The di-electron v_2 as a function of p_T in different centralities for $M_{ee} < 0.14$ GeV/c^2. The star symbol is for the measurements, and the dashed curve is for the simulated results. The charged pion v_2 [18] and neutral pion v_2 [16] are also shown. The error bars and band represent statistical and systematic uncertainties, respectively.

mass regions is in agreement with the expectations from previous measurements. For $M_{ee} < 0.14$ GeV/c^2, there is a good agreement between the measured and the simulated di-electron v_2 for each centrality interval.

4. Acknowledgements
This work was supported in part by the National Natural Science Foundation of China under Grant No. 10835005,11005103 and 11005104.

References
[1] Adams J et al (STAR Collaboration) 2005 *Nucl. Phys.* A **757** 102
[2] Rapp R and Wambach J 2000 *Adv. Nucl. phys.* **25** 1
[3] David G, Rapp R and Xu Z 2008 *Phys. Rept.* **462** 176
[4] Adare A et al (PHENIX Collaboration) 2010 *Phys. Rev.* C **81** 034911
[5] Chatterjee R et al 2007 *Phys. Rev.* C **75** 054909
[6] Llope W et al 2012 *Nucl. Instr. Meth.* A **661** S110
[7] Anderson M et al 2003 *Nucl. Instr. Meth.* A **499** 659
[8] Bichsel B 2006 *Nucl. Instr. Meth.* A **562** 154
[9] Xu Y et al 2010 *Nucl. Instr. Meth.* A **614** 28
[10] Shao M et al 2006 *Nucl. Instr. Meth.* A **558** 419
[11] Adams J et al (STAR Collaboration) 2005 *Phys. Lett.* B **616** 8
[12] Adamczyk L et al 2012 *Phys. Rev.* C **86** 024906
[13] Poskanzer A M and Voloshin S A 1988 *Phys, Rev.* C **58** 1671
[14] Abelev B I et al (STAR Collaboration) 2008 *Phys. Rev.* C **77** 054901
[15] Bai Y PhD Thesis `http://drupal.star.bnl.gov/STAR/files/Bai_Yuting.pdf`
[16] Afanasiev S et al (PHENIX Collaboration) 2009 *Phys. Rev.* C **80** 054907
[17] Abelev B I et al (STAR Collaboration) 2007 *Phys. Rev. Lett.* **99** 112301
[18] Adare A et al (PHENIX Collaboration) 2012 *Phys. Rev.* C **85** 064914

Hot Quarks 2012 IOP Publishing
Journal of Physics: Conference Series **446** (2013) 012048 doi:10.1088/1742-6596/446/1/012048

Measurement of Di-electron Continuum in Au+Au collisions at $\sqrt{s_{NN}} = 200$ GeV by PHENIX using the Hadron Blind Detector

Sky D. Rolnick for the PHENIX Collaboration
UC Riverside
E-mail: skymeson@gmail.com

Abstract. Dielectrons play an important role in heavy ion collisions since they are produced during all stages of the collision and interact only electromagnetically. The PHENIX experiment was upgraded with the Hadron Blind Detector which provides excellent background rejection by removing pairs from both partially reconstructed Dalitz decays and photon conversions. In these proceedings we will report on the results obtained from 2009 data in p+p collisions using the HBD as well as the Au+Au results obtained in 2010.

1. Introduction

Dileptons are very useful probes for studying many interesting puzzles in Heavy Ion Physics. They are produced during all stages of the collision and do not interact strongly with the medium. The dielectron continuum offers a rich and diverse region of physics to explore. It allows us to study such effects as chiral symmetry restoration or possible medium modifications of vector mesons. There are several known sources which contribute to the dielectron continuum including thermal radiation, direct photons, vector meson decays, and open charm and bottom yields. The cleanness of the dielectron measurements, compared to hadronic measurements, provides probes into multi-component flow and energy loss mechanisms and will provide checks on a variety of production mechanisms.

Previous measurements made by PHENIX in Au+Au collisions at $\sqrt{s_{NN}} = 200$ GeV have shown strong enhancement of electron pairs compared to the expected hadronic cocktail in the low mass region $m_{ee} = 0.15-0.75$ GeV$/c^2$ by a factor of $4.7 \pm 0.4(stat.) \pm 1.5(syst.)$ for minimum bias collisions. Such a large enhancement was not expected and has posed a mystery to many theorists [1]. The very low signal to background ratio of $\sim 1/200$ generates large systematics in the region of interest. To improve on this, a new measurement has been made by PHENIX utilizing the Hadron Blind Detector, which reduces the backgrounds and systematic uncertainty. In these proceedings, we will discuss the use of the Hadron Blind Detector as well as dielectron results for p+p collisions and Au+Au collisions for three centrality bins.

2. Hadron Blind Detector

The Hadron Blind Detector is a novel type of Cherenkov detector utilizing a triple GEM configuration with a CSI photo-cathode layer operated in pure CF4. The HBD has achieved the highest figure of merit of any gas Cherenkov detector to date $No = 322\ cm^{-1}$ and designed with a very low material budget of just 2.4% total radiation lengths [7]. It was successfully operated in 2009 $p+p$ data and 2010 Au+Au data sets.

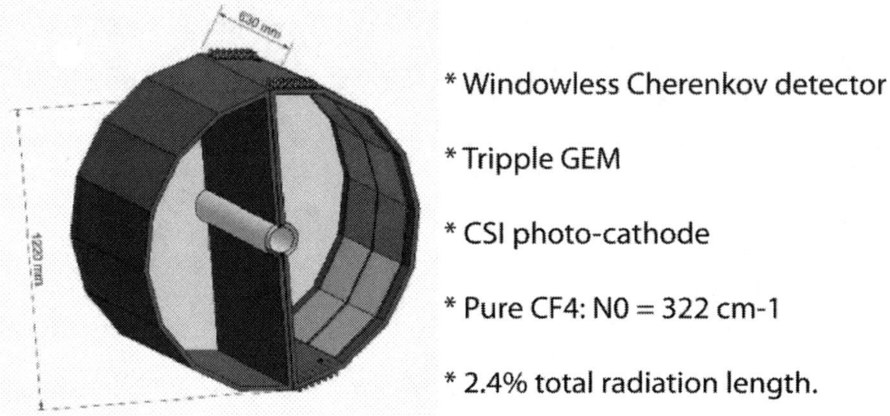

Figure 1. The HBD concept design.

The HBD was designed to work by preserving the opening angle of charged particles using the field free region of the PHENIX central arms. Electrons are tagged by the amount of Cherenkov radiation they produce and since heavier meson decays have large opening angles compared to Dalitz decays and conversions, it is possible to separate these different sources. Cherenkov light is continuously radiated along the path of the electron such that a small unfocused blob of Cherenkov light is formed on the surface of the detector. The typical photoelectron (p.e.) response is measured to be ~20 p.e. for a single electron and ~40 p.e for a double electron. In principle the difference in response should distinguish between single and double electrons.[3]

Several reconstruction algorithms have been developed and studied with a single electron tracking efficiency of ~90%.[4] The main source of background arises from partially reconstructed π^0 Dalitz decays or gamma conversion in which one electron is lost either due to acceptance or tracking inefficiencies. The HBD allows for the tagging and removal of these uncorrelated electron pairs thereby reducing the background.

3. Hadronic Cocktail

The hadronic cocktail consists of dielectron decays from known hadronic sources including Dalitz decays of π^0, η, and η' as well as direct decays from vector mesons ρ, ω, ϕ and J/ψ. The method for obtaining the cocktail has been outlined in [2] and uses a combined fit from neutral and charged pion spectra to a modified Hagedorn function. The p_T spectra of the other mesons is derived using m_T scaling of the π^0 spectra and normalized using the meson-to-pion ratio at high p_T. The momentum resolution of PHENIX tracking is taken into account using a smearing technique. The additional material of the HBD was taken into account yielding higher rates of conversion electrons which were modeled using a full Geant simulation of the PHENIX detector. The line shape of the J/ψ was also modified to account for Bremsstrahlung radiation produced by the additional material. The difference in magnetic field configurations are reflected in the acceptance of the electron pairs which is also accounted for. The heavy flavor contributions from charm and bottom are estimated using MC@NLO [5] as opposed to PYTHIA [6] since it is found to reproduce dielectron yields, especially at high p_T. The cocktail used matches very well to $p+p$ data over the whole mass range.

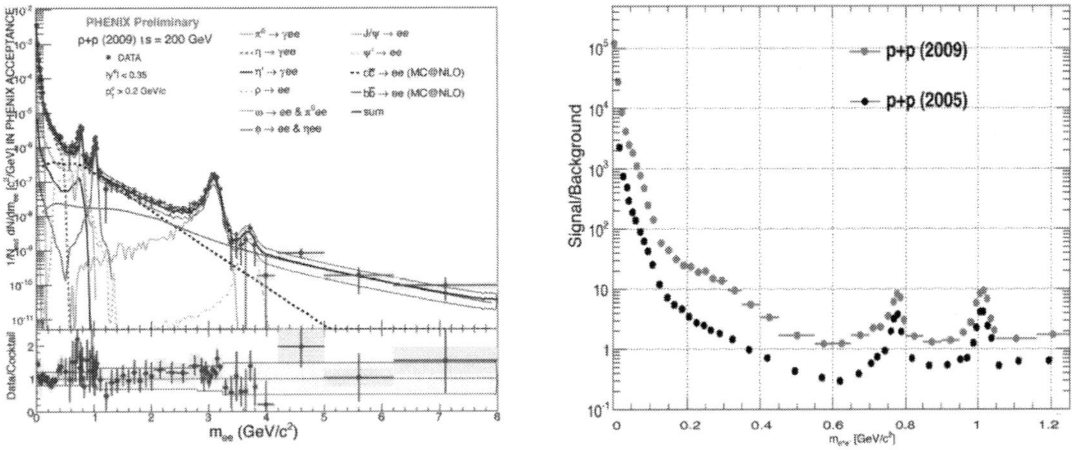

Figure 2. The left panel shows the PHENIX dielectron mass spectrum for Run9 $p+p$ overlayed with the cocktail shown by a solid line. These results show excellent agreement between data and cocktail (lower panel). The signal to background comparison for 2005 $p+p$ data (without HBD) to 2009 $p+p$ data (with HBD) is shown in the right panel.

4. Results for $p+p$ collisions

The measurements in $p+p$ collisions establish the baseline spectra without any medium effects and can be used in comparison to Au+Au results. They also provide a testing ground for understanding the HBD. The S/B ratio is improved by a factor of 5 in the low mass region by using the HBD which demonstrates the effectiveness of the HBD in background rejection.

Combinatorial background is subtracted using a like-sign subtraction technique which uses electron pairs with the same sign to estimate the correlated and uncorrelated background contributions. Examples of correlated like-sign pairs include cross pairs i.e. $\pi^0 \to \gamma\gamma \to e^+e^-e^+e^-$and jet pairs where both electrons are correlated through the jet. Differences in acceptance between like-sign and unlike-sign are corrected with a relative acceptance correction. The main advantage in using like-sign subtraction technique is that combinatorial pairs are produced at the same rate as unlike-sign foreground and the distributions will be absolutely normalized. One drawback of this technique is that the statistical precision is limited compared to mixed event techniques.

5. Results for Au+Au collisions

In Au+Au collisions, the large occupancy rate in the HBD detector presents a challenge especially in central events. The Cherenkov signal begins to get overwhelmed by other sources of radiation primarily from scintillation but also heavy ionized particles or low momentum electrons looping back into the HBD. In central events, the mean scintillation per pad in the HBD approaches $\sim$7 p.e. and when distributed over the average cluster size of 3 pads per cluster, the response from scintillation becomes comparable to the single electron response. Two techniques have been developed to reduce the background scintillation light including a mean pad subtraction based on centrality and a local background subtraction by looking at nearby pads around the clusters. Using the mean pad subtraction technique, occupancy rates can be reduced to $\sim$15% for central collisions.

Journal of Physics: Conference Series **446** (2013) 012048 doi:10.1088/1742-6596/446/1/012048

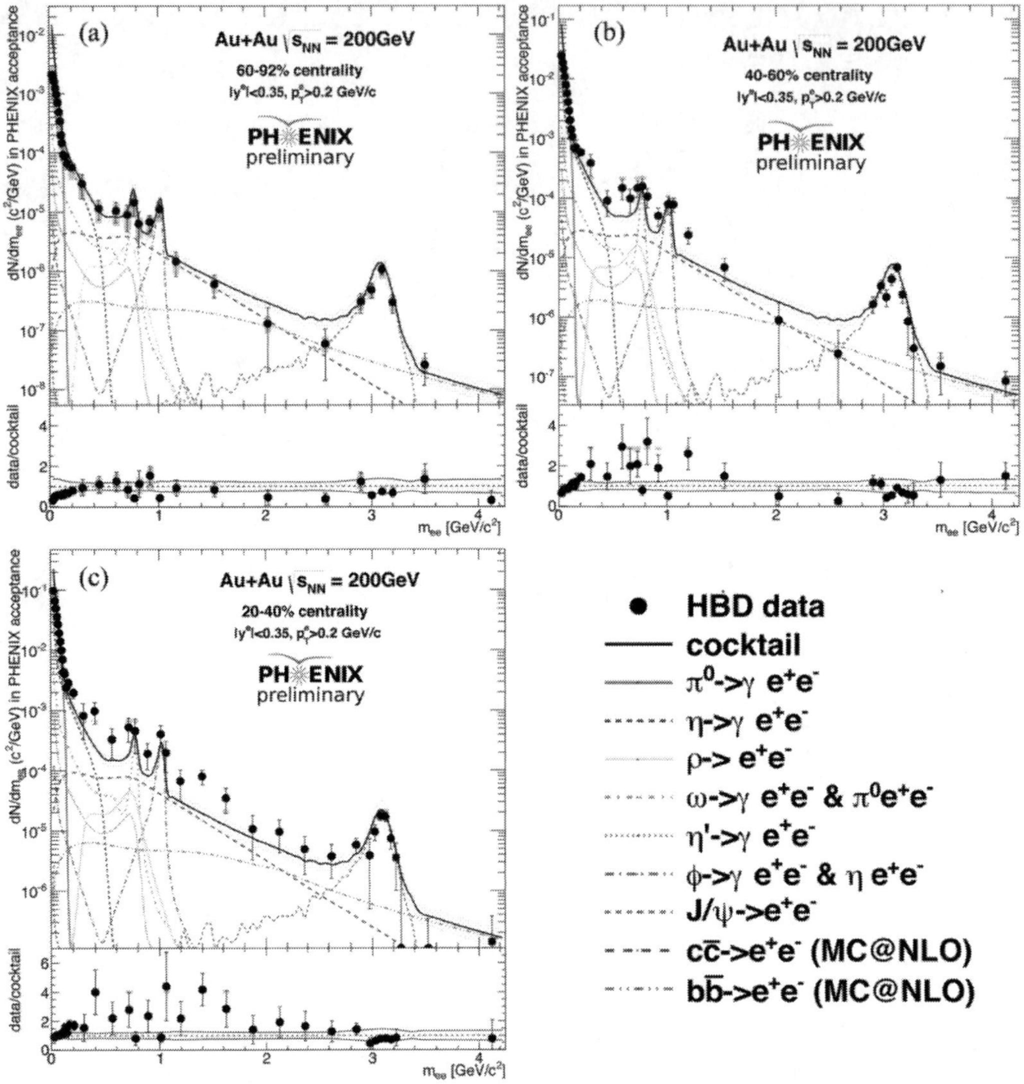

Figure 3. Dielectron mass spectra from Au+Au collisions for three centrality bins (a) 60-92%, (b) 40-60%, (c) 20-40%. The ratios to the cocktail are also shown in the lower panels.

Two neural networks were utilized to improve the electron ID and single/double electron separation. The neural networks were trained using detailed GEANT simulations. For the first stage of the analysis, a neural net selects tracks with a high probability for being an electron where studies have shown that the purity of the electron sample approaches 90%. A second neural net is used to reject contributions from double electrons in the HBD. This has been shown to remove up to 70% of double electrons reconstructed as singles by the central arm.[9]

The background subtraction technique used for Au+Au is slightly more complicated than that used for $p+p$ due to the larger combinatorial background in central events. There are two steps to remove the two types of background. First, a mixed event subtraction is used to remove the combinatorial background and then a like-sign subtraction is used to account for the correlated background. To isolate the correlated spectra from the like-sign pairs, background is subtracted using mixed events and the remaining correlated like-sign pairs are then corrected for relative acceptance, where FG12 is the unlike-sign foreground, BG12 is the unlike sign mixed

Journal of Physics: Conference Series **446** (2013) 012048 doi:10.1088/1742-6596/446/1/012048

event background, and FG(LS) and BG(LS) are the foreground and background for the like-sign respectively.

$$S = FG12 - N \times BG12 = \alpha \times (FG(LS) - N \times BG(LS))$$

The normalization is done after a pair cut efficiency is applied and a normalization factor is calculated such that the relationship $\langle BG12 \rangle = 2\sqrt{(\langle BG11 \rangle \langle BG22 \rangle)}$ holds. The results are shown in Fig. 3 for 3 different centrality bins. In Fig. 4 these results are compared to the Au+Au obtained in Run 4 before the HBD was installed.[10]

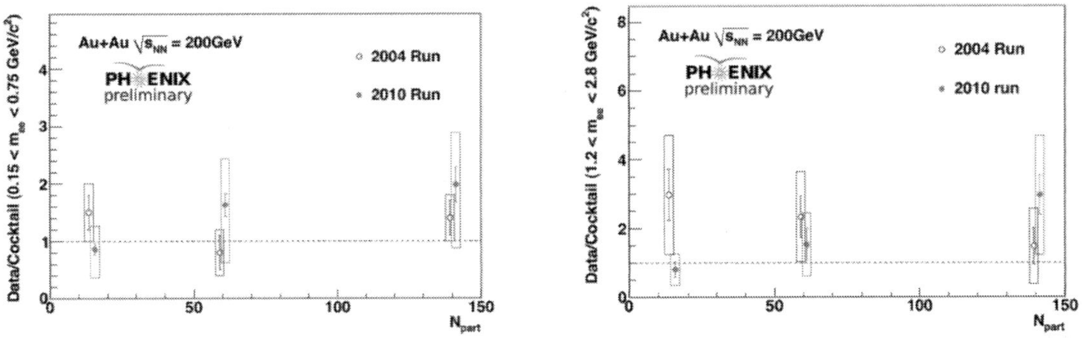

Figure 4. Comparison of Au+Au results to previous published results. We see that the latest results using the HBD are fully consistent with previous results within the systematic errors. The ratios are plotted as a function of N_{part} integrated over two different mass regions.

6. Summary

The dielectron mass spectrum has been presented for p+p and for three centrality bins of Au+Au. These results are consistent with previous published results at PHENIX. There is a hint of enhancement in the 20-40% centrality bin in the low- and intermediate-mass region. Currently this is not yet conclusive given present uncertainties and work is in progress to improve the systematics.

References

[1] C. Gale, "Electromagnetic radiation in heavy ion collisions." [arXiv:1208.2289[hep-ph]]
[2] A. Adare *et al.* [PHENIX Collaboration], Phys. Rev. C 81, 034911 (2010) [arXiv:0912.0244 [nucl-ex]].
[3] Woody *et al.* "Initial performance of the PHENIX Hadron Blind Detector at RHIC," (NSS/MIC), 2009 IEEE , pp.1002-1008, Oct. 24 2009-Nov. 1 2009
[4] QM 2011 poster by Yosuke Watanabe.
[5] S. Frixione and, B.R. Webber, JHEP 0206,029 (2002). [hep-ph/0204244]
[6] T. Sjostrand, S. Mrenna and P.Z. Skands, JHEP 0605 (2006) 026. [hep-ph/0603175]
[7] W. Anderson, *et al.* Nucl. Instr. and Meth. A (2011). [doi:10.1016/j.nima.2011.04.015]
[8] S. Afanasiev *et al.* (PHENIX), Enhancement of the dielectron continuum in $\sqrt{s_{NN}}$ = 200 GeV Au+Au collisions (2007), submitted to Phys. Rev. Lett., Preprint [arXiv:0706.3034 [nucl-ex]]
[9] "Dielectron Analysis in Au+Au Collisions at $\sqrt{s_{NN}}$ = 200 GeV with the Hadron Blind Detector." PHENIX Internal Analysis note 1040.
[10] I. Tserruya, Quark Matter 2012 proceedings. [arXiv:1211.6002 [nucl-ex]]

Journal of Physics: Conference Series **446** (2013) 012049 doi:10.1088/1742-6596/446/1/012049

Low-mass dielectron measurement in pp and Pb–Pb collisions in ALICE

Markus K. Köhler for the ALICE Collaboration
Research Division and ExtreMe Matter Institute EMMI,
GSI Helmholtzzentrum für Schwerionenforschung, Germany

E-mail: m.koehler@gsi.de

Abstract. We report on the first dielectron measurement in pp collisons at $\sqrt{s} = 7$ TeV with the ALICE detector system. The results are compared to the expected hadronic sources. The hadronic cocktail agrees to the measured dielectron continuum within statistical and systematic uncertainties. The status of the dielectron measurement in Pb–Pb collisions at $\sqrt{s_{NN}} = 2.76$ TeV is addressed.

1. Introduction

Low-mass dielectrons are an important probe for the hot and dense medium which can be created in ultrarelativistic heavy-ion collisions [1, 2]. Since leptons do not interact strongly, they carry the information from all collision stages with negligible final-state interaction. Potential modifications of the medium, such as chiral symmetry restoration, could have measurable impact on the emission pattern of dielectrons, where measurements in proton-proton (pp) collisions are important as a reference for the studies in nucleus-nucleus collisions.

ALICE [3] provides unique capabilities at the LHC for particle tracking and electron identification at low momenta. With ALICE, electrons can be identified at mid-rapidity in the central barrel ($|\eta| \lesssim 0.9$) exploiting their specific energy loss in the Inner Tracking System (ITS) and the Time Projection Chamber (TPC), with the Transition Radiation Detector (TRD), and due to their velocity as measured with the time-of-flight detector (TOF).

This writeup reports on the measurement of the production cross section of dielectrons in the mass range $0 < m_{ee} < 3.3$ GeV/c^2 for pp collisions at $\sqrt{s} = 7$ TeV. The invariant mass distribution will be compared to the expected hadronic sources.

Moreover, the status of the analysis of Pb–Pb collisions at $\sqrt{s_{NN}} = 2.76$ TeV is presented.

2. Data analysis

The results obtained in this analysis are based on data collected in 2010 for pp collisions and in 2011 for Pb–Pb collisions. The analysis was performed over the full azimuth in the electron pseudo-rapidity range $|\eta^e| < 0.8$. The transverse momentum was $p_T^e > 0.2$ GeV/c for pp collisions and $p_T^e > 0.4$ GeV/c for Pb–Pb collisions.

Electrons were identified using the combined track information from TPC and TOF. In the whole transverse momentum range TPC information was used to reject pions. In the momentum range $0.4 < p < 5$ GeV/c, kaons and protons can be separated from electrons in the TOF and were efficiently rejected.

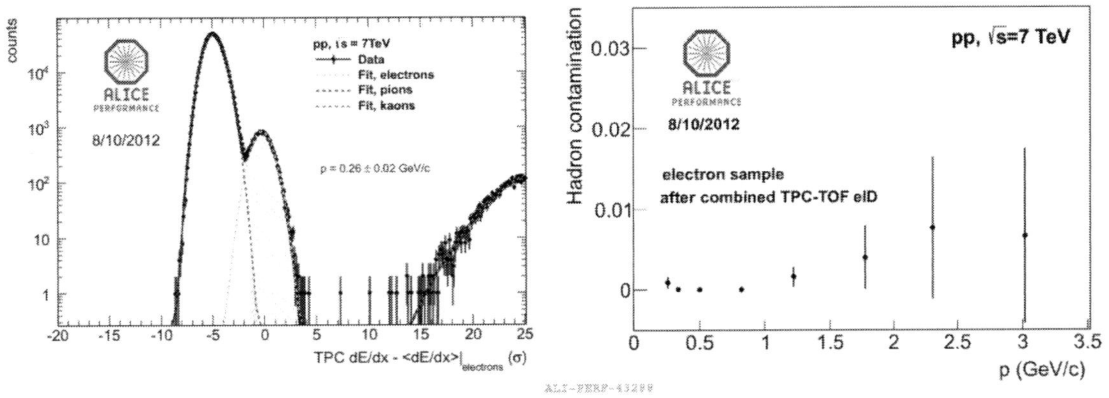

Figure 1. Estimate of the hadron contamination in the electron candidate sample. The $\mathrm{d}E/\mathrm{d}x$ distributions were fitted with Gaussians. The integral of the fits result in a contamination of at most 1%.

Figure 1 shows a data driven method for the estimation of remaining hadron contamination in the electron candidate sample. The energy loss distributions of electron candidates in the TPC were fitted by Gaussian functions, left panel of Figure 1. From the integral of the fits the hadron contamination was calculated as a function of the particle momentum, right panel of Figure 1. The hadron contamination in the electron sample is at most 1 %.

Photon conversions were rejected via their displaced vertex. Moreover, a cut was applied on the orientation of the plane spanned by the momentum of the electron and positron of a pair with respect to the magnetic field direction in the central barrel.

The measured unlike-sign spectrum is thus an overlay of the signal and combinatorial background pairs. The combinatorial background N_{+-}^{CB} was reconstructed with the same-event like-sign method. Under the assumption, that the physics signal consists only of unlike-sign pairs the combinatorial background was calculated as $N_{+-}^{CB} = 2R \times \sqrt{N_{++}N_{--}}$. Here R is a correction factor for the acceptance difference between like-sign and unlike-sign pairs, obtained from mixed event distributions, and N_{++} and N_{--} are the yields of like-sign pairs of positive and negative particles, respectively. The R factor was determined to be consistent with one within the statistical uncertainties.

The same-event like-sign method can describe the correlated background contributions from e.g. crossed pairing from double Dalitz decays at low masses or jet-like correlations at high masses. However, the drawback of the same-event like-sign method is the limited statistical accuracy of the combinatorial background compared to the mixed event method.

The unlike-sign distribution is shown together with the like-sign distribution for pp collisions in Figure 2. The raw signal was extracted by the subtraction of the like-sign distribution from the unlike-sign distribution and is corrected for efficiency. The efficiency correction was done on a track-by-track level in $(p_{\mathrm{T}}, \eta, \phi)$ bins using Monte Carlo simulations.

The expected hadronic sources of dielectrons in the mass range $0 < m_{ee} < 3.3$ GeV/c^2 were derived from p_{T} differential cross sections of π^0, η, ϕ and J/ψ measured with ALICE at mid-rapidity [4–6]. The shape of the mass distribution of dielectrons from correlated semi-leptonic decays of charmed mesons was calculated with PYTHIA [7], which was scaled to the measured charm cross section [8]. Other contributions, which were not measured, such as the η' and the ρ, were calculated by assuming m_T scaling. The detector resolution was folded into the mass spectra.

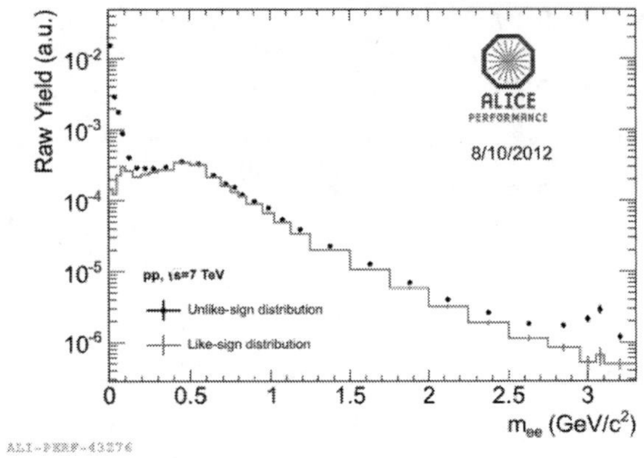

Figure 2. Invariant mass distribution of unlike-sign pairs shown together with combinatorial background pairs measured via like-sign pairs for pp collisions.

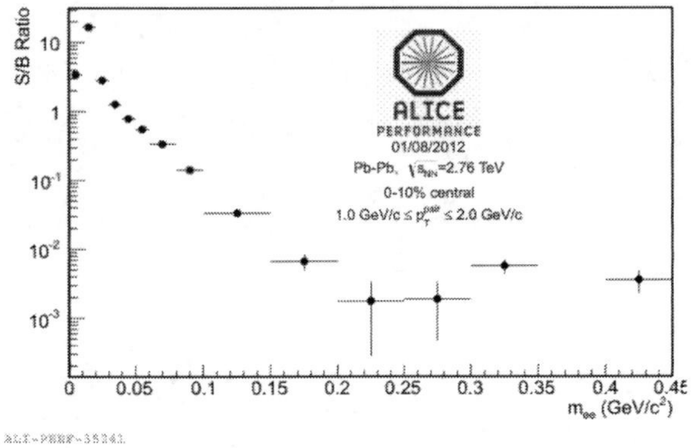

Figure 3. The S/B ratio for Pb–Pb collisions at $\sqrt{s_{NN}} = 2.76$ TeV in the low-mass region for $p_{\mathrm{T}}^{e} > 0.4$ GeV/c.

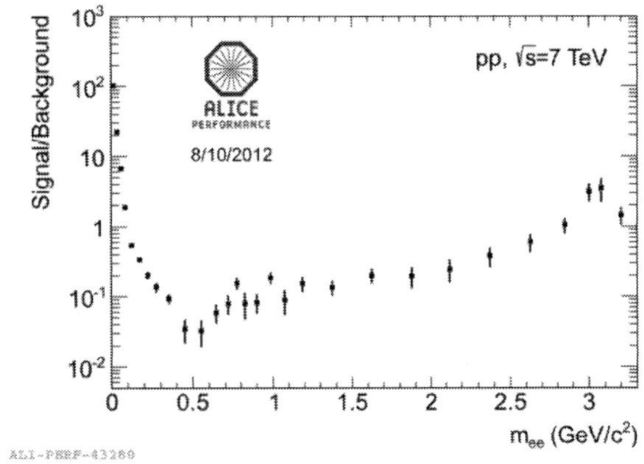

Figure 4. The S/B ratio for pp collisions in the mass range $0 < m_{ee} < 3.3$ GeV/c^2.

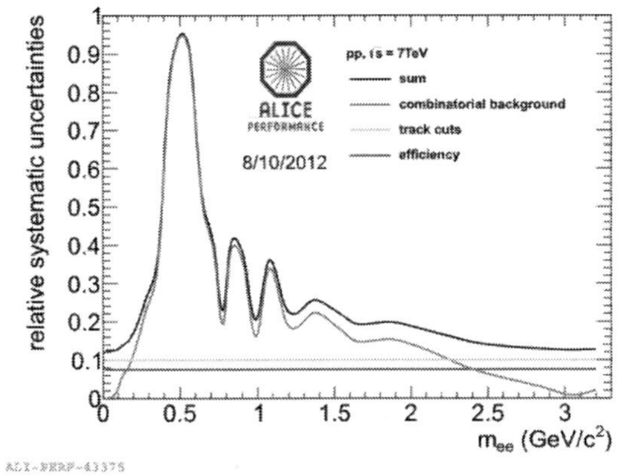

Figure 5. Sources of relative systematic uncertainties with their quadratic sum.

The analysis for Pb–Pb collisions at $\sqrt{s_{NN}} = 2.76$ TeV is currently ongoing. Figure 3 shows the signal to background ratio S/B for central (0-10%) Pb–Pb collisions in the pair transverse momentum range $1 < p_{\mathrm{T}}^{\mathrm{pair}} < 2$ GeV/c. The ratio reaches values of the order of 10^{-3}, thus the systematic evaluation of the combinatorial background is the most important step of the analysis in the Pb–Pb collision system.

3. Results

The systematic uncertainty is dominated by the combinatorial background subtraction over a wide mass range of the dielectron continuum. This uncertainty depends on the S/B ratio, shown in Figure 4. The systematic uncertainties due to track cuts and the efficiency correction contribute significantly only at very low and high masses, where the S/B ratio is large. The resulting relative systematic uncertainties are shown in Figure 5.

Figure 6 shows the comparison of the dielectron continuum with the expected hadronic sources

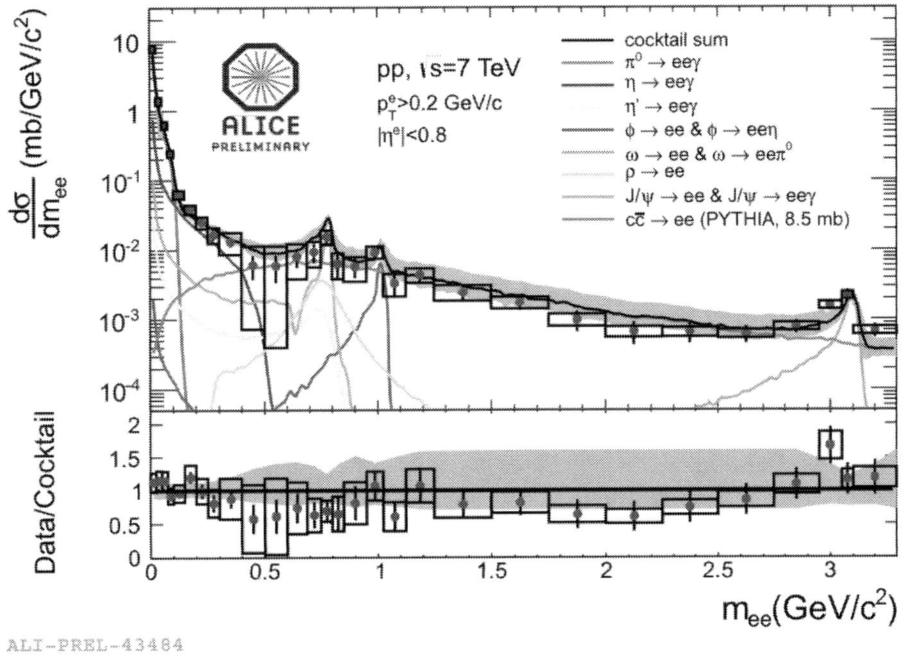

Figure 6. Background subtracted invariant mass distribution in terms of total cross section compared to the hadronic cocktail for pp collisions at $\sqrt{s}$ = 7 TeV. The boxes indicate the systematic uncertainty of the measurement, the grey band shown the systematic uncertainty of the cocktail.

for pp collisions at $\sqrt{s}$ = 7 TeV measured with the ALICE detector system. The hadronic cocktail agrees with the data within statistical and systematic uncertainties.
In the future, the proposed upgrade of the central barrel detectors [9], especially of the ITS and TPC, will help to improve the low-momentum tracking efficiency and also to detect electrons from secondary vertices from semi-leptonic heavy flavour hadron decays. This will be of special relevance for the measurement of Pb–Pb collisions.

References

[1] Rapp R, Wambach J and van Hees H (*Preprint* `arXiv:0901.3289 [hep-ph]`)
[2] Linnyk O, Cassing W, Manninen J, Bratkovskaya E L, Gossiaux P B, Aichelin J, Song T and Ko C M (*Preprint* `arXiv:1208.1279 [nucl-th]`)
[3] Aamodt K *et al.* (ALICE Collaboration) 2008 *J. Instrum.* **3** S08002
[4] Abelev B *et al.* (ALICE Collaboration) 2012 *Phys. Lett.* B **717** 162
[5] Abelev B *et al.* (ALICE Collaboration) (*Preprint* `arXiv:1208.5717 [hep-ex]`)
[6] Abelev B *et al.* (ALICE Collaboration) 2011 *Phys. Lett.* B **704** 442
[7] Sjostrand T, Mrenna S and Skands P 2006 *JHEP* **05** 026
[8] Abelev B *et al.* (ALICE Collaboration) 2012 *Journal of High Energy Physics* **1207** 116
[9] Abelev B *et al.* 2012 *CERN-LHCC-2012-012 / LHCC-I-022*

Journal of Physics: Conference Series **446** (2013) 012050 doi:10.1088/1742-6596/446/1/012050

The Physics of sPHENIX

Anne M. Sickles, for the PHENIX Collaboration
Physics Department, Brookhaven National Laboratory, Upton, NY 11973
E-mail: anne@bnl.gov

Abstract. Jet related observables have been some of the most powerful and exciting probes for understanding the matter produced in ultra-relativistic heavy ion collisions. Full jet reconstruction was begun at RHIC, and the LHC experiments have shown the power and kinematic reach of these observables. Here we discuss the sPHENIX detector and physics program which aims to bring full calorimetric based jet reconstruction to RHIC in order to explore the temperature dependence of the strongly interacting Quark Gluon Plasma.

1. Introduction
In 2010 the RHIC collaborations were charged to set out their plans for the next decade of RHIC running. After much discussion the PHENIX Collaboration decided that the most exciting heavy ion physics over the coming years was hard probe physics (jets, quarkonia and heavy flavor), but that quality measurements of these observables over the next decade were incompatible with the existing PHENIX central detectors which have small acceptance and lack hadronic calorimetry. What emerged from these considerations was a more radical plan of replacing the current central arms of PHENIX with a compact calorimeter and solenoid. This idea generated a lot of interest and the PHENIX Collaboration has recently written a proposal laying out the physics case and a planned design [1].

2. Physics of sPHENIX
The goal of the sPHENIX Upgrade to the PHENIX experiment is to make calorimetric jet measurements at RHIC in order to study the Quark Gluon Plasma in the temperature region near the critical temperature.

The discovery of the extremely low shear viscosity to entropy density ratio, η/s, established that RHIC created the QGP in a regime that was characterized by strong coupling rather than a weakly coupled gas of quarks and gluons. The η/s values required to reproduce the experimental flow measurements are small [2] and within a factor of a few of the conjectured lower quantum bound [3]. The left panel of Figure 1 shows the lower quantum bound and the perturbative calculation for $\eta/s(T)$. It is not known how η/s evolves from the near minimum value near T_c to the perturbative value at very high temperatures. Flow data are not able to constrain the temperature dependence adequately.

One interesting theoretical development is the identification of a connection between η/s and $\hat{q}$, the transverse momentum broadening per unit length of a fast parton as it traverses the QGP [4]. At weak coupling:

$$\frac{\eta}{s} = 1.25\frac{T^3}{\hat{q}} \tag{1}$$

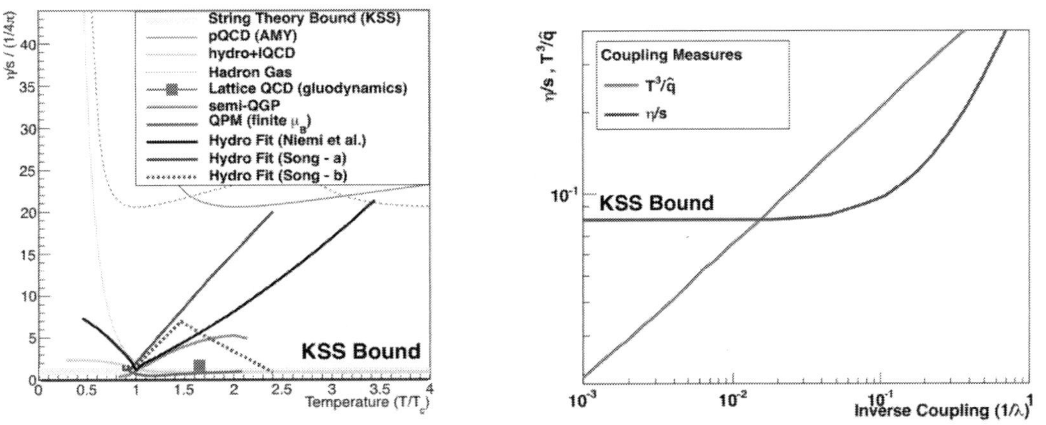

Figure 1. (left)η/s as a function of temperature. The red curves show the perturbative result from AMY [5]. The blue and black curves show parameterizations of η/s which have been used in hydrodynamical models to reproduce existing RHIC and LHC results [6, 7]. (right) $\frac{T^3}{\hat{q}}$ (red) and η/s (blue) as a function of the inverse 't Hooft coupling, λ [4].

while at strong coupling:

$$\frac{\eta}{s} \gg \frac{T^3}{\hat{q}}. \tag{2}$$

The authors of Ref. [4] state that $T^3/\hat{q}$ "is a more broadly valid measure of the coupling strength of the medium than η/s." This is illustrated in the right panel of Figure 1 which shows both $T^3/\hat{q}$ and η/s as a function of the inverse coupling. As the coupling becomes large η/s saturates at the quantum bound, $1/4\pi$, while $T^3/\hat{q}$ retains its sensitivity to further increases in the coupling strength.

The key to understanding the coupling in the QGP is to measure both η/s and $\hat{q}$, and their temperature dependences, independently. This requires high quality jet measurements at RHIC collision energies. Each collision evolves from its maximum initial temperature down as the system expands and cools. Since $\hat{q}$ depends on T^3 in pQCD the quenching is dominated by the highest temperatures in the collision. Thus, the way to study the temperature dependence is the vary the maximum initial temperature by varying the collision energy. Measurements at the LHC alone cannot do this.

3. sPHENIX Design

The jet performance desired for sPHENIX drives the design considerations. Full calorimeter coverage, both electromagnetic and hadronic is required. The detector needs to have full azimuthal coverage over $|\eta|$ <1. The currently existing Silicon Vertex Detector (VTX) is to remain for tracking, with possible additional silicon layers added at larger radii to improve the momentum resolution and improve the track finding ability.

A cut-away view of the detector is shown in Figure 2. The design is for a thin superconducting solenoid with a 2T field at a radius of 70 cm. Behind this is a silicon tungsten electromagnetic calorimeter followed by an iron scintillator hadronic calorimeter. Using tungsten as the absorber allows the electromagnetic calorimeter to be very compact; the calorimeter is about 10 cm thick. Geant 4 single particle simulations give an energy resolution of $14.2\%/\sqrt{E}$ +0.7%.

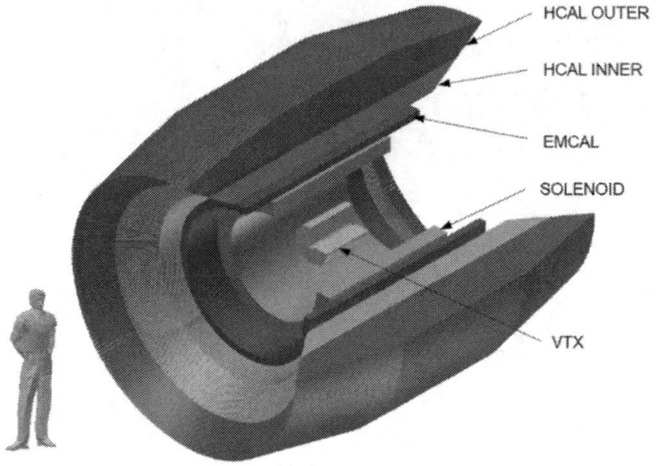

Figure 2. A cut-way view of the sPHENIX detector.

4. Jet Reconstruction Performance at RHIC

The rates for jet production at RHIC given the expected machine luminosity and expected 20 week/year runs are such that sPHENIX would have access to huge numbers of jets up to approximately 60 GeV/c for $\sqrt{s_{NN}}$= 200 GeV. Figure 3 (left) shows the pQCD rates for jet, photon and π^0 production in central Au+Au events. sPHENIX would be able to sample approximately 50B Au+Au events in a year. The numbers of jets and photons above various p_T cuts are shown in Table 1. Over 80% of the single jets that sPHENIX detects also have the opposing jet within the sPHENIX acceptance. Jet rates at $\sqrt{s_{NN}}$= 100 GeV [8] are shown in Figure 3 (right) and one 20 week run would yield 10^5 jets with $p_T > 20$ GeV/c.

For $p_T > 20$ GeV/c the yield of direct photons surpasses that of photons from π^0 decay in Au+Au collisions. As can be seen in Table 1 there are abundant direct photons at RHIC in this region. Because the photon does not interact via the strong force, it decouples from the medium after it is created. sPHENIX will be able to make measurements of both photon-jet and photon-hadron correlations to probe energy loss in this region.

	Au+Au (central 20%)	p+p	d+Au
> 20 GeV	10^7 jets 10^4 photons	10^6 jets 10^3 photons	10^7 jets 10^4 photons
> 30 GeV	10^6 jets 10^3 photons	10^5 jets 10^2 photons	10^6 jets 10^3 photons
> 40 GeV	10^5 jets	10^4 jets	10^5 jets
> 50 GeV	10^4 jets	10^3 jets	10^4 jets

Table 1. Table of jet rates at $\sqrt{s_{NN}}$= 200 GeV for different systems. Each column shows the number of jets or direct photons that would be measured within $|\eta| < 1$ in one 20 week running period.

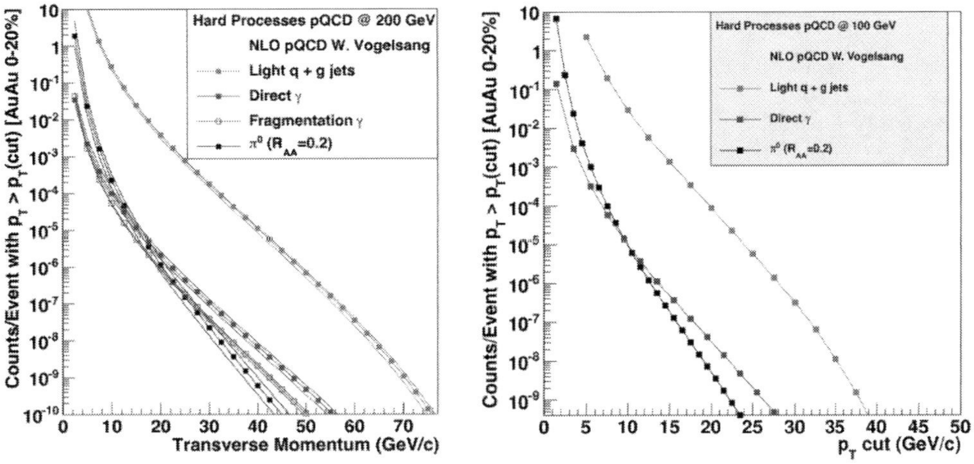

Figure 3. NLO jet, π^0 and photon rates [8] at $\sqrt{s_{NN}}$=200 GeV (left) and $\sqrt{s_{NN}}$= 100 GeV (right) as a function of the p_T cut.

Numerous studies have been done to establish the feasibility of reconstructing jets at $\sqrt{s_{NN}}$= 200 GeV in sPHENIX. A large HIJING study was done in order to evaluate the separation of true jets from fake jets (background fluctuations) [9] in an ideal calorimeter. Results for anti-k_T R = 0.2 jets are shown in Figure 4 (left). For jets with $E_T > 20$ GeV true jets dominate over fake jets. For larger jet radii the crossing point is at higher E_T, but still within the range that sPHENIX expects to have statistics for.

Dijet asymmetry measurements have been used extensively at the LHC. In heavy ion collisions, the large jet quenching decreases the fraction of symmetric (balanced) dijets and increases the fraction of unbalanced dijets. In order to estimate how well sPHENIX would be able to distinguish these scenarios we embedded PYTHIA p+p events into central HIJING events and reconstructed the jet asymmetry, A_J. We also did the same with PYQUEN events, where jet quenching is applied to PYTHIA events. The results are shown in the right panel of Figure 4. The unfolded results for both the PYTHIA and PYQUEN samples are in agreement with the initial truth asymmetry distributions.

5. sPHENIX Upgrades

As discussed above, the sPHENIX proposal in Ref. [1] includes a solenoid and electromagnetic and hadronic calorimetry. This is appropriate for jet and direct photon measurements. However, other very interesting probes, such as separated upsilon states and heavy flavor jets will require additional detectors. There are plans for additional tracking layers beyond the existing VTX and a preshower detector that will be needed for electron identification.

The physics made available by these upgrades is extremely important and the goal is to have these in place at the same time as the rest of sPHENIX. Here we highlight one example, heavy flavor jets. Heavy quarks, especially bottom, were expected to lose much less energy than light quarks due to the dead cone effect [13] suppressing gluon radiation. However, results from both RHIC and the LHC have shown evidence for substantial energy loss of both charm and bottom quarks [14, 15, 16].

If sPHENIX were to be capable of identifying heavy quark jets this would extend the p_T range of heavy quark measurements at RHIC significantly. Figure 5 shows that there are accessible rates for heavy quark production for $p_T > 30$ GeV/c. The constraints from such

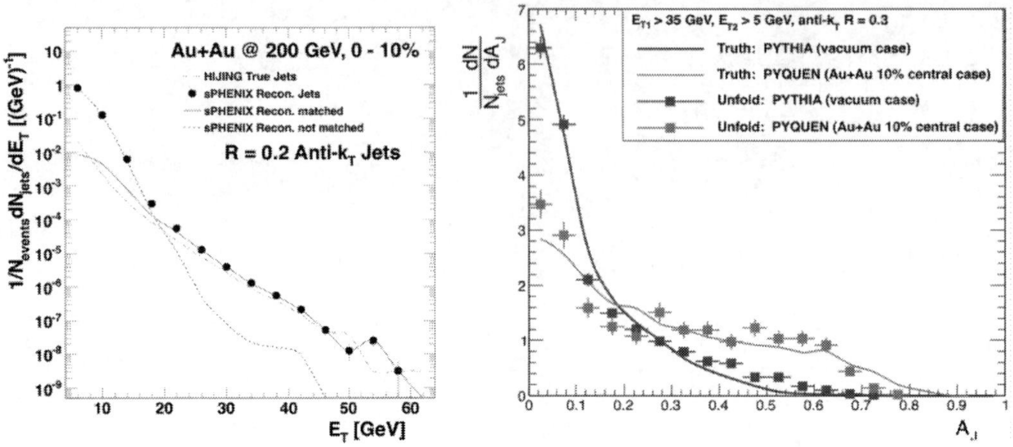

Figure 4. (left) Reconstructed single jet spectra for anti-k_T R=0.2 jets in central HIJING [10] events. The reconstructed jets are shown as points. Those reconstructed jets matched with a true jet from the HIJING event are shown with the blue curve and jets not matched with a true jet (i.e. fake jets) are shown with the black dashed curve. For jet $E_T > 20$ GeV the matched jets dominate over the unmatched jets. Figure is from Ref [9]. (right) Asymmetry, A_J, for PYTHIA [11] (blue) and PYQUEN [12] (red) for dijets embedded in central HIJING events. The truth information is shown as curves and the reconstructed and unfolded results are shown as points. Figure is from Ref. [1].

measurements would also be greatly improved due to the ability to constrain the kinematics from jet measurements which is not possible with electron or heavy meson measurements. Heavy quark jet measurements are a crucial part of understanding hard physics and measurements are necessary both at RHIC and the LHC.

6. Conclusions

The sPHENIX detector will provide the first fully calorimetric jet measurements at RHIC. These measurements are crucial to understanding the behavior of fast partons in the QGP and the properties of the plasma in a region where the coupling might be the strongest.

We have done simulations that show that anti-k_T R $= 0.2$ jets can be cleanly measured for $E_T > 20$ GeV with no additional fake jet rejection. Applying fake jet rejection techniques, already being used at the LHC [19] will decrease the jet energies and increase the jet sizes which are accessible.

The sPHENIX design exploits technological advances in the years since PHENIX was constructed. The sPHENIX design also offers the ability to add additional detectors in the central region which will enable new key physics, such as heavy quark jets and separated upsilon states. Plans are also underway to instrument the forward region with emphasis on spin, asymmetric collisions and future eRHIC running.

7. References

[1] Aidala C, Ajitanand N, Akiba Y, Akiba Y, Akimoto R *et al.* 2012 (*Preprint* 1207.6378)
[2] Luzum M and Ollitrault J Y 2012 (*Preprint* 1210.6010)
[3] Kovtun P, Son D and Starinets A 2005 *Phys. Rev. Lett.* **94** 111601 (*Preprint* hep-th/0405231)
[4] Majumder A, Muller B and Wang X N 2007 *Phys.Rev.Lett.* **99** 192301 (*Preprint* hep-ph/0703082)
[5] Arnold P B, Moore G D and Yaffe L G 2003 *JHEP* **0305** 051 (*Preprint* hep-ph/0302165)
[6] Song H, Bass S A and Heinz U 2011 *Phys.Rev.* **C83** 054912 (*Preprint* 1103.2380)

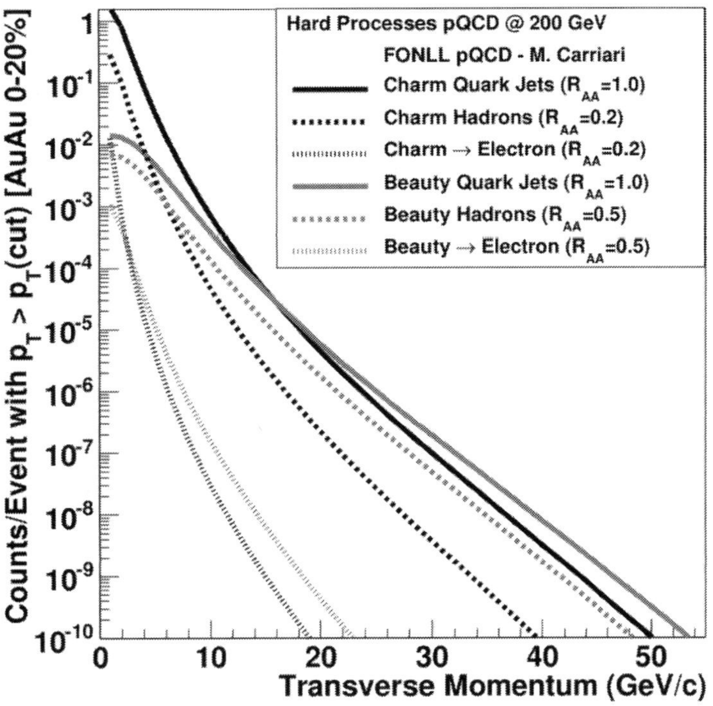

Figure 5. Fixed order next-to-leading log (FONLL) [17] results for heavy quark jet (solid curves), heavy hadron (thick dashed curves) and electron from semileptonic heavy meson decay (thin dashed curves) [18] as a function of the minimum p_T cut. Charm (black) and bottom (red) results are shown. Hadron and electron rates have been reduced by the assumed R_{AA} values shown in the figure.

[7] Niemi H, Denicol G S, Huovinen P, Molnar E and Rischke D H 2011 *Phys.Rev.Lett.* **106** 212302 (*Preprint* 1101.2442)
[8] Vogelsang W Private communication
[9] Hanks J, Sickles A, Cole B, Franz A, McCumber M *et al.* 2012 *Phys.Rev.* **C86** 024908 (*Preprint* 1203.1353)
[10] Gyulassy M and Wang X 1994 *Comput. Phys. Commun.* **83** 307 (*Preprint* nucl-th/9502021)
[11] Sjostrand T *et al.* 2001 *Comput. Phys. Commun.* **135** 238–259 (*Preprint* hep-ph/0010017)
[12] Lokhtin I and Snigirev A 2006 *Eur.Phys.J.* **C45** 211–217 (*Preprint* hep-ph/0506189)
[13] Dokshitzer Y and Kharzeev D 2001 *Phys. Lett.* **B519** 199–206 (*Preprint* hep-ph/0106202)
[14] Adare A *et al.* (PHENIX) 2007 *Phys. Rev. Lett.* **98** 172301 (*Preprint* nucl-ex/0611018)
[15] Tlusty D (STAR collaboration) 2012 (*Preprint* 1211.5995)
[16] Nguyen M (CMS Collaboration) 2012 (*Preprint* 1211.5285)
[17] Cacciari M, Nason P and Vogt R 2005 *Phys. Rev. Lett.* **95** 122001 (*Preprint* hep-ph/0502203)
[18] Cacciari M Private communication
[19] Aad G *et al.* (ATLAS Collaboration) 2012 (*Preprint* 1208.1967)

Journal of Physics: Conference Series **446** (2013) 012051 doi:10.1088/1742-6596/446/1/012051

The importance of e+A collisions at an Electron-Ion Collider

Matthew A. C. Lamont
Brookhaven National Lab, Upton, NY 11777, USA
E-mail: macl@bnl.gov

Abstract. Despite decades of high-energy nuclear physics experiments, relatively little is known about the structure of nuclei outside of the valence region. In order to explore this region, nuclear Deep Inelastic Collisions are required. The proposal to build an electron accelerator at RHIC (to form eRHIC) will be able to answer some of these outstanding questions. This paper will outline a couple of measurements that could be made at the new accelerator.

1. Introduction

Results from Deep-Inleastic Scattering (DIS) $e+p$ at HERA on the momentum-structure of the proton in the longitudinal direction were quite striking - at small Bjorken-x (the momentum fraction of the parton with respect to the hadron), the gluons were by far the most dominant parton, by over an order of magnitude [1]. In fact, linear-QCD "DGLAP" fits to the data showed that this increase showed no signs of abating. However, there is an upper bound on the total cross-section, where the hadron behaves as a "Black-disk", which is given by: $\sigma_{tot} \leq 2\pi R^2$.

One way to achieve such attenuation is though gluon saturation. Gluons themselves are self-interacting. Linear QCD equations (such as DGLAP [2] and BFKL [3]) can describe the high-gluon contribution at low-x via the splitting of high-x gluons into smaller-x gluons. To achieve the saturation required to tame the growth of the cross-section, low-x gluons can be allowed to recombine into higher-x gluons, a process described by non-linear QCD equations such as BK [4] and JIMWLK [5]. The scale at which this happens is called the saturation scale, Q_s^2 and is a function of both x and Q^2, the negative square of the momentum transfer in the collision and is small - typically a few GeV^2 for $x < 10^{-4}$. Therefore the phase-space coverage at HERA does not cover the saturation region for protons.

2. The Structure of Nuclei

The universal nature of the saturation regime, described by the "Colour Glass Condensate" (CGC) effective field theory, means that it should also occur in nuclei [6]. Existing experiments on nuclear DIS are very sparse and predominantly probe the high-x and small mass range, well away from the saturation regime. However, it is hoped that experiments at a new collider, eRHIC, will be able to study this regime. eRHIC will add an electron beam to the current RHIC hadron facility, allowing collisions for 5-30 GeV electrons with 100(250) GeV nuclei(protons) respectively [7]. Figure 1 shows the phase space coverage for e+A and polarised $e+p$ collisions (where the gluonic contribution to the proton spin at small-x will be studied) at eRHIC compared to existing measurements, as well as the saturation scale.

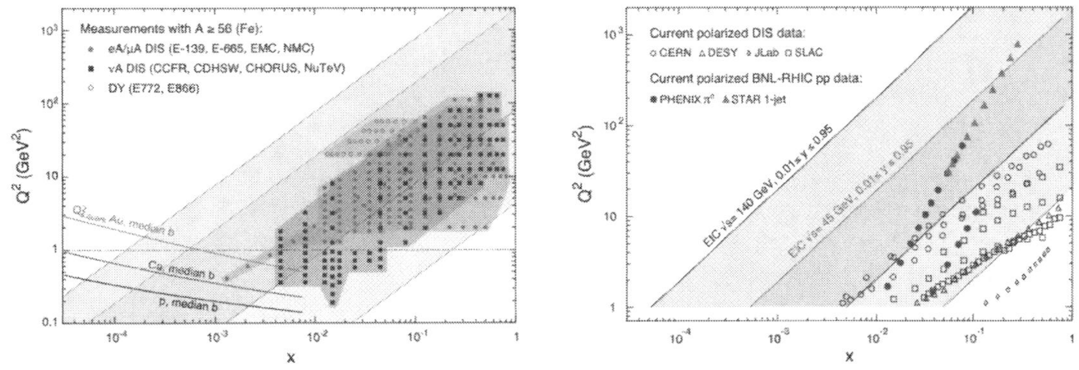

(a) (b)

Figure 1. (a) The (x,Q^2) plane coverage of existing data for $e(\nu)$+A collisions together with the coverage of eRHIC for low- and high-energy electrons ($\sqrt{s}$ = 45 and 110 GeV respectively). (b) The phase space coverage for polarised $e(p)+p$ data.

One of the important phenomena in nuclei, which will allow for the study of a saturation regime at eRHIC is the "nuclear oomph factor". That is, due to geometrical effects where a probe of a given wavelength will coherently probe all partons in the nucleus, the saturation scale in the nucleus is related to that in the proton by the nuclear radius, $A^{1/3}$:

$$(Q_s^A)^2(x) \approx A^{1/3}\left(\frac{1}{x}\right)^{\lambda} \approx A^{1/3}(Q_s^p)^2(x) \tag{1}$$

Therefore, at eRHIC , we will be able to study the nucleus in sufficient detail to answer the following questions:

(i) What is the role of strong gluon fields, parton saturation effects and collective gluon excitations in scattering off nuclei?

(ii) Can we experimentally find evidence for non-linear QCD evolution in high-energy scattering off nuclei?

(iii) What is the momentum and spatial distribution of gluons and sea quarks in nuclei?

(iv) Are there strong colour (quark and gluon density) fluctuations inside of a large nucleus? How does the nucleus respond to the propagation of a colour charge through it?

The following sections will describe two of the measurements which have been identified to answer some of these questions. For descriptions of other measurements, see elsewhere [8].

3. Inclusive Structure Functions

The reduced cross-section of a DIS collision can be written in terms of structure functions, as given by:

$$\sigma_r(x, Q^2) = F_2(x, Q^2) - \frac{y}{Y+}F_L(x, Q^2) \tag{2}$$

where F_2 and F_L represent the (anti)quark and gluon momentum distributions respectively, y is a measure of the inelasticity of the collision and $Y^+ = 1+(1\text{-y})^2$. Inelasticity is described in terms of x, Q^2 and s, the centre-of-mass energy via the relation: $y = \frac{Q^2}{sx}$. In order to measure F_L well, it is imperative to take data at a number of energies. As this didn't happen at HERA, the gluon contribution was extracted via the scaling violation of a fit to the F_2 data.

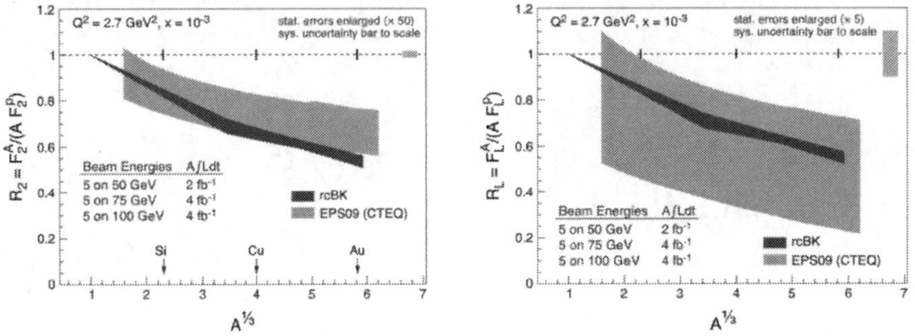

Figure 2. (a) The ratio of F_2 in nuclei to that in protons. Both theoretical and experimental uncertainties are shown. (b) The ratio of F_L in nuclei to that in protons. Note that both the theoretical and experimental uncertainties are larger for F_L than F_2.

In order to study how well eRHIC will be able to measure the F_2 and F_L distributions for different nuclei, Monte-Carlo studies have been performed using the PYTHIA event generator and nuclear PDFs. For ease of viewing, Figure 2 plots the $F_2(F_L)$ for each nucleus divided by the $F_2(F_L)$ in a proton. Two theoretical curves are shown on the plot, rcBK is a non-linear QCD model [9] and EPS09 is a linear QCD model [10]. EPS09 performs a full error calculation and this is represented by the shaded grey band in each figure. The rcBK simulation does not perform a full error calculation, rather the width of the band comes from two different assumptions of the A dependence of the saturation scale. These figures show clearly that whilst the existing nuclear data outlined in Figure 1 can place some constraints on the quark (F_2) distributions, there are very few constraints on the gluons (F_L). Also shown on Figure 2 are the statistical and systematic errors obtained from the Monte-Carlo, scaled to a total cross-section of 10 fb^{-1} for each nucleus, corresponding to 6 months running per species. The statistical uncertainties are enlarged so that they can be seen and are placed at unity.

4. Di-hadron Correlations

One of the important measurements to come from RHIC was the suppression of di-hadron correlations, back-to-back in azimuth, in central Au+Au collisions at mid-rapidity whilst at the same time, no suppression was observed in p+p and d+Au collisions [11]. Di-hadron correlations were used as a proxy for jets in high-multiplicity A+A collisions and the interpretation of this result was that high-momentum particles have their energy quenched as they pass through a de-confined medium. Further experiments went on to show that if these measurements were performed at forward rapidities, then suppression was observed already in the d+Au collisions [12]. This is a striking result and has so far only been explained by models incorporating saturation effects [13]. This is one of the first pieces of evidence of saturation at small-x in nuclei.

One of the disadvantages of p(d)+A collisions is that there is no direct correlation between the rapidity and the x of the collision. Therefore, it is important to also perform these measurements in e+A collisions, where the x can be measured precisely through a reconstruction of the scattered electron. Di-hadron correlations are important as they give information on not just the gluon distribution, but on multi-gluon correlations and theories predict that there will be a similar suppression of correlations in e+A correlations as was observed in d+A correlations. This is presented in Figure 3 (a) which shows the suppression of the away-side correlation ($\Delta\phi=\pi$) for e+p, e+Ca and e+Au respectively, where the suppression level increases for larger nuclei [14].

Figure 3 (b) shows what can be achieved experimentally. Again, Monte-Carlo simulations, this time using PYTHIA along with DPMJET-III, were used to simulate 10 fb^{-1} of saturation

Journal of Physics: Conference Series **446** (2013) 012051 doi:10.1088/1742-6596/446/1/012051

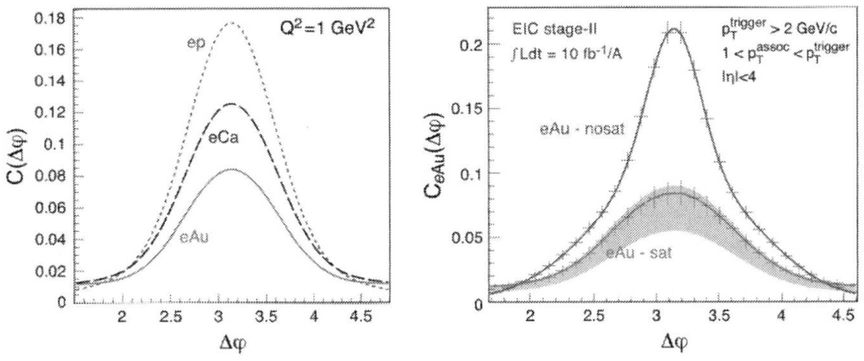

Figure 3. (a) The away-side di-hadron correlation for different nuclei from a saturation model. (b) the away-side di-hadron correlation for e+Au from a Monte Carlo model with and without saturation, including the estimated experimental uncertainties for 6 months running.

and non-saturation e+Au collisions respectively. The statistical and systematic uncertainties are shown on the plot and reveal that at eRHIC, it will be possible to differentiate between the two scenarios.

5. Conclusions

Studies with Monte-Carlo simulations have shown that by performing e+A collisions at eRHIC, with electron energies ranging from 5 to 30 GeV, we are able to answer some of the most important outstanding questions in nuclear physics, namely, to understand the spatial and momentum distributions (longitudinal and transverse) of quarks and gluons at low-x. In fact, due to the many assumptions that have to be made in interpreting hadron-hadron interactions, such a facility is the *only* place where these questions can be answered unambiguously.

References

[1] H1 and ZEUS Collaborations, F. D. Aaron *et al.*, JHEP **01**, 109 (2010).
[2] Y. L. Dokshitzer, Sov. Phys. JETP **46**, 641 (1977).
V. Gribov and L. Lipatov, Sov. J. Nucl. Phys. **15**, 438 (1972).
G. Altarelli and G. Parisi, Nucl. Phys **B126**, 298 (1977).
[3] I. Balitsky and L. Lipatov, Sov. J. Nucl. Phys. **28**, 822 (1978).
E. Kuraev, L. Lipatov and V. S. Fadin, Sov. Phys. JETP **45**, 199 (1977).
[4] I. Balitsky, Nucl. Phys. **B463**, 99 (1996).
Y. V. Kovchegov, Phys. Rev. **D60**, 034008 (1999).
Y. V. Kovchegov, Phys. Rev. **D61**, 074018 (2000).
[5] E. Iancu, A. leonidov and L. D. McLerran, Phys. Lett. **B510**, 133 (2001).
J. Jalilian-Marian, A. Kovner, A.Leonidov and H. Weigert, Phys. Rev. **D59**, 014014 (1998).
[6] E. Iancu, A. Leonidov and L. D. McLerran, Phys. Lett. **B510**, 133 (2001).
[7] V. N. Litvinenko *et al.*, ArXiv: 1109.2819.
[8] Gluons and the quark sea at high energies: Distributions, polarization and tomography., Eds D. Boer, M. Diehl, R. Milner, R. Venugopalan, W. Vogelsang, BNL-96164-2011, INT-PUB-11-034, JLAB-THY-11-1373 (arXiv:1108.1713).
T. Toll (these proceedings).
[9] J. L. Albacete, N. Armesto, J. G. Milhano and C. A. Salgado, Phys. Rev. **D80**, 034031, (2009).
[10] K. J. Eskola, J. Paukkunen and C. A. Salgado, JHEP **04**, 65 (2009).
[11] STAR Collaboration, J. Adams *et al.*, Phys. Rev. Lett. **91**, 072304 (2003).
[12] STAR Collaboration, E. Braidot, Nucl. Phys. **A854**, 168 (2011).
[13] J. L. Albacete and C. Marquet, Phys. Rev. Lett. **105**, 162301 (2010).
[14] F. Dominguez, C. Marquet, B-W Xiao and F. Yuan, Phys. Rev. **D83**,10500 (2001).
F. Dominguez, B-W Xiao and F. Yuan, Phys. Lett. **106**,022301 (2001).

Journal of Physics: Conference Series **446** (2013) 012052 doi:10.1088/1742-6596/446/1/012052

Physics performance studies for the ALICE inner tracker upgrade

Johannes Stiller for the ALICE Collaboration

Physikalisches Institut, University of Heidelberg, Im Neuenheimer Feld 226, 69120 Heidelberg

E-mail: `stiller@physi.uni-heidelberg.de`

Abstract. During the second long shutdown of the LHC in 2018, the ALICE Collaboration plans to install an upgrade of the ALICE Inner Tracking System (ITS) in the central barrel with seven layers of silicon detectors starting at 2.2 cm radial distance from the interaction region and a material budget as low as 0.3 % radiation length per layer. A single-hit resolution of 4 μm and a readout rate capability of up to 50 kHz in Pb–Pb collisions will allow new and unique measurements in the heavy-quark sector, i.e. charm and beauty. Using detailed Monte Carlo simulations of pp and Pb–Pb collisions we study the performance for heavy-flavor detection with an upgraded ITS in the following benchmark analyses: Charm meson and baryon production, i.e. $D^0 \rightarrow K^-\pi^+$ and $\Lambda_c^+ \rightarrow pK^-\pi^+$, and beauty meson and baryon production, i.e. displaced vertices of $B^+ \rightarrow \overline{D}^0\pi^+$ and $\Lambda_b \rightarrow \Lambda_c^+\pi^-$.

1. Introduction

A Large Ion Collider Experiment (ALICE) [1] is designed to study strongly interacting matter at extreme energy densities in high-energy nucleus-nucleus (AA) collisions at the CERN Large Hadron Collider (LHC). The individual subsystems of the detector are arranged in a central barrel and a forward muon spectrometer, which are specifically designed to study Pb–Pb as well as pp and p-Pb collisions as reference. The main tracking devices in the central barrel are, in the order from closest to furthest away from the interaction region, the Inner Tracking System (ITS), the Time Projection Chamber (TPC), the Transition Radiation Detector (TRD) and the Time Of Flight (TOF) system. At central rapidity and down to low transverse momenta (p_T), ALICE has unique particle identification (PID) capabilities as well as high tracking and vertexing precision. These allow for a detailed characterization of the Quark-Gluon Plasma (QGP). Within this scope the light quark sector is extensively studied by ALICE. For example, the charged particle production in Pb–Pb collisions was determined via the nuclear modification factor (R_{AA}) and it was observed that the suppression of high-p_T particles strongly depends on event centrality [2]. For the first time, higher harmonic anisotropic flow was measured for charged particles in Pb–Pb collisions [3], giving insight into initial spatial anisotropy and fluctuations. Further, a large enhancement in the strange baryon/meson-ratio was observed [4], which may be explained by coalescence models. In addition, ALICE has performed several measurements in the heavy-flavor sector. A strong suppression of D meson production in Pb–Pb at LHC energies was measured for the first time down to $p_T > 2$ GeV/c [5]. In comparison, models predict a hierarchy in the energy loss, $\Delta E_g > \Delta E_c > \Delta E_b$, for example due to the dead cone effect [6]. Furthermore, data from Pb–Pb collisions indicates a non-zero elliptic flow of D mesons [7], which suggests that charm quarks might participate in the collective expansion of the system. However, to achieve deeper insight, especially within the scope of rare heavy quark production, high-statistics, precision measurements at very low p_T are required.

2. Future LHC and ALICE upgrade strategy

The so-called *Phase 0* of the LHC schedule was completed in the beginning of 2013. After the first long shutdown from 2013 to 2014, in which the ALICE detector completion is scheduled, the *Phase 1* program will then start in 2015 until 2017. In this time period, a total of 1 nb^{-1} of Pb–Pb at $\sqrt{s_{NN}} = 5.5$ TeV and further pp and p–Pb reference data are planned to be collected. In the second long shutdown from 2017 until 2018, significant detector upgrades are planned. Beginning in 2018 with the *Phase 2,* the LHC will increase its luminosity with Pb beams up to an interaction rate of about 50 kHz, which corresponds to an instantaneous luminosity of 6×10^{27} $cm^{-2}s^{-1}$. With the high-rate upgrade a total amount of data of more than 10 nb^{-1} (Pb–Pb) and more than 6 pb^{-1} (pp reference) will be available in ALICE. The ALICE upgrade concept [8] aims for new high-precision AA measurements of rare probes at low p_T. Among these are measurements of D mesons to zero p_T, charm and beauty baryons (accessible for the first time) and the measurement of beauty via displaced $D^0 \rightarrow K\pi$ (accessible for the first time) to almost zero p_T. The latter is limited by the statistical uncertainty obtained, which is much larger than the systematic uncertainty from theory-driven methods. Within this concept, the ALICE upgrade entails a new, high-resolution and low-material ITS. This detector yields significant improvements for vertexing and tracking at low p_T, namely of the pointing resolution (up to a factor of 5 better) as well as the tracking efficiency and the momentum resolution. Further, for low-p_T heavy-flavor measurements, a continuous readout is planned at a rate of up to 50 kHz for Pb–Pb and several MHz for pp collisions, with High Level Trigger data compression based on charge clusters stemming from particle tracks. This is necessary because of the very high trigger rates for the most interesting signals, as background candidates that pass the selections would also fire the trigger. In case of a conventional trigger and due to the small signal-over-background ratio (S/B) of many signals, a minimum p_T threshold would be required to stay at affordable rates, i.e. for Λ_c^+ with $p_T > 2$ GeV/c gives a rate of 16 kHz. Such a cut at trigger level prevents the measurement at very low p_T. Besides a new ITS readout, an upgrade of the readout systems of most of the detectors of the central barrel to a pipelined readout is foreseen, including a complete replacement of the TPC readout chambers with gas electron multipliers (GEMs) at the endcaps.

3. ITS upgrade concept

The rareness and the low S/B ratio of the desired signals at low p_T pose strong constraints on the design of the upgraded ITS [9]. In order to achieve the required improvement of the impact parameter resolution, the radial position of the first layers of silicon detectors will be moved closer to the interaction point (39 mm $\rightarrow$ 22 mm). Depending on the final design, the total material budget X/X_0 per layer will be reduced from presently 1.14 % to as low as 0.3 %. The overall pixel size improves from currently 50 μm (rϕ) $\times$ 425 μm (z) to O(20 μm $\times$ 20 μm) or O(50 μm $\times$ 50 μm) for monotlithic and state-of-the-art hybrid pixels respectively. An improved tracking efficiency and p_T resolution at low p_T is achieved by adding an additional detection layer (7 in total) and increasing the pixel granularity. The radial extension increases to 22 – 430 mm. In addition it will be possible to quickly insert or remove the detector for yearly maintenance. Two design options have been studied for the upgraded ITS, either consisting of seven layers of pixel detectors or of three inner layers of pixel detectors and four outer layers of strip detectors. The first option yields better standalone tracking efficiency and p_T resolution, whereas the second design has better standalone PID capabilities. Here, only the first option will be addressed.

4. Physics performance of the upgraded ITS

The reconstruction of $D^0 \rightarrow K\pi$ decays, which was performed on Pb–Pb data with good precision, is used as a benchmark to quantify the improved performance of the new ITS. In order to estimate the ITS upgrade performance, a so-called *Hybrid* simulation approach was used. This is a simple scaling of the residuals of impact parameters ($d_{0,r\phi}$, $d_{0,z}$) and transverse momentum based on their true values from Monte Carlo (MC) simulations. The scaling factors are the corresponding ratios of the "upgrade to current" resolution for these parameters.

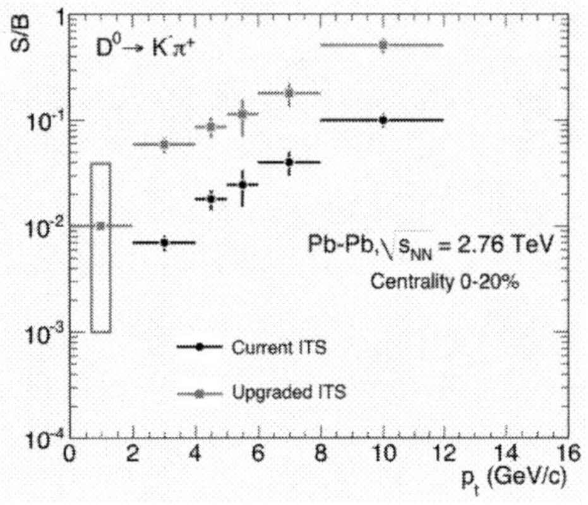

Figure 1. Comparison of the D^0 signal-over-background ratio for current and upgraded ITS [9].

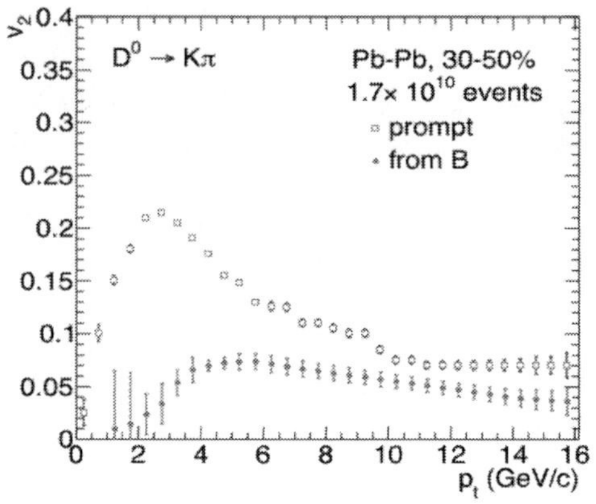

Figure 2. Estimated statistical uncertainty on v_2 of prompt and secondary D^0 mesons for $L_{int} = 10\ nb^{-1}$ [9].

As visible in Fig. 1, the S/B ratio of the measurement improves substantially due to the improved separation of the signal topology from the background. In both scenarios, the same selection cuts were applied and a similar amount of signal was selected. In the range $0 < p_T < 2$ GeV/c the expected signal and background yields were scaled from pp FONLL predictions at 2.76 TeV, as the D meson productions have never been measured in central Pb–Pb collisions in this p_T range. The spread of the R_{AA} (varied between 0.3 and 1) and of the FONLL prediction are the major sources of the rather large uncertainty. The cut efficiencies were determined using MC simulations. Scaling to an integrated luminosity of 10 nb^{-1}, a statistical significance above one hundred is measured at any p_T, which leads to higher precision and extended p_T range of the measurements of v_2 and R_{AA}. The former is shown in Fig. 2 for prompt D and secondary D from B, whose relative contributions are determined from the analysis of D^0 displacement from the primary vertex. Here, v_2 was calculated based on the in-plane and out-of-plane yields with respect to the Event Plane (EP) direction, neglecting higher harmonics. Only statistical uncertainties are shown, as most of the systematic uncertainties cancel in the ratio. Further, first performance studies on the D_s^+ production in Pb–Pb collisions with the upgrade ITS show, that the measurements of R_{AA} and v_2 would largely benefit from the improved tracking precision and high statistics provided by the *high-rate* upgrade of the ALICE central barrel. Another benchmark case is the measurement of charmed baryons ($\Lambda_c^+ \rightarrow pK^-\pi^+$), which strongly relies on the excellent ALICE PID capabilities with TPC and TOF. However, a Λ_c^+ signal was not observed in Pb–Pb data due to the large combinatorial background. Here, because of the short mean proper decay length of the Λ_c^+, $c\tau \approx 60\ \mu$m, very high spatial resolution is needed in order to cleanly identify the secondary vertex. With the improved detector the most effective cut variables are the $\cos(\theta_p)$, the decay length and the requirement of a minimum p_T of the three decay tracks. The pointing angle θ_p describes the angle between the straight connection line of primary and secondary vertex and the direction of the reconstructed momentum vector of the decaying particle. As shown in Fig. 3, the relative statistical uncertainty improves significantly for all p_T. Here, three different scenarios are considered, namely the current and upgraded ITS with *no high-rate* capability (0.1 nb^{-1}) and the upgraded ITS with *high-rate* capability (10 nb^{-1}). In the *high-rate* scenario, Λ_c production should be measurable down to $p_T = 2$ GeV/c. Based on this new measurement, for example the enhancement $(\Lambda_c^+/D)_{Pb-Pb}/(\Lambda_c^+/D)_{pp}$ can be determined, as shown in Fig. 4. The points are drawn on a line (dashed), which captures the trend and magnitude of the Λ/K_S^0 ratio. Systematic uncertainties, which mainly come from feed-down from Λ_b decays, are displayed as boxes. Two model calculations are presented to indicate the expected sensitivity of the baryon to meson ratio in the measurement. The tagging of Λ_c^+ also allows for the full reconstruction of $\Lambda_b \rightarrow \Lambda_c^+\pi^-$.

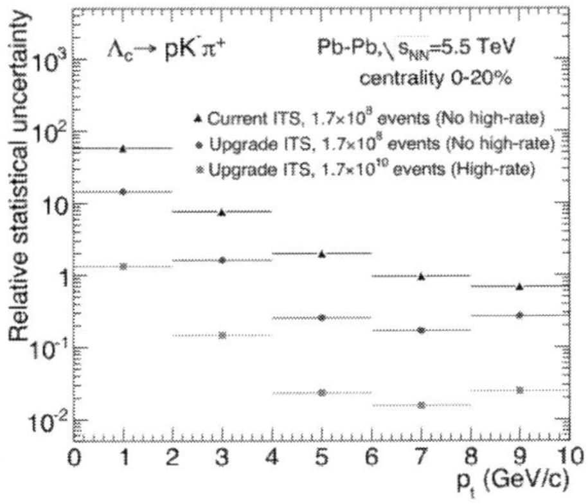

Figure 3. Comparison of statistical precision for different cases of ITS performance and integrated luminosity [9].

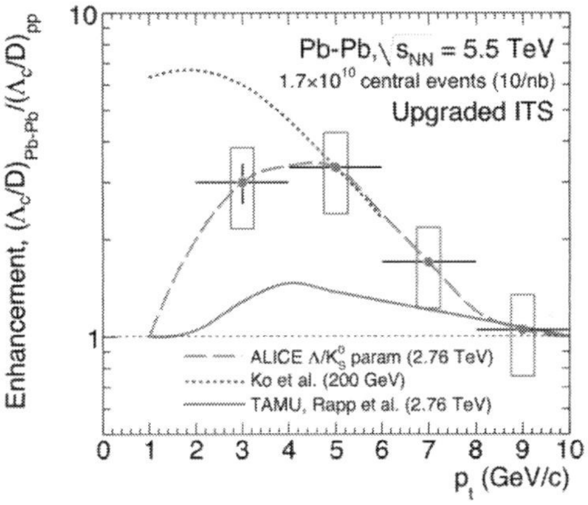

Figure 4. Enhancement of the Λ_c^+/D ratio in central Pb–Pb with respect to pp collisions for $L_{int} = 10$ nb^{-1} [9].

5. Conclusion and outlook

The ALICE experiment shows outstanding performance in studying strongly interacting matter. For a deeper understanding of, e.g., heavy-flavor energy loss mechanisms, azimuthal anisotropy and in-medium hadronization, the ALICE upgrade concept foresees to build a new, high-resolution, low-material ITS in the second long shutdown of the LHC. As a fundamental component of the ALICE upgrade concept, this detector will open the door for a set of new high precision measurements of rare probes at very low p_T. Among these are the measurement of open charm mesons down to zero p_T, charmed baryons, beauty via displaced $D^0 \rightarrow K\pi$ from B decays and low-mass di-leptons. In addition, extensive studies are ongoing for the full reconstruction of $B^+ \rightarrow \overline{D}^0\pi^+(\overline{D}^0 \rightarrow K^+\pi^-)$ and $\Lambda_b \rightarrow \Lambda_c^+\pi^-(\Lambda_c^+ \rightarrow pK\pi)$.

References

[1] ALICE Collaboration, *ALICE: Physics Performance Report, Volume I*, J.Phys. **G30**, 1517(2004).
[2] ALICE Collaboration, arXiv:1208.2711v1 [hep-ex] (2012).
[3] ALICE Collaboration, Phys.Rev.Lett. **107**, 032301 (2011), arXiv:1105.3865v2 [nucl-ex]
[4] I. Belikov (ALICE Collaboration), arXiv:1109.4807v2 [hep-ex] (2011).
[5] ALICE Collaboration, JHEP **09**, 112 (2012), arXiv:1203.2160v4 [nucl-ex].
[6] Y. Dokshitzer and D. Kharzeev, Phys. Lett. B **519**, 199 (2001), doi:10.1016/S0370-2693(01)01130-3.
[7] G. Luparello, these proceedings.
[8] ALICE Collaboration, CERN-LHCC-2012-012 (2012).
[9] ALICE Collaboration, CERN-LHCC-2012-013 (2012).

Journal of Physics: Conference Series **446** (2013) 012053 doi:10.1088/1742-6596/446/1/012053

Understanding the nuclear initial state with an electron ion collider

Tobias Toll
Brookhaven National Laboratory, Upton, NY, USA
E-mail: ttoll@bnl.gov

Abstract. In these proceedings I describe how a future electron-ion collider will allow us to directly measure the initial spatial distribution of gluons in heavy ions, as well as its variance ("lumpiness") in exclusive diffraction. I show the feasibility of such a measurement by means of simulated data from the novel event generator Sartre.

1. Introduction

Measurements over the last decade in heavy ion collision experiments at RHIC indicate the formation of a strongly coupled plasma of quarks and gluons (sQGP). This sQGP appears to behave like a "near-perfect liquid" with a ratio of the shear viscosity to entropy density (η/s) approaching $1/4\pi$ [1, 2, 3, 4]. Despite the significant insight that the sQGP is a strongly correlated nearly perfect liquid, little is understood about how the system is created. The largest uncertainty in our understanding of the evolution of a heavy-ion collision comes from our limited knowledge of the initial conditions, such as the spatial distribution of gluons in the nuclei as well as how it fluctuates around its mean. For example, different assumptions about the nuclear initial distributions give differences up to factors of two for the obtained η/s value [5, 6]. Measurements of the initial gluon distributions with existing machines is only possible indirectly and with large uncertainties. The study of diffractive events in electron-ion collisions would allow one to directly measure the initial condition of the colliding ions providing its spatial gluon distributions as well as the underlying fluctuations ("lumpiness"). In fact, exclusive diffractive eA events are the only way to study the initial spatial distributions and shed light on these fundamental questions.

Much of the content in these proceedings are discussed in more detail in the EIC white paper [7] and in [8].

2. Diffraction in eA

When an electron and an ion collide, they do so electromagnetically via the exchange of a virtual photon, which interacts with a constituent parton in the ion. At a given collision energy s, such collisions are characterized by the virtuality of the photon Q^2 and the longitudinal momentum fraction of the participating parton x. In a *diffractive* eA event, the electron collides with the ion leaving the nucleus intact, via exchange of a color neutral virtual pomeron as well as the virtual photon. This gives a couple of more characteristic kinematic variables, namely $t = (p - p')^2$, which is the square four-momentum difference between the incoming and outgoing ions, and $x_{\mathbb{P}}$ which is the longitudinal momentum fraction of the ion carried by the pomeron. Experimentally, these colorless exchanges produce a rapidity gap, i.e. an angular region in the detector without hadronic activity. The interaction with the nucleus is either elastic or inelastic, and in the latter case the nucleus subsequently radiates a photon or breaks up into color neutral fragments. When it is elastic, the event is called **coherent** and when it is inelastic, the event is called **incoherent**.

3. Imaging gluons in space

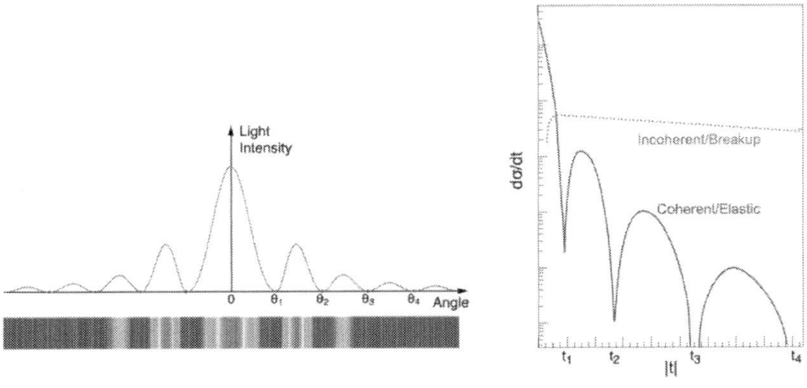

Figure 1. Left: Diffractive pattern of light scattering off a target. Right: A similar pattern is seen in coherent eA scattering as a function of t. The incoherent curve is directly proportional to the variance of the initial states in the target nucleus probed at different values of t.

When light is scattered inelastically against a target, the scattered light shows a diffractive pattern with regions of more and less intense light. These regions are distributed in angle such that the angles of intensity minima $\theta_i \propto 1/kR$, where k is the wave number and R is the size of the target (see Fig. 1). There is thus direct information about the size of the target in the scattered light. Similarly, the spectrum of the elastic cross-section with respect to the hadronic momentum transfer t in diffractive eA events is related to the transverse spatial distribution of the gluons in the ion through a Fourier transform. According to the Good-Walker picture [9], the incoherent cross-section is a direct measure of the lumpiness of the gluons in the ion: $\frac{d\sigma}{dt} \propto \left\langle |\mathcal{A}|^2 \right\rangle - |\langle \mathcal{A} \rangle|^2$, where $\mathcal{A}$ is the quantum mechanical amplitude of the event, and the average is taken over spatial distributions of gluons in the nucleus. Note that the second term on the R.H.S. is the coherent part of the cross-section and the first term is the total (incoherent and coherent) cross-section.

In order to access t in these events, the complete final state has to be measured, which is in general experimentally impossible since some of the ion remnants will always be lost down the beam-pipe. It is only possible to measure t in exclusive processes, such as vector-meson production or Deeply Virtual Compton Scattering (DVCS), where t can be reconstructed by the four-momenta of the produced particle, as well as the scattered electron.

3.1. The dipole model and Sartre

A novel Monte Carlo event generator called Sartre [8] has been constructed to simulate exclusive vector meson and DVCS events. It is based on an extension of the dipole model bSat [10, 11] from ep to eA. In the dipole model, the virtual photon splits up into a quark-antiquark pair – a color dipole. This dipole then interacts with the ion via one or many color-less two gluon exchanges. The dipole fragments into a vector meson, or annihilates into a real photon (DVCS). The bSat model contains an explicit dependence on the impact parameter b. The amplitude in bSat contains two main parts: 1. The wave-overlap, i.e. the overlap of the wave-functions of the incoming virtual photon and the outgoing vector meson. 2. The dipole cross-section, i.e. the hard interaction between the dipole and the ion. These parts are modelled in coordinate space,

Journal of Physics: Conference Series **446** (2013) 012053 doi:10.1088/1742-6596/446/1/012053

and in order to express the amplitude in momentum space, which corresponds to what can be measured, they have to be Fourier transformed. The dipole cross-section for eA in bSat is:

$$\frac{\mathrm{d}\sigma_{q\bar{q}}^{(A)}}{\mathrm{d}^2\mathbf{b}}(x,r,\mathbf{b}) = 2\Big[1-\exp\Big(-\frac{\pi^2}{2N_C}r^2\alpha_S(\mu^2)xg(x,\mu^2)\sum_{i=1}^{A}T(|\mathbf{b}-\mathbf{b}_i|)\Big)\Big].$$

where $xg(x,\mu^2)$ is the DGLAP gluon density taken at a scale $\mu^2 = 4/r^2 + \mu_0^2$, where r is the dipole size, μ_0^2 is the starting scale in the evolution, and $N_C = 3$ is the number of colours. The function $T(b)$ is a Gaussian profile of a nucleon in impact parameter space, and $\mathbf{b}_i$ is the position of each nucleon. $\sum_{i=1}^{A} T(|\mathbf{b}-\mathbf{b}_i|)$, is thus the sum of the contributions from all nucleons in the ion at a given impact parameter $\mathbf{b}$. The positions $\mathbf{b}_i$ are distributed according to the Woods-Saxon potential. Many of these configurations of nucleons are then averaged over for the first and second moments of the amplitude such that both the incoherent and coherent parts of the cross-section can be calculated.

There is also a linearized version of bSat, called bNonSat[10] implemented in Sartre. The dipole cross-section in bNonSat is:

$$\frac{\mathrm{d}\sigma_{q\bar{q}}^{(A)}}{\mathrm{d}^2b} = \frac{\pi^2}{N_C}r^2\alpha_s(\mu^2)xg(x,\mu^2)\sum_{i=1}^{A}T(|\mathbf{b}-\mathbf{b}_i|) \tag{1}$$

In bSat there are numerous pomerons exchanges between the dipole and the nucleus, which causes the cross-section to saturate for small x and large dipole sizes. In bNonSat there is only one pomeron exchanged, giving no such saturation.

The amplitude of the event is a Fourier transform from coordinate- to momentum space t, of the convolution of the wave overlap and dipole cross-section. By squaring this amplitude one obtain the cross-section as a function of t. By looking at its mean and variance with respect to the spatial gluon distribution we can access its coherent and incoherent parts.

3.2. Resulting cross-sections

In Fig. 2 the differential cross-section with respect to t, $\mathrm{d}\sigma/\mathrm{d}t$ is shown, for both J/ψ- and ϕ-meson production, for both dipole models. The coherent cross-section clearly exhibits the typical diffractive pattern, and the incoherent cross-section, which is proportional to the lumpiness of the nucleus, is also shown.

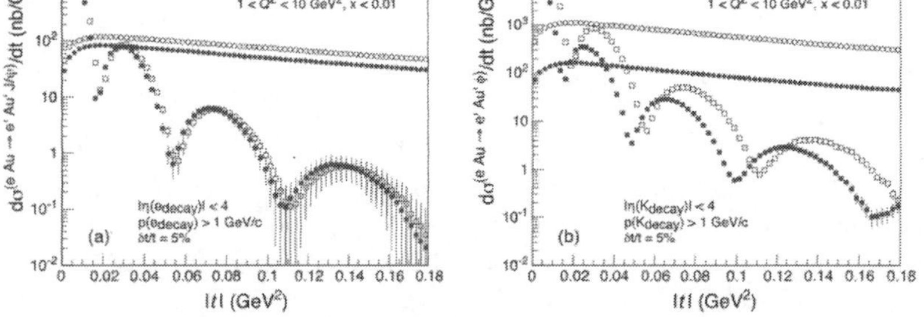

Figure 2. Differential distributions with respect to t for exclusive J/ψ (left) and ϕ (right) for coherent and incoherent events. Both bSat and bNonSat models are shown.

Following [12], one can regain the impact-parameter distribution of gluons by performing a Fourier transform on the amplitude. The amplitude can be obtained by taking the square root of the cross-section, switching its sign in every second minimum. The Fourier transform of this modified amplitude $\sqrt{\mathrm{d}\sigma_{\mathrm{coherent}}/\mathrm{d}t}|_{\mathrm{mod}}$ is:

$$F(b) = \frac{1}{2\pi}\int_0^\infty \mathrm{d}\Delta\Delta J_0(\Delta b)\sqrt{\frac{\mathrm{d}\sigma_{\mathrm{coherent}}}{\mathrm{d}t}(\Delta)}\Bigg|_{\mathrm{mod}}, \tag{2}$$

where $\Delta = \sqrt{-t}$ and J_0 is a Bessel function. The function $F(b)$ is a function of impact-parameter only. For bNonSat, $F(b)$ is directly proportional to the input Woods-Saxon density function, while for bSat the relation is more complex.

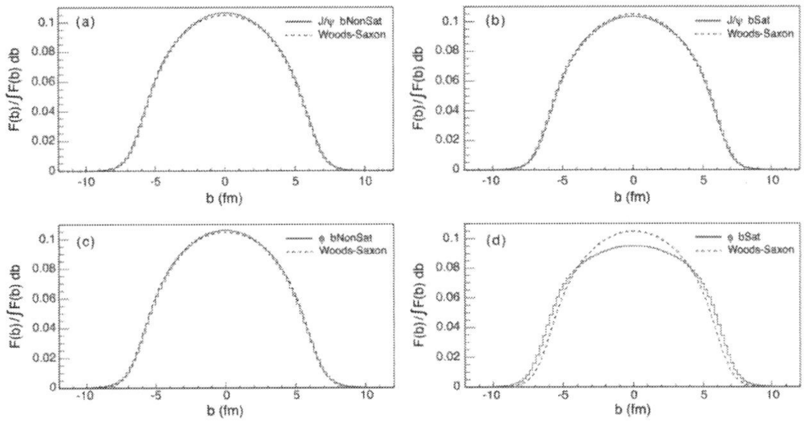

Figure 3. The Fourier transforms obtained from the distributions in Fig. 2 for J/ψ-mesons in (a) and (b) and ϕ-mesons in (c) and (d). Both bSat and bNonSat results are shown with error bands (too small to be seen). The input Woods-Saxon distribution is shown as a reference.

Figure 3 shows the resulting Fourier transforms of the coherent curves in Fig. 2, using the range where $|t| < 0.36$ GeV^2. The obtained distributions have been normalized to unity. As a reference the original input distribution is shown, which is the Woods-Saxon function normalized to unity. The bNonSat curves for ϕ- and J/ψ-meson production reproduce the shape of the input distribution perfectly as is expected since the bNonSat amplitude is directly proportional to the input distribution. For bSat, the shape of the J/ψ curve also reproduces the input distribution, while the ϕ curve does not. This is not surprising, as the size of the J/ψ meson's wave function is much smaller than the ϕ meson's. Therefore, the J/ψ is better suited for probing the transverse structure of the nucleus. As a sign of the robustness of the method, the resulting curves include error-bands (too small to see) coming from the statistical uncertainties in Fig. 2, and they do not require a large range in t; $|t| < 0.2$ gives a good convergence for the shape.

The results in Figure 3 provide a strong indication that the EIC and the LHeC will be able to obtain the nuclear spatial gluon distribution from the measured coherent t-spectrum from exclusive J/ψ production in eA, in a model independent fashion.

4. References

[1] K. Adcox *et al.* [PHENIX Collaboration], Nucl. Phys. A **757**, 184 (2005) [nucl-ex/0410003].
[2] J. Adams *et al.* [STAR Collaboration], Nucl. Phys. A **757**, 102 (2005) [nucl-ex/0501009].
[3] B. B. Back, M. D. Baker, M. Ballintijn, D. S. Barton, B. Becker, R. R. Betts, A. A. Bickley and R. Bindel *et al.*, Nucl. Phys. A **757**, 28 (2005) [nucl-ex/0410022].
[4] I. Arsene *et al.* [BRAHMS Collaboration], Nucl. Phys. A **757**, 1 (2005) [nucl-ex/0410020].
[5] Z. Qiu and U. W. Heinz, Phys. Rev. C **84**, 024911 (2011) [arXiv:1104.0650 [nucl-th]].
[6] S. Esumi [PHENIX Collaboration], J. Phys. G **38**, 124010 (2011) [arXiv:1110.3223 [nucl-ex]].
[7] A. Deshpande *et al.*, arXiv:1212.1701 [nucl-ex].
[8] T. Toll, T. Ullrich [arXiv:1211.3048 [hep-ph]].
[9] M. L. Good, W. D. Walker, Phys. Rev. **120**, 1857-1860 (1960).
[10] H. Kowalski and D. Teaney, Phys. Rev. D **68**, 114005 (2003) [hep-ph/0304189].
[11] H. Kowalski, L. Motyka, G. Watt, Phys. Rev. **D74**, 074016 (2006). [hep-ph/0606272].
[12] S. Munier, A. M. Stasto and A. H. Mueller, Nucl. Phys. B **603**, 427 (2001) [hep-ph/0102291].

Muon Physics in ALICE: The MFT Upgrade Project

Antonio Uras for the ALICE MFT Working Group
IPNL, Université Claude Bernard Lyon-I and CNRS-IN2P3, Villeurbanne, France
E-mail: antonio.uras@cern.ch

Abstract. The ALICE experiment is dedicated to the study of the quark gluon plasma in heavy-ion collisions at the CERN LHC. The Muon Forward Tracker (MFT) is under consideration by the ALICE experiment to be part of its program of detector upgrades to be installed during the LHC Long Shutdown 2 (LS2) planned for 2018. Designed as a silicon pixel detector added in the Muon Spectrometer acceptance ($-4.0 < \eta < -2.5$) upstream of the hadron absorber, the MFT will allow a drastic improvement of the measurements that are presently done with the Muon Spectrometer and, in addition, will give access to new measurements that are not possible with the present Muon Spectrometer setup. Motivations and preliminary results are discussed here, concerning the measurement of prompt and displaced charmonia, open heavy flavors, and low mass dimuons in central Pb–Pb collisions at $\sqrt{s_{NN}} = 5.5$ TeV.

1. Introduction

Lepton-pair measurements always played a key-role in high energy nuclear physics: leptons arrive to the detectors almost unaffected, being not sensitive to the strong color field dominating inside the QCD matter, thus providing an ideal tool to probe the whole evolution of nuclear collisions. The ALICE experiment at the CERN LHC measures single-lepton and lepton-pair production both at central rapidity, in the electron channel, and at forward rapidity, in the muon channel. Identification and measurement of muons in ALICE are performed in the Muon Arm [1], covering the pseudo-rapidity region $-4 < \eta < -2.5$. Starting from the nominal interaction point (IP) the Muon Arm is composed of the following elements: a hadron absorber allowing a reliable muon identification; a dipole magnet providing a field integral of 3 Tm; a set of five tracking stations, each one composed of two cathode pad chambers with a spatial resolution of about $\sim 100\ \mu$m in the bending direction; an iron wall, which absorbs the residual secondary hadrons emerging from the front absorber; the muon trigger system, consisting of 4 resistive plate chambers, with a time resolution of about 2 ns. The ALICE experiment has already an intense physics program based on muon measurements, developed along three main directions: study of quarkonia production [2, 3], of open heavy flavor production [4, 5], of low mass dimuons [6]. In these proceedings we present the Muon Forward Tracker upgrade for the ALICE Muon Arm: after addressing the motivations of the project, a survey of the main results from the physics performances studies is taken.

2. Motivations for a Muon Forward Tracker in ALICE

Despite the remarkable results already obtained, the current ALICE muon physics program suffers from several limitations, basically due to the multiple scattering induced on the muon tracks by the hadron absorber. The details of the vertex region are completely smeared out: in particular, this prevents us to disentangle open charm and open beauty production without making assumptions relying on physics models, and makes it impossible to distinguish prompt and displaced J/ψ production. In addition, only very limited possibilities are given to reject muons coming from semi-muonic decays of pions and kaons, representing an important background both in single muon and dimuons analyses, in particular at low masses and low $p_{\rm T}$. Finally, the degradation of the kinematics, imposed by the presence of the hadron absorber, plays a crucial role in determining the mass resolution for the resonances, especially at low masses. To overcome these limitations, and better exploit the unique kinematic range accessible by the ALICE Muon Arm at the LHC, the Muon Forward Tracker (MFT) was proposed in the context of the ALICE upgrade plans, to take place in the years 2017/2018 during the LHC Long Shutdown 2. The MFT is a silicon pixel detector added in the Muon Spectrometer acceptance ($-4.0 < \eta < -2.5$) upstream of the hadron absorber. The basic idea, motivating the integration of the MFT in the ALICE setup, is the possibility to match the extrapolated muon tracks, coming from the tracking chambers *after* the absorber, with the clusters measured in the MFT planes *before* the absorber. With a correct match between muon tracks and MFT clusters, the muon tracks will gain enough pointing accuracy to permit a reliable measurement of their offset with respect to the primary vertex of the interaction. As realistic simulations show, correct matching rates as large as 60 % and offset resolutions of $\approx$ 100 μm are expected already for muons with $p_{\rm T} \approx 1$ GeV/c. For $p_{\rm T} \approx 3$ GeV/c, less than 10 % of the muons are found to be wrongly matched and the offset resolution stays below 70 μm.

3. Charmonia

The interest of measuring charmonia production in heavy-ion collisions at low $p_{\rm T}$, mainly comes from the possibility to understand the interplay between suppression/recombination mechanisms in the deconfined medium. From this point of view, the MFT is called to give two main contributions: (i) reducing the uncertainties on the extraction of the ψ' signal; (ii) allowing a reliable disentanglement between prompt and displaced charmonia production.

The main limitation affecting the measurement of the ψ' signal with the current Muon Arm setup comes from the poor signal over background (S/B) ratio, ranging from 0.2 % to 0.5 % according to the $p_{\rm T}$ range. When comparing these values to the expected systematic uncertainty on the evaluation of the shape and the normalization of the background, estimated from the analysis of the current J/ψ data to be $\sim$ 0.1 %, one can easily conclude that the only way to improve the uncertainty on the extraction of the ψ' signal is to increase the S/B. The introduction of the MFT in the Muon Arm setup goes exactly in this direction: thanks to its capability to access the details of the vertex region, it makes it possible to operate quality cuts on the dimuon sample, rejecting at the same time: (i) muon pairs in which at least one muon is produced within the hadron absorber, and (ii) muon pairs whose muons are not recognized to share a common origin. Improvement by a factor 3 to 10 is found for the S/B and the systematic errors, depending on the $p_{\rm T}$; moreover, although the improvement of the S/B is obtained by means of a sensible statistics loss, both the significance and the statistical errors on the isolated signal (after background subtraction) are preserved or even slightly improved. The comparison between the expected statistical and systematic uncertainties, in the two scenarios without and with the MFT, is shown in Figure 1: we present the results in the form of uncertainties on the nuclear modification factor R_{AA} of the ψ', under the hypothesis $R_{AA} = 0.3$.

Concerning the measurement of the J/ψ production from beauty mesons, it should be remarked that this kind of measurement is simply not accessible with the current Muon Arm setup. The MFT will make it feasible thanks to the possibility to disentangle prompt and displaced J/ψ vertices: the separation between the two components is performed on a statistical basis by means of the analysis of the distributions of the pseudo-proper decay length of the dimuons falling within the J/ψ mass window, defined as $l_{J/\psi} = L_{xy} m_{J/\psi}/p_{\rm T}$, where L_{xy} is the most probable transverse decay length of the b-hadron, in the laboratory frame. This latter is evaluated as $L_{xy} \approx \frac{\hat{u}^T \cdot \vec{r}}{\hat{u}^T \cdot \hat{u}}$, where $\vec{r}$ represents the transverse displacement vector between the

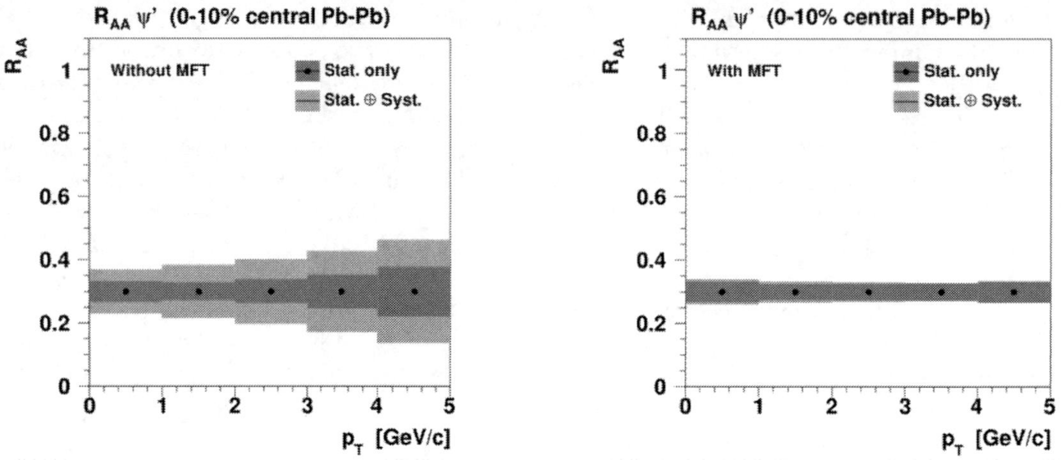

Figure 1. Expected uncertainties on the measurement of the nuclear modification factor R_{AA} of the ψ', in the two scenarios without (left panel) and with (right panel) the MFT. The scenario $R_{AA}(\psi') = 0.3$ is considered here. Red, smaller boxes account for statistical uncertainties alone; green, larger ones represent the sum in quadrature of statistical and systematic uncertainties.

primary vertex and the one of the $\mu^+\mu^-$ pair, and $\hat{u}$ is the unit vector in the direction of the J/ψ momentum. Technically, the analysis is performed by decomposing the measured pseudo-proper decay length into the three components coming from background dimuons, prompt and displaced J/ψ. The normalization of the background component being conveniently fixed by means of a simultaneous fit on the dimuon mass spectrum, we are finally left with the superposition of the pseudo-proper decay length distributions from prompt and displaced J/ψ. The feasibility of the measurement depends therefore on the detector performance in distinguishing the two contributions: preliminary investigations suggest that the ratio between displaced and prompt production can be measured with an error smaller than 10 % for $p_T > 1$ GeV/c, while uncertainties as large as 20 % are expected for $p_T < 1$ GeV/c, due to the smaller S/B, the less favorable displaced/prompt ratio, and the less significantly different decay length distributions.

4. Open Heavy Flavors

Heavy flavor production is an important probe of the hot and dense medium produced in ultra-relativistic heavy-ion collisions. Due to their large masses, charm and beauty quarks are produced in the initial hard scattering, which makes them a precious source of information on the early stages of the collision; on the other hand, thanks to their long lifetimes, heavy-flavored hadrons can experience the further stages of the evolution of the medium.

Measurement of open heavy flavor production via the analysis of the offset distributions, for both single muons and correlated muon pairs, represents a primary item and a natural calling for the MFT physics. It should be remarked that the open heavy flavor measurements in single muons, possible with the current Muon Arm setup, do not allow a separation between the open charm and open beauty components. In addition, they rely on the analysis of the p_T distributions, implying a dependence either on the assumed models or, for the analysis of the Pb–Pb data, on the extrapolation of the available measurements to the rapidity range of the Muon Arm: in this latter case, no trustworthy result can be obtained below $p_T \approx 4$ GeV/c, because of the uncertainties on the background subtraction. In a scenario including the MFT, the measurement is performed on the basis of the analysis of the muons' offset: indeed, offset distributions can be decomposed – with rather limited model assumptions – into the three expected components coming from the background, the open charm and the open beauty processes. Preliminary results have shown that this procedure is reliable down to $p_T \approx 2$ GeV/c

in the single muons analysis: below this limit, the similarities between the offset distributions of the open beauty and background components only allow the open charm to be properly measured; it should be noticed, however, that a complementary measurement of beauty production down to zero p_T will be provided by the analysis of displaced J/ψ cited above. Similar investigations are ongoing for the analysis of the muon pairs populating the dimuon mass continuum between the ϕ and J/ψ resonances, and between the ψ' and Υ resonances.

5. Low Mass Dimuons

The major interest in studying low-mass-dimuon production is represented by the possibility to study the bulk properties and the space-time evolution of the hot and dense QCD matter formed in ultra-relativistic heavy-ion collisions, revealing microscopic properties such as the relevant degrees of freedom and the hadronic excitation spectrum in the deconfined medium. The main experimental challenge in pursuing this research, comes from the necessity to subtract with high accuracy the large combinatorial background coming from semi-muonic decays of pions and kaons, in order for the signals of interest to be properly identifiable among the rich superposition of sources populating the low mass dimuon spectrum.

Here, the MFT is expected to be doubly helpful. Firstly, since we deal with prompt signals, the possibility to apply tight cuts on the offset of the muons and dimuons will enhance the S/B by factors as large as 10, without any loss of significance. Secondly, the much improved mass resolution – by factors larger than 3, thanks to the precise measurement of the opening angle of muon pairs – will enable us to precisely identify the 2-body decays of the narrow light resonances, which could be either subtracted to isolate the signals in the underlying continuum, or taken as the object of dedicated investigations.

6. Conclusions

The muon physics program of the ALICE experiment is constrained by the intrinsic limitations of the current setup of the Muon Arm. In this report we discussed how the addition of a silicon Muon Forward Tracker in the acceptance of the Muon Spectrometer should overcome these limitations increasing the physics potential of the muon measurements in ALICE. In particular, the rejection power gained against muon pairs from the combinatorial background will allow us to dramatically improve the uncertainties associated to the measurement of the ψ' charmonium state, while the identification of displaced and prompt J/ψ production will be made possible by means of the measurement of the pseudo-proper decay length distributions. The enhanced pointing accuracy gained by the muon tracks will also give us the possibility to disentangle open charm and beauty production on the basis of the analysis of the single muon and dimuon offset distributions, starting from $p_T > 2$ GeV/c. Low mass dimuons, finally, will enormously profit from the significantly improved rejection power for the combinatorial background, as well as from the much better mass resolution available for the low mass narrow resonances.

References

[1] The ALICE Collaboration, *The ALICE experiment at the CERN LHC* JINST **3**, S08002 (2008)
[2] The ALICE Collaboration, *J/ψ suppression at forward rapidity in Pb–Pb collisions at $\sqrt{s_{NN}} = 2.76$ TeV* Phys. Rev. Lett. **109**, 072301 (2012)
[3] The ALICE Collaboration, *J/ψ polarization in pp collisions at $\sqrt{s} = 7$ TeV* Phys. Rev. Lett. **108** 082001 (2012)
[4] The ALICE Collaboration, *Production of muons from heavy flavor decays at forward rapidity in pp and Pb–Pb collisions at $\sqrt{s_{NN}} = 2.76$ TeV* Phys. Rev. Lett. **109**, 112301 (2012)
[5] The ALICE Collaboration, *Heavy flavor decay muon production at forward rapidity in proton–proton collisions at $\sqrt{s} = 7$ TeV* Phys. Lett. B **708**, 265 (2012)
[6] The ALICE Collaboration, *Light vector meson production in pp collisions at $\sqrt{s} = 7$ TeV* Phys. Lett. B **710**, 557 (2012)

Journal of Physics: Conference Series **446** (2013) 012055 doi:10.1088/1742-6596/446/1/012055

Strange and Multi-Strange Particle Production in ALICE

D.D. Chinellato[1] **for the ALICE Collaboration**

[1] University of Houston, Physics Department, 617 Science & Research Building 1, Houston, TX 77204

E-mail: david.dobrigkeit.chinellato@cern.ch

Abstract. The production of strange and multi-strange hadrons in proton-proton (pp) and lead-lead (Pb-Pb) collisions is studied with the ALICE experiment at the CERN LHC. These particles are reconstructed via their weak decay topologies, exploiting the tracking and particle identification capabilities of ALICE. Measurements of central rapidity yields of Λ, Ξ^- and Ω^- baryons, their antiparticles and K^0_S mesons are presented as a function of transverse momentum for Pb-Pb collisions at $\sqrt{s_{NN}} = 2.76$ TeV. They are compared to those observed in pp collisions as well as to results from lower energy nucleus-nucleus measurements.

PACS Numbers: 25.40.Ep, 25.75.-q, 25.75.Dw

1. Introduction

The study of strangeness production in both proton-proton (pp) and lead-lead (Pb-Pb) collisions contributes to the understanding of particle production and hadrochemistry in general. This is not only due to the fact that there is no net strangeness content in the colliding systems but also because strange quarks are the lightest and most abundantly produced among the higher generation quarks.

In proton-proton collisions, the strange hadron measurements serve not only as a benchmark for the heavy-ion collision measurements but may also offer an insight into the different production mechanisms at play. At low transverse momenta (p_T), soft interactions are expected to be the dominant source of particle production, while at high p_T it is expected that interactions with high transferred momenta, i.e. those that can be described well with perturbative QCD (pQCD), take over as the most significant production mechanism.

In heavy-ion collisions, it has been proposed that strange quarks may be produced thermally in the high energy density medium formed [1]. In this scenario, the final yields would exhibit a *strangeness enhancement*. This is defined as a higher yield of strange and multi-strange hadrons in Pb-Pb collisions when compared to production rates measured in proton-proton collisions with an appropriate scaling for the volume, such as the number of participant nucleons (N_{part}).

2. Data Analysis

Strange and multi-strange hadrons are reconstructed by their weak decay topologies in a sample of about 130×10^6 pp collisions at $\sqrt{s} = 7$ TeV and about 30×10^6 Pb-Pb collisions at $\sqrt{s_{NN}} = 2.76$ TeV, both collected in 2010. K^0_S, Λ and $\bar{\Lambda}$ are reconstructed by combining two charged tracks into V-shaped decays (the so-called V0 topology) while Ξ^-, $\bar{\Xi}^+$, Ω^- and $\bar{\Omega}^+$ are reconstructed by further combining one Λ or $\bar{\Lambda}$ candidate with a third track (the so-called cascade decay topology). The V0 and cascade decay candidates are formed only if the tracks involved satisfy certain geometrical restrictions to ensure consistency with the expected topology. These initial selection criteria are thus called topological selections. Furthermore, the energy loss for ionization of the TPC gas is used as a particle identification criterion for all daughter tracks, reducing background substantially. An upper limit on the actual lifetime selection is also imposed in terms of the PDG $c\tau$ of each particle for the V0 analysis and serves to remove secondaries from material interactions. In addition, a competing decay rejection in Armenteros-Podolanski parameter space is applied for selecting K^0_S candidates in the Pb-Pb analysis, where only candidates satisfying $|\alpha_{arm}| < 5\ p_{\mathrm{T}}^{arm}$ are accepted into the K^0_S analysis. In this equation, $\alpha_{arm} = (p_{||}^+ - p_{||}^-)/(p_{||}^+ + p_{||}^-)$, where $p_{||}$ is the momentum component of the positive or negative daughter particle parallel to the momentum of the parent, and p_{T}^{arm} is the momentum component perpendicular to the parent momentum given in GeV/c. This selection has been shown to consistently remove background without creating any false peaks and acts to select decays that are more symmetric in phase space. Finally, in the case of multi-strange baryons, candidates with an invariant mass close to the Ξ mass under the appropriate daughter mass hypothesis are rejected for the Ω analysis.

Signal extraction is performed by sampling the background in areas adjacent to the invariant mass peaks and subtracting this background from the peak region. The background areas are p_{T} dependent to account for the variation of resolution with momentum. For the Λ and $\bar{\Lambda}$ analysis, we subtract the contribution to the Λ signal coming from Ξ decays by first computing a feeddown matrix that relates the Ξ production in a given momentum range to the secondary Λ detection at any Λ p_{T}. Once this matrix is calculated from Monte Carlo simulations using the same selections as in the real data analysis, we scale the production rates in simulation by the data to obtain the Λ signal that needs to be subtracted from the measurement. Feeddown from neutral Ξ is roughly equal to that from charged Ξ; together they account for about 20% of the raw measured Λ signal in both pp and Pb-Pb collisions.

In order to compute detection efficiency, we use PYTHIA [2] (for pp) or HIJING [3] (for Pb-Pb) event generators and subsequently propagate the produced particles through the full ALICE geometry with GEANT3 [4, 5]. When computing this efficiency for Pb-Pb collisions and also for the multi-strange in pp, the Monte Carlo productions were enriched with injected particles in order to reduce the statistics required for proper determination of detection efficiencies.

Systematic uncertainties are evaluated point-by-point for the topological selections, signal extraction and track quality cuts. Other sources of systematic uncertainties such as normalization, feeddown correction, material budget, GEANT3 (anti-)proton cross-section, TPC particle identification, proper lifetime selection and competing decay rejection are taken to be p_{T} independent.

3. Measurements in Proton-Proton Collisions

The K^0_S, Λ and $\bar{\Lambda}$ efficiency- and feeddown-corrected spectra in pp collisions at $\sqrt{s} = 7$ TeV are shown in Fig. 1. The $\bar{\Lambda}/\Lambda$ ratio is compatible with unity throughout the measured transverse momentum range of 0.4 - 10.0 GeV/c. For the extraction of the integrated production rates, we fit the spectra with Tsallis functions, as done previously in lower energy pp strangeness analyses in ALICE [6]. While a 15% extrapolation at low p_{T} is required for the computation of the Λ yields, only a negligible extrapolation is needed for the K^0_S yield.

The total production rates of K^0_S, Λ and $\bar{\Lambda}$ measured by ALICE are obtained in the inelastic event (INEL) category and are observed to be 20% lower than the rates measured by CMS for non-single-diffractive (NSD) events [7]. This matches the INEL and NSD cross-section ratio $\sigma_{\mathrm{NSD}}/\sigma_{\mathrm{INEL}}$ that has been measured to be of about 80% [8], so this difference is consistent

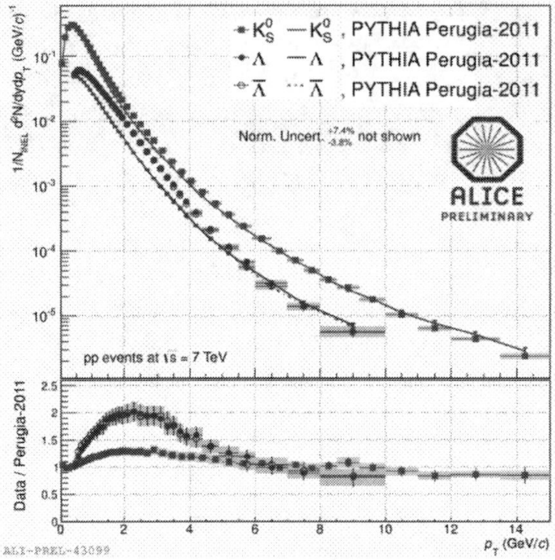

Figure 1. Λ, $\bar{\Lambda}$ and K^0_S transverse momentum spectra in pp collisions at $\sqrt{s}$ = 7 TeV, represented with points, compared to Pythia P2011, represented with lines. The bottom panel shows the ratio of data to calculation.

with the expectation that there is a negligible strange particle production at mid-rapidity in single-diffractive events compared to NSD events.

We compare the measured transverse momentum spectra to PYTHIA Perugia-2011 [9, 10]. This tune is chosen because it provides the best description of multi-strange baryon production [11] while simultaneously also being tuned to the charged-particle multiplicity measured in pp collisions at 7 TeV. It can be seen in Fig. 1 that, while the calculation reproduces the data above 6 GeV/c, at lower p_T the differences are much larger, although not as large as those observed for multi-strange baryons [11].

4. Measurements in Heavy-Ion Collisions

Figure 2 shows the K^0_S and Λ transverse momentum spectra in Pb-Pb collisions at $\sqrt{s_{NN}} = 2.76$ TeV. $\bar{\Lambda}$ spectra are not drawn, however the $\bar{\Lambda}/\Lambda$ ratio is compatible with unity. We observe that the ratio Λ/π is constant with centrality, as seen in the left panel of Fig. 3, which means chemistry does not significantly change with centrality. A more thorough evaluation of the hadrochemistry in heavy-ion collisions is currently under development by combining these measurements and others, such as of other, non-strange particles and resonances.

We have also measured transverse momentum spectra for Ξ^-, $\bar{\Xi}^+$, Ω^- and $\bar{\Omega}^-$, thus enabling the study of strangeness enhancement for $|S| = 2$ and $|S| = 3$ cases. This can be seen in Fig. 3, right panel, where yields are divided by $N_{\rm part}$, the number of participating nucleons provided by a Glauber model, to account for the system size in nuclear collisions in the pp comparison. In this plot, we see an increasing enhancement with both increasing strangeness content as well as with increasing centrality. When compared to lower energy measurements done at SPS [12, 13] and at RHIC [14], the enhancement is seen to decrease with increasing beam energy, continuing the trend first observed at the SPS. However, since the Λ/π ratio has been shown to be constant with centrality and, additionally, the charged-particle multiplicity does not scale with $N_{\rm part}$ [15], other alternatives to the $N_{\rm part}$ scaling are currently being studied to quantify and factor

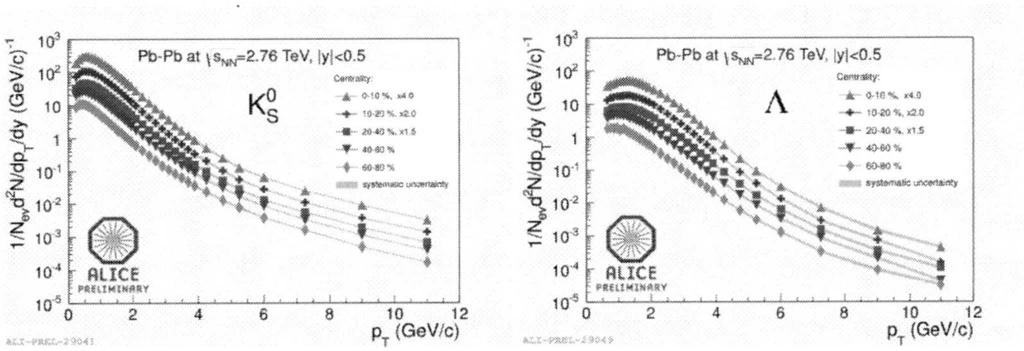

Figure 2. p_T spectra of K^0_S (left) and Λ (right) in Pb-Pb collisions for 5 centrality intervals.

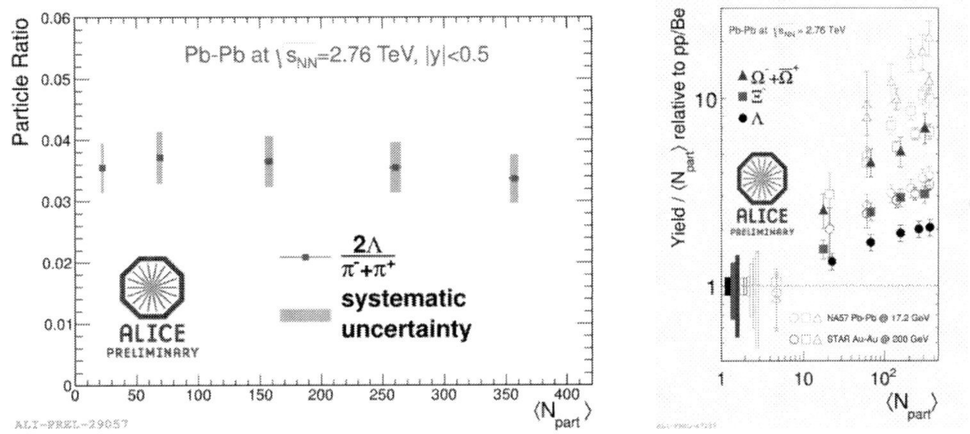

Figure 3. Left panel: Λ/π ratio according to centrality. Right panel: Strangeness enhancement for Λ, Ξ^- and Ω^- shown with $\langle N_{\text{part}}\rangle$ scaling and compared to lower energy measurements from NA57 and STAR.

out the effect of a general increase in particle multiplicity and that of an actual strangeness enhancement.

References

[1] Rafelski J and Müller B 1982 *Phys. Rev. Lett.* **48** 1066–1069
[2] Sjostrand T, Mrenna S and Skands P Z 2006 *JHEP* **05** 026
[3] Wang X N and Gyulassy M 1991 *Phys. Rev. D* **44**(11) 3501–3516
[4] Brun R *et al.* 1985 *CERN Data Handling Division DD/EE/841*
[5] Brun R *et al.* 1994 *CERN Program Library LongWrite- up, W5013*
[6] Aamodt K *et al.* 2011 *Eur. Phys. J. C* **71** 1594
[7] Khachatryan V *et al.* (CMS) 2011 *JHEP* **2011**(5) 1–40
[8] Abelev B *et al.* (ALICE Collaboration) 2012 (*Preprint* arXiv:1208.4968v1)
[9] Skands P Z 2010 *Phys. Rev.* **D82** 074018
[10] Field R 2010 (*Preprint* arXiv:1010.3558v1)
[11] Abelev B *et al.* 2012 *Phys. Lett. B* **712** 309 – 318
[12] Antinori F *et al.* (NA57 Collaboration) 2006 *J. Phys. G* **32** 427
[13] Antinori F *et al.* (NA57 Collaboration) 2010 *J. Phys. G* **37** 045105
[14] Abelev B I *et al.* (STAR Collaboration) 2008 *Phys. Rev. C* **77**(4) 044908
[15] Aamodt K *et al.* (ALICE Collaboration) 2011 *Phys. Rev. Lett.* **106**(3) 032301

Journal of Physics: Conference Series **446** (2013) 012056 doi:10.1088/1742-6596/446/1/012056

Hadronic resonances at ALICE

A G Knospe (for the ALICE Collaboration)

Department of Physics, The University of Texas at Austin, 1 University Station C1600, Austin, TX 78712-0264 USA

E-mail: anders.knospe@cern.ch

Abstract. Measurements of the ratios of hadronic resonance yields to non-resonance yields can be used to study the properties of the hadronic phase of high-energy heavy-ion collisions. A change in resonance masses or widths could be an indication of chiral symmetry restoration. Measurements of resonances in proton-proton collisions provide an important baseline for measurements in heavy-ion collisions as well as data for tuning QCD-inspired particle production models. The ALICE collaboration has measured the $K^*(892)^0$ and $\phi(1020)$ resonances in Pb–Pb collisions at $\sqrt{s_{NN}} = 2.76$ TeV and the $K^*(892)^0$, $\phi(1020)$, and $\Sigma^*(1385)^{\pm}$ resonances in pp collisions at $\sqrt{s} = 7$ TeV. These measurements – including transverse momentum spectra, ratios to non-resonances, masses, and widths – are discussed and compared to theoretical predictions.

1. Introduction

These proceedings report measurements of resonances in pp collisions at $\sqrt{s} = 7$ TeV and Pb–Pb collisions at $\sqrt{s_{NN}} = 2.76$ TeV. The study of resonance production yields in pp collisions provides an important baseline for comparison with results in Pb–Pb collisions. Resonance spectra in pp collisions can also be used to tune the predictions of QCD-inspired event generators such as PYTHIA [1] and PHOJET [2].

In nucleus-nucleus collisions, resonances can be used to study the properties of the hadronic phase and to search for the signatures of chiral symmetry restoration in the medium. Resonances may be produced through regeneration in the hadronic phase and the observable resonance signal may be reduced through the re-scattering of the decay daughters [3–6]. Statistical models [4, 7, 8] can be used to predict the ratios of resonance yields to the yields of non-resonances as functions of the chemical freeze-out temperature T_{ch} and the time Δt between chemical and thermal freeze-out; simultaneous measurements of at least two resonance-to-non-resonance ratios (*e.g.*, K^{*0}/K and $\Lambda^*(1520)/\Lambda$) may allow constraints to be placed on T_{ch} and Δt. It is expected that chiral symmetry will be restored at approximately the same critical temperature as the phase transition to the QGP [9]. Resonances that interact with and decay inside the medium (when chiral symmetry is at least partially restored) may be shifted off-shell and may also have shorter lifetimes than observed in vacuum [10–12]. It is expected [13, 14] that it would be easiest to observe the signatures of chiral symmetry restoration in the range $2 < p_T < 10$ GeV/c at LHC energies, due to regeneration and re-scattering at low p_T and the time dilation of decays for high p_T.

The ALICE detector is described in [15]. The analyses described in these proceedings use the Inner Tracking System (ITS, for tracking and vertex finding), the Time Projection Chamber (TPC, for tracking and particle identification through energy loss), the Time-of-Flight

(TOF) detector (for particle identification), and the V0 detector (which provides minimum-bias triggers).

2. Resonances in pp Collisions at $\sqrt{s} = 7$ TeV

The analyses of the K^{*0} and ϕ mesons[1] in pp collisions at $\sqrt{s} = 7$ TeV are described in detail in [16]. The K^{*0} (ϕ) meson is reconstructed in 80 (60) M minimum-bias pp collisions via the decay channel $K^{*0} \rightarrow \pi^{\pm}K^{\mp}$ ($\phi \rightarrow K^{-}K^{+}$) which has a branching ratio of 0.666 (0.489) [17]. The decay products of the K^{*0} are identified using the TOF; the TPC is used if no TOF signal is present. The decay products of the ϕ are identified using the TPC, with the addition of TOF information if a signal is present. The invariant-mass distributions of unlike-charge πK and KK pairs (with $|y_{\mathrm{pair}}| < 0.5$) in the same event are constructed. Two methods are used to estimate the combinatorial background. First, the background is extracted from the distributions of like-charge pairs in the same event. Second, each decay product is paired with unlike-charge tracks from a different event (with a similar primary vertex z position and charged track multiplicity). The like-charge (event mixing) method is the default used for the K^{*0} (ϕ) analysis. The background-subtracted K^{*0} (ϕ) invariant-mass distributions are fitted with a Breit-Wigner (Voigtian[2]) peak added to a 2nd-order polynomial (which describes the residual background). The product of the efficiency and acceptance is extracted from 60 M Monte-Carlo collisions. The systematic uncertainties in the corrected spectra account for variations due to changing the track selection cuts, the combinatorial background, the mass range of the fit, and the order of the residual-background polynomial. The corrected transverse-momentum spectra cover the ranges $0 < p_{\mathrm{T}} < 6$ GeV/c for the K^{*0} and $0.4 < p_{\mathrm{T}} < 6$ GeV/c for the ϕ. These spectra are fitted with Lévy-Tsallis functions [18], which are used to extract the mean transverse momentum $\langle p_{\mathrm{T}} \rangle$ and to estimate the ϕ meson yield for low p_{T}. The ratios of the p_{T}-integrated K^{*0} and ϕ yields to the yields of charged kaons are calculated.

The $\Sigma^{*\pm}$ and $\overline{\Sigma}^{*\pm}$ baryons are reconstructed in 211 M minimum-bias pp collisions in the rapidity range $|y| < 0.8$ via the decay channel $\Sigma^{*\pm} \rightarrow \pi^{\pm}\Lambda \rightarrow \pi^{\pm}\pi^{-}p$. The two steps of the decay process have a combined branching ratio of 0.56 [17]. The Λ baryons are identified by cuts on their decay topology and invariant mass, and additional topological cuts are applied the $\pi\Lambda$ pairs. The combinatorial background is constructed using the event mixing technique (each Λ is paired with a π from 5 other events). The residual background is parametrized by the sum of a 3rd-order polynomial and a Gaussian, which accounts for correlated $\pi\Lambda$ pairs from $\Lambda^{*}(1520)$ decays. The product of the efficiency and acceptance is calculated using simulated pp collisions. The systematic uncertainties from sources including the track selection cuts, topological cuts, combinatorial background normalization, residual background parametrization, and the material budget in the simulations have been evaluated.

The corrected K^{*0}, ϕ, and $\frac{1}{2}(\Sigma^{*+} + \overline{\Sigma}^{*-})$ spectra are shown in Figure 1 and compared to predictions from PYTHIA and PHOJET. The PYTHIA Perugia 2011 tune describes the K^{*0} spectrum and the ϕ spectrum for $p_{\mathrm{T}} > 3$ GeV/c well, while the Perugia 0 tune underestimates the spectra. PHOJET and the PYTHIA ATLAS-CSC tune describe the meson spectra for $p_{\mathrm{T}} > 1$ GeV/c. The PYTHIA D6T tune deviates from the measured spectra for $p_{\mathrm{T}} > 2$ GeV/c. The $\Sigma^{*\pm} + \overline{\Sigma}^{*\mp}$ spectra are well described by the PYTHIA ATLAS-CSC tune for $p_{\mathrm{T}} > 2$ GeV/c, while other models under-predict the data.

3. Resonances in Pb–Pb Collisions at $\sqrt{s_{\mathrm{NN}}} = 2.76$ TeV

The K^{*0} (ϕ) meson is reconstructed in 8.2 (9.5) M Pb–Pb collisions. The decay products are identified using TPC energy loss measurements. For the K^{*0} (ϕ) meson, the combinatorial

[1] K^{*0} refers to both the $K^{*}(892)^{0}$ and its antiparticle; masses are omitted from all particle symbols
[2] A Breit-Wigner function convoluted with a Gaussian, which accounts for the detector resolution

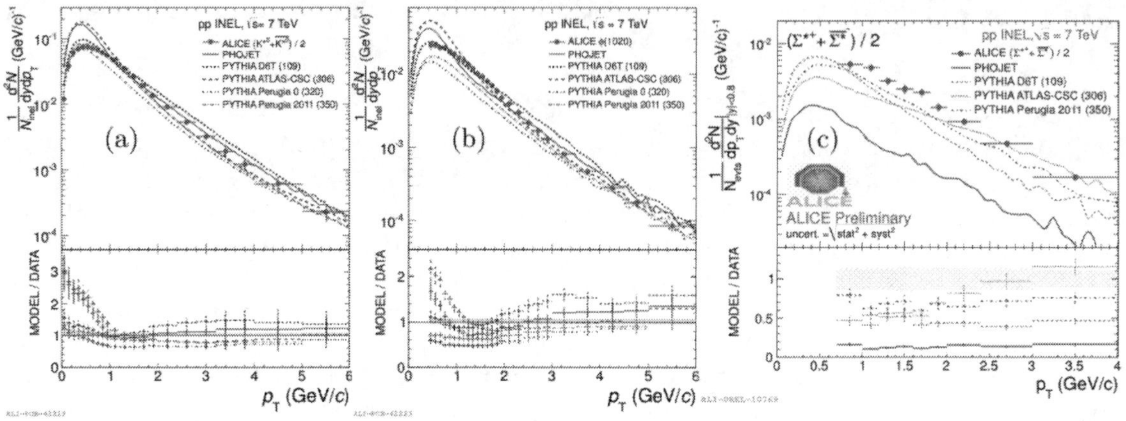

Figure 1. Resonance spectra in pp collisions at $\sqrt{s} = 7$ TeV compared to predictions from event generators: K^{*0} (a), ϕ (b), $\frac{1}{2}(\Sigma^{*+} + \overline{\Sigma}^{*-})$ (c).

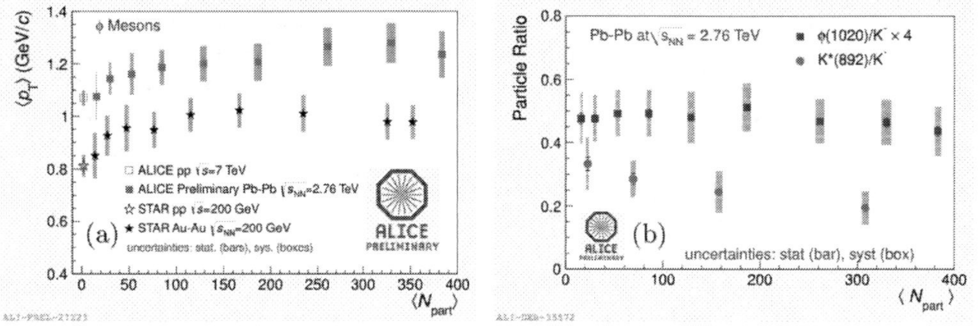

Figure 2. (a): Mean transverse momentum for ϕ mesons in pp and A–A collisions measured using the ALICE detector and the STAR detector [19, 20]. (b): The K^{*0}/K^- and ϕ/K^- ratios measured in Pb–Pb collisions at $\sqrt{s_{NN}} = 2.76$ TeV as functions of $\langle N_{part} \rangle$.

background is calculated using the like-charge (event mixing) method, the residual background is parametrized using a 1st- (2nd-)order polynomial, and the peaks are fit using a Breit-Wigner function. The systematic uncertainties in the corrected spectra account for variations due to changing the track selection cuts, particle identification cuts, the combinatorial background, the mass range of the fit, the order of the residual-background polynomial, and the method used to extract the particle yield (integrating the histogram and subtracting the residual background fit or integrating the peak fit itself). The corrected spectra are fitted with Boltzmann-Gibbs Blast-Wave functions [21]; the ϕ meson fits are used to estimate the yields for $p_T < 0.5$ GeV/c. The values of $\langle p_T \rangle$ and the ratios of the p_T-integrated K^{*0} and ϕ yields to the yields of charged kaons [22] are calculated.

The masses and widths of the K^{*0} and ϕ mesons were extracted from the Breit-Wigner peak fits. The K^{*0} mass deviates from the vacuum value by about 10 MeV/c^2 for $p_T < 2$ GeV/c. Similar deviations in the K^{*0} mass are observed in pp collisions at $\sqrt{s} = 7$ TeV, suggesting that the deviations in Pb–Pb collisions are instrumental effects (K^{*0} peaks from GEANT simulations are being studied to verify this). Furthermore, the deviation occurs in a momentum range where the fraction of regenerated resonances (which would have vacuum properties) would be large and the signal would be heavily modified by re-scattering [13, 14]. The K^{*0} width is consistent with the vacuum value. The ϕ mass (width) deviates from the vacuum value by 0.5 (1-2) MeV/c^2. When ϕ peaks from GEANT simulations are fitted using the same procedure, similar deviations from the vacuum value are observed.

The values of $\langle p_{\mathrm{T}} \rangle$ for ϕ mesons in pp and nucleus-nucleus collisions at RHIC and LHC energies are shown as functions of $\langle N_{\mathrm{part}} \rangle$ in Figure 2a. The $\langle p_{\mathrm{T}} \rangle$ values measured in pp collisions ($\langle N_{\mathrm{part}} \rangle = 2$) at $\sqrt{s} = 7$ TeV are consistent with the values measured in peripheral Pb–Pb collisions at $\sqrt{s_{\mathrm{NN}}} = 2.76$ TeV. Higher $\langle p_{\mathrm{T}} \rangle$ values are observed for K^{*0} and ϕ in A–A collisions at LHC energies than at RHIC [19, 20] energies; increases in $\langle p_{\mathrm{T}} \rangle$ from RHIC to LHC energies are also observed [22] for charged pions and kaons and appear to be due to increased radial flow in A–A collisions at LHC energies.

The ratios of the p_{T}-integrated yields $\mathrm{K}^{*0}/\mathrm{K}^-$ and ϕ/K^- in Pb–Pb collisions at $\sqrt{s_{\mathrm{NN}}} = 2.76$ TeV are shown as functions of $\langle N_{\mathrm{part}} \rangle$ in Figure 2b. While the ϕ/K ratio is independent of centrality, the K^{*0}/K ratio decreases for more central collisions (which is an indication that re-scattering effects become larger in central Pb–Pb collisions). The statistical thermal model of Ref. [7] has been used to predict particle yields in central Pb–Pb collisions at $\sqrt{s_{\mathrm{NN}}} = 2.76$ TeV, assuming a temperature of 164 MeV. The model prediction (without re-scattering) of $\phi/\mathrm{K}^- = 0.128$ is consistent with the measured value, while the prediction of $\mathrm{K}^{*0}/\mathrm{K}^- = 0.31$ is larger than the measured value. From RHIC [19] to LHC energies, the K^{*0}/K and ϕ/K ratios are independent of $\sqrt{s_{\mathrm{NN}}}$; the ϕ/K ratio is independent of the collision system.

4. Conclusions

The spectra of the K^{*0}, ϕ, $\Sigma^{*\pm}$, and $\overline{\Sigma}^{*\pm}$ resonances in pp collisions at $\sqrt{s} = 7$ TeV have been compared to predictions by event generators. The PYTHIA Perugia 2011 tune describes the K^{*0} spectrum and portions of the ϕ spectrum well; the PHOJET model and the PYTHIA D6T and ATLAS-CSC tunes describe only portions of the K^{*0} and ϕ spectra. The PYTHIA ATLAS-CSC tune describes the $\frac{1}{2}(\Sigma^{*\pm} + \overline{\Sigma}^{*\mp})$ spectra for $p_{\mathrm{T}} > 2$ GeV/c. The values of $\langle p_{\mathrm{T}} \rangle$ in Pb–Pb collisions at $\sqrt{s_{\mathrm{NN}}} = 2.76$ TeV are observed to be higher than measured at RHIC, consistent with previous observations of larger radial flow at LHC energies. The $\mathrm{K}^{*0}/\mathrm{K}^-$ ratio decreases for more central Pb–Pb collisions, suggesting that re-scattering effects are larger in central collisions.

References

[1] Sjöstrand T, Mrenna S and Skands P 2006 *J. High Energy Phys.* **5** 026 1–581
[2] Engel R and Ranft J 1996 *Phys. Rev.* D **54** 4244–62
[3] Bleicher M and Stöcker H 2004 *J. Phys.* G **30** S111–8
[4] Markert C, Torrieri G and Rafelski J 2002 *Proc. of PASI 2002* (*Preprint* hep-ph/0206260)
[5] Vogel S and Bleicher M 2005 *Proc. of Nucl. Phys. Winter Meeting 2005 in Bormio* (*Preprint* nucl-th/0505027v1)
[6] Bleicher M and Aichelin J 2002 *Phys. Lett.* B **530** 81–7
[7] Andronic A, Braun-Munzinger P and Stachel J 2011 *J. Phys.* G **38** 124081
[8] Torrieri G and Rafelski J 2001 *Phys. Lett.* B **509** 239–45
[9] Petreczky P 2007 *Nucl. Phys.* A **785** 10–7
[10] Rapp R, Wambach J and van Hees H 2009 *Preprint* 0901.3289v1
[11] Brown G E and Rho M 2002 *Rhys. Rep.* **363** 85–171
[12] Brodsky S J and de Teramond G F 1988 *Phys. Rev. Lett.* **60** 1924–7
[13] Bleicher M *et al* 1999 *J. Phys.* G **25** 1859–96
[14] Markert C, Bellwied R and Vitev I 2008 *Phys. Lett.* B **669** 92–7
[15] Aamodt K *et al* (ALICE Collaboration) 2008 *J. Inst.* **3** No. S08002 i–245
[16] Abelev B *et al* (ALICE Collaboration) 2012 *CERN-PH-EP-2012-221* (*Preprint* 1208.5717v2)
[17] Beringer J *et al* (Particle Data Group) 2012 *Phys. Rev.* D **86** 010001
[18] Tsallis C 1988 *J. Stat. Phys.* **52** 479
[19] Abelev B I *et al* (STAR Collaboration) 2009 *Phys. Rev.* C **79** 064903
[20] Adams J *et al* (STAR Collaboration) 2005 *Phys. Lett.* B **612** 181–9
[21] Schnedermann E, Sollfrank J and Heinz U 1993 *Phys. Rev.* C **48** 2462–75
[22] Abelev B *et al* (ALICE Collaboration) 2012 *CERN-PH-EP-2012-230* (*Preprint* 1208.1974v1)

Journal of Physics: Conference Series **446** (2013) 012057 doi:10.1088/1742-6596/446/1/012057

Vector meson production in pp collision at $\sqrt{s} = 7$ TeV and at $\sqrt{s} = 2.76$ TeV measured with the ALICE detector

Ester Anna Rita Casula
Università di Cagliari and INFN, Sezione di Cagliari, Italy
E-mail: ester.anna.rita.casula@cern.ch

Abstract. Strangeness enhancement is one of the possible signatures of the Quark Gluon Plasma formation and can be accessed through the measurement of ϕ meson production with respect to ρ and ω mesons. Vector mesons can be detected through their decays into muon pairs with the ALICE muon spectrometer.
We present transverse momentum spectra of ϕ and $\rho + \omega$ mesons in the rapidity range $2.5 < y < 4$ in p-p collisions at $\sqrt{s} = 7$ TeV, as well as the absolute production cross sections at $\sqrt{s} = 7$ TeV and at $\sqrt{s} = 2.76$ TeV. We also show the $\phi/(\rho + \omega)$ ratio in Pb-Pb collisions at $\sqrt{s_{NN}} = 2.76$ TeV.

1. Introduction

Low mass meson (ρ, ω, ϕ) production provides key information on the hot and dense state of strongly interacting matter produced in high-energy heavy ion collisions. Among them, strangeness enhancement can be accessed through the measurement of ϕ meson production, while the measurement of the ρ spectral function can be used to reveal in-medium modifications of hadron properties close to the QCD phase boundary [1]. Vector meson production in pp collisions provides a reference for these studies and it is interesting by itself, since it can be used to tune particle production models in the LHC energy range.
We analyzed the vector mesons [2] produced in the rapidity range $2.5 < y < 4$ at the ALICE experiment [3] and detected through their decays into muon pairs.
The ALICE muon spectrometer is composed of a front hadron absorber, a set of cathode pad chambers (five stations, each one composed of two chambers) for the track reconstruction in a dipole field, two stations of two resistive plate chambers (RPC) for the muon trigger and an iron wall acting as a muon filter. The muon trigger is fired when at least three of the four RPC planes give a signal compatible with a tracklet in the muon trigger system. A minimum bias trigger, independent from the muon trigger and based on a set of forward scintillators and on a silicon pixel detector placed in the vertex region, was used to evaluate the integrated luminosity.

2. Analysis in pp collisions at $\sqrt{s} = 7$ TeV and at $\sqrt{s} = 2.76$ TeV

Data were collected in pp collisions in 2010 at $\sqrt{s} = 7$ TeV and in 2011 at $\sqrt{s} = 2.76$ TeV. Muon tracks were selected asking that the tracks reconstructed in the tracking stations matched

the ones in the trigger chambers (single muon p_{T} trigger threshold ~ 0.5 GeV/c) and that their rapidity was in the range $2.5 < y_\mu < 4$. Muon pairs were selected requiring that the dimuon rapidity was inside the interval $2.5 < y_{\mu\mu} < 4$.
The combinatorial background in the opposite sign dimuon mass spectrum was subtracted using the event mixing technique. The S/B ratio at the ϕ peak is $\sim$1 at 7 TeV and $\sim$2 at 2.76 TeV. The mass spectrum, shown in Fig. 1 for dimuon $p_{\mathrm{T}} > 1$ GeV/c, was described as a superposition of light meson decays into muon pairs, with an additional contribution coming from charm and beauty semi-muonic decays. Low-mass resonances shapes come from a Monte Carlo simulation with a parametric generator [2], while open charm and beauty have been generated using a parametrization of PYTHIA. The cut on dimuon p_{T} is due to the fact that for $p_{\mathrm{T}} < 1$ GeV/c the background subtraction is not under sufficient control.
The numbers of ϕ and ρ+ω were extracted through the fit of the dimuon invariant mass spectrum with the aforementioned sources, using the full data sample corresponding to an integrated luminosity of ~ 85 nb^{-1}. We obtained $N_\phi = (3.20 \pm 0.15) \cdot 10^3$ and $N_{\rho+\omega} = (6.83 \pm 0.20) \cdot 10^3$ at 7 TeV, and $N_\phi = 350 \pm 42$ and $N_{\rho+\omega} = 801 \pm 40$ at 2.76 TeV. We used the full data sample to extract the ϕ and $\rho + \omega$ p_{T} distributions as well.

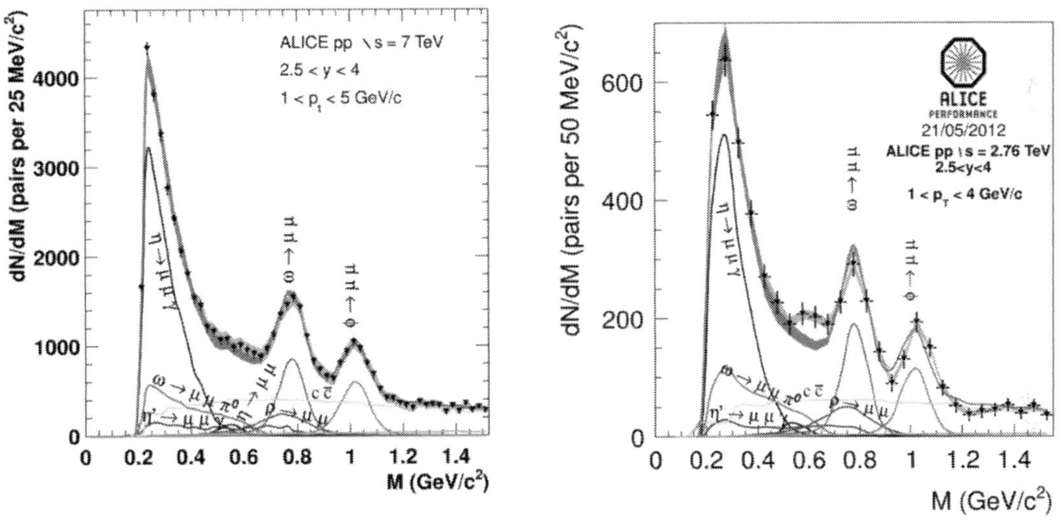

Figure 1. Fit to the dimuon invariant mass spectrum for $p_{\mathrm{T}} > 1$ GeV/c, at $\sqrt{s} = 7$ TeV (left) and at $\sqrt{s} = 2.76$ TeV (right). Blue band: systematic uncertainty from background subtraction. Red band: uncertainty in the relative normalization of the sources.

At 7 TeV, part of the data was not collected with the minimum bias trigger in parallel with the muon trigger. For this fraction, the integrated luminosity could not be measured and the cross sections were determined with the remaining sub-sample, corresponding to an integrated luminosity $\mathrm{L_{INT}} = 55.7 \pm 2.8(\mathrm{syst})$ nb^{-1}, while the cross section at 2.76 TeV was determined from a sample with $\mathrm{L_{INT}} = 17.6 \pm 0.5(\mathrm{syst})$ nb^{-1}, in both cases through the formula $\sigma_\phi = \frac{N_\phi^c}{BR(\phi\to\mu^+\mu^-)} \frac{1}{\mathrm{L_{INT}}}$, where N_ϕ^c is the measured number of ϕ mesons corrected for efficiency and acceptance.
We obtained $\sigma_\phi(2.5 < y < 4, 1 < p_{\mathrm{T}} < 5\ \mathrm{GeV}/c) = 0.940 \pm 0.084(\mathrm{stat}) \pm 0.078(\mathrm{syst})$ mb at 7 TeV [2], and $\sigma_\phi(2.5 < y < 4, 1 < p_{\mathrm{T}} < 4\ \mathrm{GeV}/c) = 0.587 \pm 0.070(\mathrm{stat}) \pm 0.045(\mathrm{syst})$ at 2.76 TeV. The main sources of systematic uncertainty are those on the luminosity (5% at 7 TeV and 3% at 2.76 TeV), the tracking efficiency (3% and 4% respectively) and the muon trigger efficiency

(4% and 5% respectively). We have also taken into account the uncertainty on the background subtraction (2% at 7 TeV and 1 % at 2.76 TeV) and the one on the ϕ branching ratio into dileptons (1%).

In order to extract the ω cross section, the ρ and ω contributions must be disentangled, leaving the ρ normalization as an additional free parameter in the fit to the dimuon mass spectrum. The result of the fit for $1 < p_T < 5$ GeV/c gives $\sigma_\rho/\sigma_\omega = 1.15 \pm 0.20(\text{stat}) \pm 0.12(\text{syst})$. The ω production cross section, calculated from this ratio, is $\sigma_\omega(2.5 < y < 4, 1 < p_T < 5 \text{ GeV}/c) = 5.28 \pm 0.54(\text{stat}) \pm 0.49(\text{syst})$ mb.

In Fig. 2, left side, the p_T-differential cross section $d^2\sigma_\phi/dp_T dy$ at 7 TeV is shown (black triangles). Data are here compared with some commonly used models as PHOJET [5] and PYTHIA [6] (Perugia-0 [7], Perugia-11 [8], ATLAS-CSC [9] and D6T [10] tunes): PYTHIA Perugia-0 and PYTHIA Perugia-11 underestimate the data, while the other models reproduce the data.

In Fig. 2, right side, the p_T-differential cross section $d^2\sigma_\omega/dp_T dy$ at 7 TeV is shown (black triangles). Comparison with the models shows that all the PYTHIA tunes reproduce the p_T slope, even if D6T overestimate the data, while PHOJET gives a slightly harder spectrum.

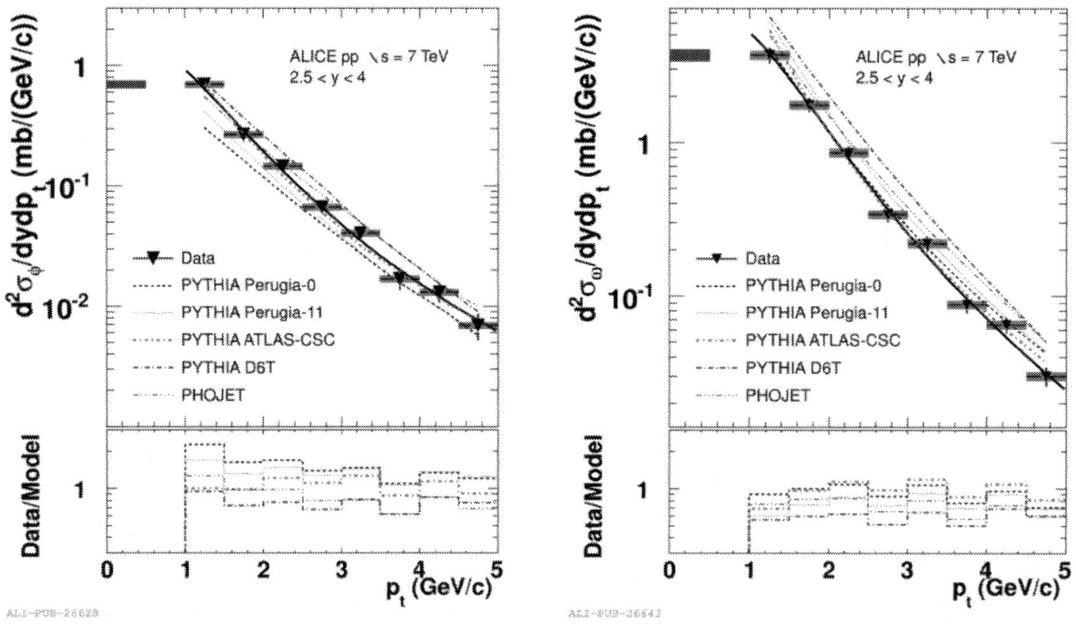

Figure 2. ϕ (left) and ω (right) differential cross sections compared with PHOJET and several tunes of PYTHIA

3. Analysis in Pb-Pb collisions at $\sqrt{s} = 2.76$ TeV

Data in Pb-Pb collisions were collected in 2011 at $\sqrt{s_{NN}} = 2.76$ TeV. The selections applied were the same as in pp analysis, with an additional cut on the single muon p_T at 0.85 GeV/c to reduce the background. The number of opposite sign muon pairs was $\sim 2 \cdot 10^6$.

Combinatorial background was evaluated with the event mixing technique also in Pb-Pb analysis; a cut on dimuon p_T at 2 GeV/c was applied too, because for dimuon $p_T < 2$ GeV/c the acceptance for ϕ, ρ and ω was close to 0, a consequence of the fact that the single muon p_T trigger threshold in Pb-Pb collisions was set at ~ 1 GeV/c. S/B ratio at the ϕ peak is ~ 0.1 for

most central collisions, increasing up to ~ 3 for peripheral collisions.
In Fig. 3, left side, the dimuon invariant mass spectrum for dimuon $p_T > 2$ GeV/c is shown. The other processes are here described as an empirical continuum that describes the sources of the low mass resonances as well as other characteristic sources which may be present in heavy-ion collisions, such as thermal dileptons.
In Fig. 3, right side, the $N_\phi/N_{\rho+\omega}$ ratio, corrected for the acceptance, is shown as a function of the number of participating nucleons N_{part}. The Pb-Pb values obtained in four different centrality classes are compared to the value obtained in pp analysis at 2.76 TeV.
$N_\phi/N_{\rho+\omega}$ increases from pp to Pb-Pb; within the uncertainty, no centrality dependence is observed. Systematic uncertainties are due to the uncertainty in the branching ratio of ϕ into dileptons, to the variations related to p_T cut on single muons (we have tried three different cuts at 0.7, 0.85 and 1 GeV/c), to the description of the correlated background (we have also used Crystal Ball functions to describe ϕ and $\rho+\omega$ peaks, with an exponential plus a constant, or an exponential plus a Landau to describe the other processes) and to the uncertainty on the mass resolution, which is a fixed parameter in the fit.

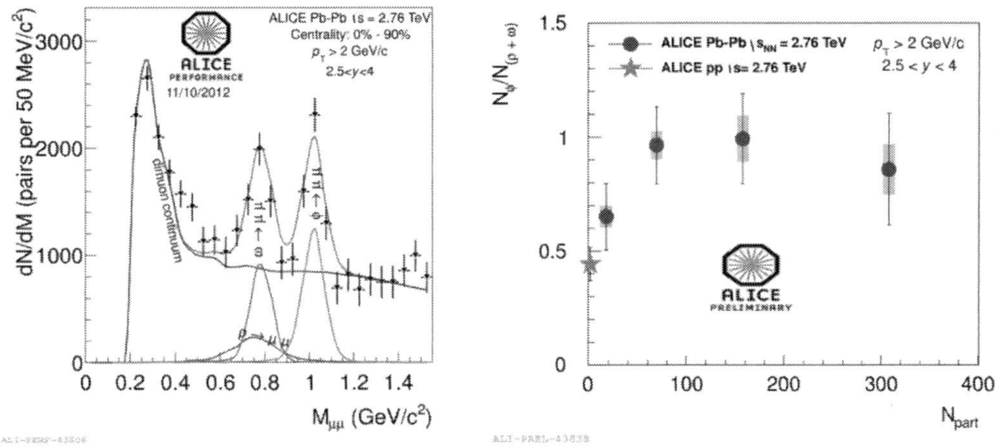

Figure 3. Dimuon invariant mass spectrum integrated in centrality for Pb-Pb collisions at $\sqrt{s}_{NN} = 2.76$ TeV (left) and $N_\phi/N_{\rho+\omega}$ ratio as a function of the number of participating nucleons N_{part} (right)

In conclusion, we measured both integrated and p_T-differential cross sections of ϕ and ω in pp collisions at $\sqrt{s} = 7$ TeV and integrated and p_T-differential cross sections of ϕ at $\sqrt{s} = 2.76$ TeV. In Pb-Pb collisions work is in progress to measure nuclear modification factor of ϕ mesons.

References

[1] P. Braun-Munzinger, J. Wambach, *Rev. Mod. Phys. 81* (2009) 1031.
[2] B. Abelev et al. (ALICE collaboration), *Phys. Lett. B* 710 (2012) 557-568.
[3] K. Aamodt et al. (ALICE collaboration), *JINST* 3 (2008) S08002.
[4] K. Oyama et al. (ALICE collaboration), *CERN-Proceedings-2011-001* 39 (2011).
[5] R. Engel, *Z. Phys. C* 66 (1995) 203; R. Engel, J. Ranft, *Phys. Rev. D* 54 (1996) 4244.
[6] T. Sjöstrand et al., *J. High Energy Phys.* 0605 (2006) 026.
[7] P.Z. Skands, *arXiv:0905.3418* (2009).
[8] P.Z. Skands, *Phys. Rev. D* 82 (2010) 074018.
[9] C. Buttar et al., *Acta Phys. Pol. B* 35 (2004) 433.
[10] R. Field, *Acta Phys. Pol. B* 39 (2008) 2611.

Journal of Physics: Conference Series **446** (2013) 012058 doi:10.1088/1742-6596/446/1/012058

Coherent J/ψ photoproduction with the ALICE experiment at the LHC

Jaroslav Adam

Faculty of Nuclear Sciences and Physical Engineering
Czech Technical University in Prague

E-mail: jaroslav.adam@cern.ch

Abstract. Results from the ALICE experiment on coherent J/ψ photoproduction in ultra-peripheral Pb-Pb collisions at $\sqrt{s_{NN}} = 2.76$ TeV are presented. The $J/\psi \to \mu^+\mu^-$ channel has been used to measure the cross section in the rapidity interval $-3.6 < y < -2.6$. The photoproduction cross section has been compared to several theoretical models. The best agreement between data and theoretical predictions is found for models which include nuclear gluon shadowing.

1. Introduction

Heavy vector meson photoproduction is a good probe of the nuclear gluon distribution at low Bjorken-x, because the photoproduction cross section is directly related to the gluon distribution. In the case of J/ψ photoproduction at LHC energies in the forward rapidity $-3.6 < y < -2.6$, the Bjorken-x takes values around $x \approx 10^{-2}$ or $x \approx 10^{-5}$, depending on which nucleus emitted the photon. At the gamma-nucleon level, the photonuclear reactions are described via the two-gluon exchange mechanism.

An experimental opportunity to produce such high energy photonuclear reactions are the so called ultra-peripheral heavy-ion collisions (UPC), which are defined as collisions with impact parameter b larger than sum of the nucleus radii. In this case, hadronic interactions are supressed and the interaction is mediated by the electromagnetic field. The field can be considered as a flux of virtual photons $\mathrm{d}N/\mathrm{d}k$, where k is the photon energy. Photoproduction cross section is then given by $\sigma_X = \int \mathrm{d}k(\mathrm{d}N/\mathrm{d}k)\sigma_X^{\gamma}(k)$, where $\sigma_X^{\gamma}(k)$ is the photonuclear cross section [1].

2. Physics of ultra-peripheral collisions

To calculate the spectrum of virtual photons created by the electromagnetic field, the Weizsäcker-Williams method can be used. In this approach, Lorentz contracted electromagnetic field is decomposed into a frequency spectrum of virtual quanta applying the Fourier transformation. Assuming a particle with charge q, moving at velocity β in minimal distance b_{min}, the virtual photons spectrum is given by the following equation [2]

$$I(\omega) = \frac{2}{\pi}\frac{q^2}{c}\beta^2[xK_0(x)K_1(x) - \frac{\beta^2}{2}x^2(K_1^2(x) - K_0^2(x))], \quad (1)$$

where K_0 and K_1 are modified Bessel functions and $x = \omega b_{min}/c\beta\gamma$.

The general cross section formula for J/ψ photoproduction in Pb-Pb collisions consists of two contributions with the opposite rapidity sign, because each lead ion can serve either as a photon source or as target. Denoting N_γ as the photon flux and y as the J/ψ rapidity in gamma-nucleus center of mass system, the cross section takes the form

$$\frac{\mathrm{d}\sigma(PbPb \to J/\psi PbPb)}{\mathrm{d}y} = N_\gamma(y)\sigma_{\gamma Pb \to J/\psi Pb}(y) + N_\gamma(-y)\sigma_{\gamma Pb \to J/\psi Pb}(-y). \quad (2)$$

Gamma-nucleus cross section $\sigma_{\gamma Pb \to J/\psi Pb}$ depends on the square of the gluon PDF.

3. The ALICE experiment

For this analysis, the forward muon spectrometer, the VZERO scintillators and the Zero-Degree calorimeters were used [3].

The VZERO counters consist of scintillator elements and cover the pseudorapidity $2.8 < \eta < 5.1$ and $-3.7 < \eta < -1.7$ for VZERO-A and VZERO-C, respectively. The VZERO detector is used for triggering and for event selection. Two hadronic Zero-Degree Calorimeters (ZDC) are placed at 116 m at both sides of the interaction point. They are used to detect neutrons and protons in the forward direction.

The muon spectrometer covers the pseudorapidity interval $-4.0 < \eta < -2.5$. It consists of five tracking MWPC stations and four RPC trigger plates placed in the magnetic field created by a 3 Tm dipole magnet. Between the interaction point and the spectrometer a 10 interaction length thick composite absorber is placed.

4. Data analysis

The available data sample consists of $3.16\text{x}10^6$ events with integrated luminosity 55 μb^{-1} taken in 2011 Pb-Pb collisions at $\sqrt{s_{NN}} = 2.76$ TeV. In order to select UPC events the special trigger was set up, requiring (i) single muon trigger above a 1 GeV/c p_T threshold, (ii) at least one hit in VZERO-C detector and (iii) no hits in the VZERO-A detector (opposite side to the muon arm). The luminosity of data sample was measured using the UPC trigger scalers from the cross section of the reference trigger classes, which was obtained by the van der Meer scan method.

The offline event selection criteria for the $J/\psi \to \mu^+\mu^-$ analysis required coherent ultra-peripheral processes signatures. In order to ensure an exclusive reaction only two tracks in the muon spectrometer in an otherwise empty detector were required.

The signatures are (i) neutron ZDC signal < 6 TeV, (ii) VZERO timing compatible with crossing beams and (iii) two reconstructed tracks in the muon arm with opposite charges. The second group of analysis cuts ensures the muon track quality, this selection consists of (i) momentum dependent DCA (distance between the vertex and the track) to reject multiple scattering in the absorber and (ii) both tracks pseudorapidity $-3.7 < \eta_{1,2} < -2.5$.The last group of analysis requirements select reconstructed candidate having (i) rapidity $-3.6 < y < -2.6$, (ii) transverse momentum $p_T < 0.3$ GeV/c and (iii) invariant mass $2.8 < M_{inv} < 3.4$ GeV/c^2.

Following the above criteria, 117 J/ψ candidates remained.

5. Signal extraction

5.1. Dimuon invariant mass

The invariant mass corresponding to dimuon candidates which satisfied the above selection criteria is shown in Figure 1 in the interval $2.2 < M_{inv} < 4.6$ GeV/c^2. We can clearly see the J/ψ peak above the background mainly coming from $\gamma\gamma \to \mu^+\mu^-$ QED process. The fit is performed using the Crystal Ball function [4] for the signal and an exponential function for the background. The J/ψ yield obtained from the corresponding fit is $N_{yield} = 96 \pm 12$(stat) ± 6(syst). The exponential slope parameter is in very good agreement with the STARLIGHT MC expectation (-1.4 ± 0.2 GeV^{-1}c^2 for data and -1.39 ± 0.01 GeV^{-1}c^2 for MC) [5].

5.2. Physics processes involved

There are 4 main processes contributing to the exclusive J/ψ production decaying in the $\mu^+\mu^-$ channel.

The coherent J/ψ production is our signal of interest consisting of transverse polarized, low p_T J/ψ. In this case the photon couples coherently to all nucleons. The incoherent production occurs when photon couples to a single nucleon. J/ψ produced by this mechanism have higher p_T and are suppressed by the $p_T < 0.3$ GeV/c cut.

The ψ', produced by ultra-peripheral processes, may contribute to the signal by its decay $\psi' \rightarrow J/\psi + \mathrm{X}$. The corresponding simulations include a ψ' sample generated with the STARLIGHT MC [5] while the decays were simulated using PYTHIA [6].

The last source of background is the QED $\gamma\gamma \rightarrow \mu^+\mu^-$ interaction.

Figure 2 (right) shows dimuon p_T distribution with the fit performed using MC templates for each of the contributing processes. The number of coherent candidates can be extracted by the equation $N^{coh}_{J/\psi} = \frac{N_{yield}}{1+f_I+f_D}$. Using the yield, the incoherent fraction $f_I = 0.12^{+0.14}_{-0.04}$ (incoherent J/ψ contribution) and the feed-down $f_D = 0.11 \pm 0.06$ (J/ψ from ψ' decay), we obtain $N^{coh}_{J/\psi} = 78 \pm 10(\mathrm{stat})^{+7}_{-11}(\mathrm{syst})$ candidates. The contributions f_I and f_D are obtained from the fits on the p_T distribution (Figure 2).

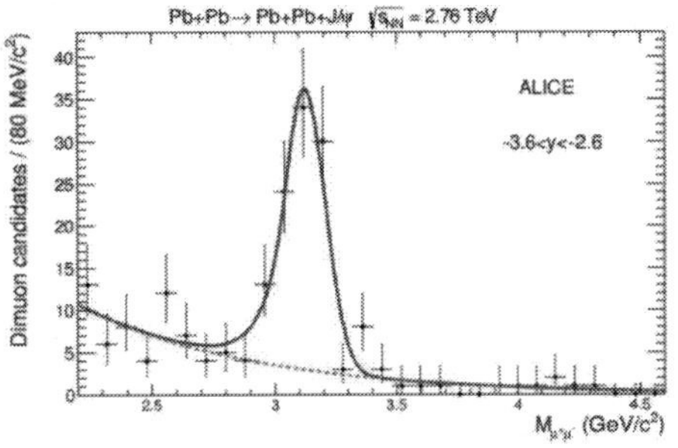

Figure 1. Invariant mass distribution for events satisfying the selection criteria.

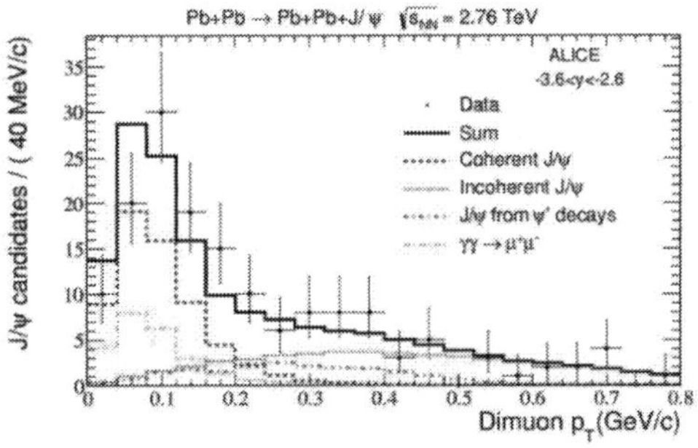

Figure 2. Dimuon p_T distribution for the accepted events with extended p_T range.

6. Acceptance and efficiency correction

In order to calculate the detection efficiency, the STARLIGHT generator has been used to provide coherent and incoherent J/ψ and $\gamma\gamma \rightarrow \mu^+\mu^-$ samples. The generated events were folded with the detector Monte Carlo simulation using GEANT3.

The correction $(\mathrm{Acc} \times \varepsilon)_{J/\psi}$ is defined as the ratio of accepted over generated candidates and the resulting values are $(\mathrm{Acc} \times \varepsilon)_{J/\psi} = 16.6\%$ and 14.3% for coherent and incoherent productions respectively.

7. Coherent J/ψ differential cross section

The following equation 3 provides cross section calculation using $J/\psi \rightarrow \mu^+\mu^-$ branching ratio BR, number of coherent J/ψ and $\gamma\gamma \rightarrow \mu^+\mu^-$ candidates $N^{coh}_{J/\psi}$ and $N_{\gamma\gamma}$, the corresponding corrections $\mathrm{Acc} \times \varepsilon$, the $\gamma\gamma \rightarrow \mu^+\mu^-$ QED cross section $\sigma_{\gamma\gamma}$ and rapidity interval $\Delta y = 1$ in present case.

$$\frac{\mathrm{d}\sigma^{coh}_{J/\psi}}{\mathrm{d}y} = \frac{1}{\mathrm{BR}(J/\psi \rightarrow \mu^+\mu^-)} \cdot \frac{N^{coh}_{J/\psi}}{N_{\gamma\gamma}} \cdot \frac{(\mathrm{Acc} \times \varepsilon)_{\gamma\gamma}}{(\mathrm{Acc} \times \varepsilon)_{J/\psi}} \cdot \frac{\sigma_{\gamma\gamma}}{\Delta y} \tag{3}$$

The QED continuum pair production cross section $\sigma_{\gamma\gamma}$ has been used for normalization, so that the $\sigma^{coh}_{J/\psi}$ formula (3) is independent of luminosity and of trigger efficiency, but the theoretical uncertainty in $\sigma_{\gamma\gamma}$ contributes 20% to the coherent cross section systematic uncertainty.

The coherent J/ψ differential cross section in $-3.6 < y < -2.6$ and $p_T < 0.3$ GeV/c is $\mathrm{d}\sigma^{coh}_{J/\psi}/\mathrm{d}y = 1.00 \pm 0.18(\mathrm{stat})^{+0.24}_{-0.26}(\mathrm{syst})$ mb.

8. Theoretical predictions

The measured coherent differential cross section of J/ψ photoproduction is shown in Figure 3, together with several model predictions [7]. The uncertainty is given by the quadratic sum of the statistical and systematic errors. Figure 4 provides the ratio of the cross sections between two rapidity bins, taking advantage that some parts of the systematic errors cancel. The ratio of the cross sections is $R = 1.36 \pm 0.36(\mathrm{stat}) \pm 0.19(\mathrm{syst})$.

The best agreement is observed for partonic models AB-EPS08, AB-EPS09 and RST-LTA, which include nuclear gluon shadowing.

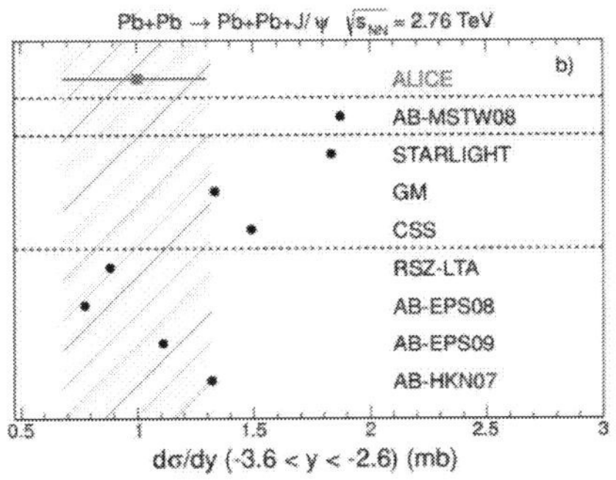

Figure 3. Coherent J/ψ cross section.

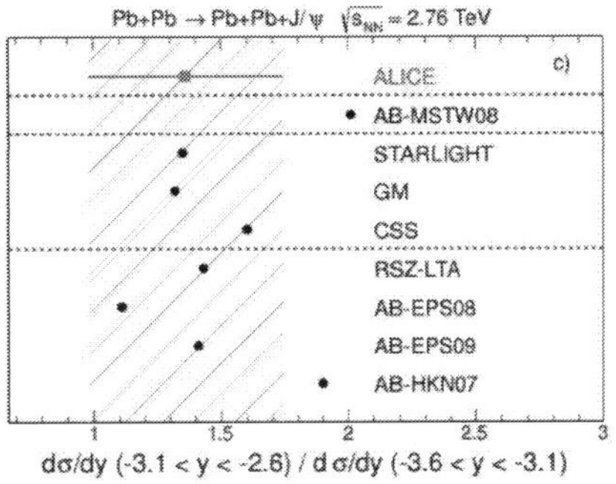

Figure 4. Ratio of the cross sections between two rapidity bins.

9. Conclusions

The ALICE Collaboration made the first LHC measurement on exclusive J/ψ photoproduction in Pb-Pb collisions at $\sqrt{s_{NN}} = 2.76$ TeV [7], the cross section in the forward rapidity interval $-3.6 < y < -2.6$ was estimated as $\mathrm{d}\sigma^{coh}_{J/\psi}/\mathrm{d}y = 1.00 \pm 0.18(\mathrm{stat})^{+0.24}_{-0.26}(\mathrm{syst})$ mb. The value is in agreement with partonic models including nuclear gluon shadowing AB-EPS08, AB-EPS09 and RST-LTA.

Acknowledgements

This work was supported by grant LK11209 of MŠMT ČR.

References

[1] A. J. Baltz et al. Phys. Rept. 458 (2008) 1
[2] Jackson J. D., Classical Electrodynamics 2nd Ed., John Wiley & Sons, New York (1975)
[3] K. Aamodt et al. [ALICE Collaboration], JINST 3 (2008) S08002
[4] J. E. Gaiser, Ph.D. thesis, SLAC-R-255, 1982
[5] STARLIGHT website, http://starlight.hepforge.org/
[6] T. Sjöstrand, Comput. Phys. Commun. 82 (1994) 74
[7] B. Abelev et al. [ALICE Collaboration], Phys. Lett. B 718 (2013) 1273 - 1283

AUTHOR INDEX

Adam, J 244
Aiola, S 19
Armesto, N 40
Bartz, S P 80
Bass, S A 149
Bazavov, A 44
Beck, H 54
Bouras, I 93
Campbell, J M 58
Cao, S 149
Casula, E A R 240
Chen, G 89
Chinellato, D D 232
Coleman-Smith, C E 32
Colonna, M 76
Connors, M 36
Cui, X 197
Dainese, A 141
Ding, L 173
El, A 93
Fasel, M 161
Ficnar, A 7
Filip, P 137
Florent, A 193
Fochler, O 93, 185
Fries, R J 11, 89
Gavin, S 121
Greco, V 76, 105
Greiner, C 93, 185
Gulbrandsen, K 113
Gyulassy, M 7
Horvat, S P 72
Hu, Z 181
Innocenti, G M 169
Kapusta, J I 80
Kim, Y 28
Knospe, A G 236
Köhler, M K 206
Lamont, M A C 216
Lardeux, A 177
Lee, Y-J 1
Lisa, M A 58
Lohner, D 117
Luparello, G 165
Luzum, M 84
Ma, H 40
Martinez, M 40
Mas, A 189
Mehtar-Tani, Y 40
Mizher, A J 68
Mohapatra, S 101
Moschelli, G 121
Müller, B 149
Niemi, H 93
Noronha, J 7
Oldag, E W 97
Pandit, Y 50
Plumari, S 76, 105
Pokharel, R 121
Puglisi, A 105
Qin, G-Y 149
Reed, R 23
Rodriguez, R 11
Rolnick, S D 201
Romita, R 153
Salgado, C A 40
Scardina, F 76, 105
Sickles, A M 210
Skokov, V 133
Sorensen, P 62
Stiller, J 220
Teaney, D 109
Thomas, D 157
Timmins, A R 129
Toll, T 224
Uphoff, J 185
Uras, A 228
Vajzer, M 15
Velicanu, D 125
Xu, Z 93, 185
Yan, L 109